Scientific Advances in STEM: Synergies to Achieve Success. 2nd Edition

Scientific Advances in STEM: Synergies to Achieve Success. 2nd Edition

Editors

Yadir Torres Hernández
Ana María Beltrán Custodio
Manuel Félix Ángel

MDPI • Basel • Beijing • Wuhan • Barcelona • Belgrade • Manchester • Tokyo • Cluj • Tianjin

Editors
Yadir Torres Hernández
University of Seville (US)
Spain

Ana María Beltrán Custodio
University of Seville (US)
Spain

Manuel Félix Ángel
Universidad de Sevilla
Spain

Editorial Office
MDPI
St. Alban-Anlage 66
4052 Basel, Switzerland

This is a reprint of articles from the Topic published online in the open access journals *Materials* (ISSN 1996-1944), *Polymers* (ISSN 2073-4360), *Foods* (ISSN 2304-8158), *Sensors* (ISSN 1424-8220), and *Sustainability* (ISSN 2071-1050) (available at: https://www.mdpi.com/topics/stem_2nd).

For citation purposes, cite each article independently as indicated on the article page online and as indicated below:

LastName, A.A.; LastName, B.B.; LastName, C.C. Article Title. *Journal Name* **Year**, *Volume Number*, Page Range.

ISBN 978-3-0365-5193-7 (Hbk)
ISBN 978-3-0365-5194-4 (PDF)

Cover image courtesy of Sergio Muñoz

© 2022 by the authors. Articles in this book are Open Access and distributed under the Creative Commons Attribution (CC BY) license, which allows users to download, copy and build upon published articles, as long as the author and publisher are properly credited, which ensures maximum dissemination and a wider impact of our publications.

The book as a whole is distributed by MDPI under the terms and conditions of the Creative Commons license CC BY-NC-ND.

Contents

About the Editors vii

Preface to "Scientific Advances in STEM: Synergies to Achieve Success. 2nd Edition" ix

Paloma Trueba, Carlos Navarro, Mercè Giner, José A. Rodríguez-Ortiz, María José Montoya-García, Ernesto J. Delgado-Pujol, Luisa M. Rodríguez-Albelo and Yadir Torres
Approach to the Fatigue and Cellular Behavior of Superficially Modified Porous Titanium Dental Implants
Reprinted from: *Materials* 2022, 15, 3903, doi:10.3390/ma15113903 1

Ana M. Beltrán, Mercè Giner, Ángel Rodríguez, Paloma Trueba, Luisa M. Rodríguez-Albelo, Maria Angeles Vázquez-Gámez, Vanda Godinho, Ana Alcudia, José M. Amado, Carmen López-Santos and Yadir Torres
Influence of Femtosecond Laser Modification on Biomechanical and Biofunctional Behavior of Porous Titanium Substrates
Reprinted from: *Materials* 2022, 15, 2969, doi:10.3390/ma15092969 17

Paula Navarro, Alberto Olmo, Mercè Giner, Marleny Rodríguez-Albelo, Ángel Rodríguez and Yadir Torres
Electrical Impedance of Surface Modified Porous Titanium Implants with Femtosecond Laser
Reprinted from: *Materials* 2022, 15, 461, doi:10.3390/ma15020461 35

Ricardo Chávez-Vásconez, Sheila Lascano, Sergio Sauceda, Mauricio Reyes-Valenzuela, Christopher Salvo, Ramalinga Viswanathan Mangalaraja, Francisco José Gotor, Cristina Arévalo and Yadir Torres
Effect of the Processing Parameters on the Porosity and Mechanical Behavior of Titanium Samples with Bimodal Microstructure Produced via Hot Pressing
Reprinted from: *Materials* 2022, 15, 136, doi:10.3390/ma15010136 49

Johar Amin Ahmed Abdullah, Mercedes Jiménez-Rosado, Antonio Guerrero and Alberto Romero
Gelatin-Based Biofilms with Fe_xO_y-NPs Incorporated for Antioxidant and Antimicrobial Applications
Reprinted from: *Materials* 2022, 15, 1966, doi:10.3390/ma15051966 67

Pablo Sánchez-Cid, José Fernando Rubio-Valle, Mercedes Jiménez-Rosado, Víctor Pérez-Puyana and Alberto Romero
Effect of Solution Properties in the Development of Cellulose Derivative Nanostructures Processed via Electrospinning
Reprinted from: *Polymers* 2022, 14, 665, doi:10.3390/polym14040665 81

María Alonso-González, Manuel Felix and Alberto Romero
Rice Bran-Based Bioplastics: Effects of Biopolymer Fractions on Their Mechanical, Functional and Microstructural Properties
Reprinted from: *Polymers* 2022, 14, 100, doi:10.3390/polym14010100 97

Ismael Santana, Manuel Félix, Antonio Guerrero and Carlos Bengoechea
Processing and Characterization of Bioplastics from the Invasive Seaweed *Rugulopteryx okamurae*
Reprinted from: *Polymers* 2022, 14, 355, doi:10.3390/polym14020355 111

Luis Concha, Ana Luiza Resende Pires, Angela Maria Moraes, Elizabeth Mas-Hernández, Stefan Berres and Jacobo Hernandez-Montelongo
Cost Function Analysis Applied to Different Kinetic Release Models of *Arrabidaea chica* Verlot Extract from Chitosan/Alginate Membranes
Reprinted from: *Polymers* 2022, 14, 1109, doi:10.3390/polym14061109 . . . . **129**

Manuel Felix, Antonio Guerrero and Cecilio Carrera-Sánchez
Optimization of Multiple $W_1/O/W_2$ Emulsions Processing for Suitable Stability and Encapsulation Efficiency
Reprinted from: *Foods* 2022, 11, 1367, doi:10.3390/foods11091367 . . . . **143**

Julia Martín, Carmen Mejías, Marina Arenas, Juan Luis Santos, Irene Aparicio and Esteban Alonso
Occurrence of Linear Alkylbenzene Sulfonates, Nonylphenol Ethoxylates and Di(2-ethylhexyl)phthalate in Composting Processes: Environmental Risks
Reprinted from: *Sustainability* 2022, 14, 186, doi:10.3390/su14010186 . . . . **161**

Raquel Cañete and M. Estela Peralta
ASDesign: A User-Centered Method for the Design of Assistive Technology That Helps Children with Autism Spectrum Disorders Be More Independent in Their Daily Routines
Reprinted from: *Sustainability* 2022, 14, 516, doi:10.3390/su14010516 . . . . **173**

Wei Feng, Ling Zhao and Yue Chen
Research on Collaborative Innovation Mode of Enterprise Group from the Perspective of Comprehensive Innovation Management
Reprinted from: *Sustainability* 2022, 14, 5304, doi:10.3390/su14095304 . . . . **201**

Javier Antonio Guerra, Juan Ignacio Guerrero, Sebastián García, Samuel Domínguez-Cid, Diego Francisco Larios and Carlos Leon
Design and Evaluation of a Heterogeneous Lightweight Blockchain-Based Marketplace
Reprinted from: *Sensors* 2022, 22, 1131, doi:10.3390/s22031131 . . . . **221**

Manuel Peña, Félix Biscarri, Enrique Personal and Carlos León
Decision Support System to Classify and Optimize the Energy Efficiency in Smart Buildings: A Data Analytics Approach
Reprinted from: *Sensors* 2022, 22, 1380, doi:10.3390/s22041380 . . . . **243**

Robin Cabeza-Ruiz, Luis Velázquez-Pérez, Alejandro Linares-Barranco and Roberto Pérez-Rodríguez
Convolutional Neural Networks for Segmenting Cerebellar Fissures from Magnetic Resonance Imaging
Reprinted from: *Sensors* 2022, 22, 1345, doi:10.3390/s22041345 . . . . **269**

About the Editors

Yadir Torres Hernández

Yadir Torres Hernández. Interests: design and manufacture of porous materials; surface modification (physical and chemical); biofuntional (osseointegration, cells, and bacterial response) and tribo-mechanical (instrumented micro-indentation, fracture, fatigue, scratch resistance, and wear) behavior; biomaterials; tool materials (cemented carbides, cermet's, and multi-layered: alumina-zirconia, WC-Co/WC-Co, and Cermet/WC-Co); powder metallurgy (conventional and space-holder technique).

Ana María Beltrán Custodio

Ana María Beltrán Custodio. Interests: active packaging; bioplastics; electrospinning; interfaces; injection moulding, mixing; proteins; rheology; food by-products.

Manuel Félix Ángel

Manuel Félix Ángel. Interests: design; nano-structure and chemical studies by scanning-transmission electron microscopy techniques; biomaterials.

Preface to "Scientific Advances in STEM: Synergies to Achieve Success. 2nd Edition"

1. Introduction and Scope

The *Escuela Politécnica Superior* (EPS) from Universidad de Sevilla (US) is a higher education engineering center that aims to provide specific training and research in each area of knowledge from Science, Techology, Engineering and Maths (STEM), as well as the generation and transfer of knowledge. To be at the forefront of innovation, research plays a key role, and, at this center, there are researchers working in all STEM areas, not only Maths and Physics but also Electrical, Electronic, Materials, and Chemistry Engineering.

Among the different research activities promoted by the center, the EPS-US annually organizes a workshop on Research, Development and Innovation, which reached its 8th edition in 2021. Advances in research in several fields of Science and Engineering are presented at this event, involving the participation of students of all levels (from BSc to PhD), as well as academic and research staff not only from this center, but also participants from different universities located in seven countries. The multidisciplinary nature of this center has led to the establishment of collaborations among research groups from different disciplines, the combination of scientific knowledge from basic with applied research, and the use of different research facilities. Fundamental science helps us to understand phenomenological foundations, while applied science focuses on products and technological developments, highlighting the need to transfer knowledge to society and the industrial sector. This is also an opportunity for transfering knowledge, since industrial PhD students also participate and companies can present their demands and establish collaborations with research groups.

This book collects articles published on the special topic *Scientific Advances in STEM: Synergies to Achieve Success, 2nd Edition*, including publications from the journals *Materials*, *Polymers*, *Foods*, *Sustainability* and *Sensors*, all published by MDPI. These papers cover selected cutting-edge research hosted at the EPS-US and presented at the 8th Symposium on Research, Development and Innovation of the academic center, which also covers the collaboration with national and international institutions. In fact, some of these works are framed within strategic research areas developed during recent years, presenting their progress in each edition of this workshop, as it was corroborated by their publication of the results in the first edition of this topic [1]. The editors wish to disseminate the multidisciplinary research carried out under the institutions of the EPS-US, contributing to establishing new collaborations among research groups.

2. Contributions

This book collects a total of 16 research works from the STEM field. As mentioned previously, they were published in five MDPI journals. Six contributions are collected from *Materials*, analysing different issues. This was the case for the analysis of fatigue and cellular performance of novel superficially treated porous titanium dental implants fabricated using conventional powder metallurgy and space holder techniques [2]. Additionally, in the field of biomaterials, Beltrán et al. [3] evaluated the texture and surface roughness of porous titanium samples manufactured by the space holder powder metallurgy technique superficially modified by the Yb-doped laser. The area of titanium-based biomaterials was also complemented by studies aimed to characterise the macroporosity inherent in the manufacturing process, including the effect of the femtosecond laser treatment of the surface of titanium discs [4], as well as the influence of processing parameters on the bimodal microstructure in biomedical implants [5].The journal *Materials* also included a paper on the formation of gelatin-based biofilms containing Fe_xO_y NPs [6]. These materials exhibited antioxidant and antimicrobial properties.

Related to material science, four contributions are included from *Polymers* and are focused on the development of bio-based materials. With this aim, the authors explored the development of cellulose-derived nanostructures (processed by electrospinning) [7] as well as the development of bioplastics by injection molding, using undervalued raw materials for this goal (rice bran and invasive seaweed) [8,9]. Moreover, this field of knowledge was enriched by simulation in terms of cost function analysis applied to arrabidaea chica verlot extracts [10].

This book also includes a contribution from *Foods*, where the manufacture of double emulsions was described by Felix et al. [11]. These emulsions were intended to be carriers of future food products with specific functionalities.

Three contributions are included from *Sustainability*. Martin et al. [12] analysed the presence of linear alkylbenzene sulfonates, nonylphenol ethoxylates and Di(2-ethylhexyl)phthalate in composting processes. They elucidated the environmental risks that it may involve. Furthermore, Cañete et al. [13] proposed new strategies to improve family and work conciliation with children suffering from psychological pathologies. Eventually, Feng et al. [14] conducted comprehensive research on innovation management in collaborative enterprise groups.

This book is finally complemented by three contributions from *Sensors*. In one of them, Guerra et al. [15] proposed a low-level blockchain marketplace to share the power of participants as requested. Slightly related to this topic, Peña et al. [16] suggested a system to classify and optimize energy input in smart buildings using an analytics approach. Eventually, convolutional neural networks to automatically segment cerebellar fissures using brain magnetic resonance imaging were investigated by Cabeza-Ruíz et al. [17].

3. References

1. *Scientific Advances in STEM: From Professor to Students*; MDPI, 2021; ISBN 978-3-0365-1776-6.

2. Trueba, P.; Navarro, C.; Giner, M.; Rodríguez-Ortiz, J.A.; Montoya-García, M.J.; Delgado-Pujol, E.J.; Rodríguez-Albelo, L.M.; Torres, Y. Approach to the Fatigue and Cellular Behavior of Superficially Modified Porous Titanium Dental Implants. *Materials* **2022**, 15, 3903.

3. Beltrán, A.M.; Giner, M.; Rodríguez, Á.; Trueba, P.; Rodríguez-Albelo, L.M.; Vázquez-Gámez, M.A.; Godinho, V.; Alcudia, A.; Amado, J.M.; López-Santos, C.; et al. Influence of Femtosecond Laser Modification on Biomechanical and Biofunctional Behavior of Porous Titanium Substrates. *Materials* **2022**, 15, 2969.

4. Navarro, P.; Olmo, A.; Giner, M.; Rodríguez-Albelo, M.; Rodríguez, Á.; Torres, Y. Electrical Impedance of Surface Modified Porous Titanium Implants with Femtosecond Laser. *Materials* **2022**, 15, 461.

5. Chávez-Vásconez, R.; Lascano, S.; Sauceda, S.; Reyes-Valenzuela, M.; Salvo, C.; Mangalaraja, R.V.; Gotor, F.J.; Arévalo, C.; Torres, Y. Effect of the Processing Parameters on the Porosity and Mechanical Behavior of Titanium Samples with Bimodal Microstructure Produced via Hot Pressing. *Materials* **2021**, 15, 136.

6. Abdullah, J.A.A.; Jiménez-Rosado, M.; Guerrero, A.; Romero, A. Gelatin-Based Biofilms with FexOy-NPs Incorporated for Antioxidant and Antimicrobial Applications. *Materials* **2022**, 15, 1966.

7. Sánchez-Cid, P.; Rubio-Valle, J.F.; Jiménez-Rosado, M.; Pérez-Puyana, V.; Romero, A. Effect of Solution Properties in the Development of Cellulose Derivative Nanostructures Processed via Electrospinning. *Polymers* **2022**, *14*, 665.

8. Alonso-González, M.; Felix, M.; Romero, A. Rice Bran-Based Bioplastics: Effects of Biopolymer Fractions on Their Mechanical, Functional and Microstructural Properties. *Polymers* **2021**, *14*, 100.

9. Santana, I.; Félix, M.; Guerrero, A.; Bengoechea, C. Processing and Characterization of Bioplastics from the Invasive Seaweed Rugulopteryx okamurae. *Polymers* **2022**, *14*, 355.

10. Concha, L.; Resende Pires, A.L.; Moraes, A.M.; Mas-Hernández, E.; Berres, S.; Hernandez-Montelongo, J. Cost Function Analysis Applied to Different Kinetic Release Models of Arrabidaea chica Verlot Extract from Chitosan/Alginate Membranes. *Polymers* **2022**, *14*, 1109.

11. Felix, M.; Guerrero, A.; Carrera-Sánchez, C. Optimization of Multiple W1/O/W2 Emulsions Processing for Suitable Stability and Encapsulation Efficiency. *Foods* **2022**, *11*, 1367.

12. Martín, J.; Mejías, C.; Arenas, M.; Santos, J.L.; Aparicio, I.; Alonso, E. Occurrence of Linear Alkylbenzene Sulfonates, Nonylphenol Ethoxylates and Di(2-ethylhexyl)phthalate in Composting Processes: Environmental Risks. *Sustainability* **2021**, *14*, 186.

13. Cañete, R.; Peralta, M.E. ASDesign: A User-Centered Method for the Design of Assistive Technology That Helps Children with Autism Spectrum Disorders Be More Independent in Their Daily Routines. *Sustainability* **2022**, *14*, 516.

14. Feng, W.; Zhao, L.; Chen, Y. Research on Collaborative Innovation Mode of Enterprise Group from the Perspective of Comprehensive Innovation Management. *Sustainability* **2022**, *14*, 5304.

15. Guerra, J.A.; Guerrero, J.I.; García, S.; Domínguez-Cid, S.; Larios, D.F.; León, C. Design and Evaluation of a Heterogeneous Lightweight Blockchain-Based Marketplace. *Sensors* **2022**, *22*, 1131.

16. Peña, M.; Biscarri, F.; Personal, E.; León, C. Decision Support System to Classify and Optimize the Energy Efficiency in Smart Buildings: A Data Analytics Approach. *Sensors* **2022**, *22*, 1380.

17. Cabeza-Ruiz, R.; Velázquez-Pérez, L.; Linares-Barranco, A.; Pérez-Rodríguez, R. Convolutional Neural Networks for Segmenting Cerebellar Fissures from Magnetic Resonance Imaging. *Sensors* **2022**, *22*, 1345.

Yadir Torres Hernández, Ana María Beltrán Custodio, and Manuel Félix Ángel
Editors

Article

Approach to the Fatigue and Cellular Behavior of Superficially Modified Porous Titanium Dental Implants

Paloma Trueba [1], Carlos Navarro [2], Mercè Giner [3,*], José A. Rodríguez-Ortiz [1], María José Montoya-García [4], Ernesto J. Delgado-Pujol [1], Luisa M. Rodríguez-Albelo [1] and Yadir Torres [1]

[1] Departamento de Ingeniería y Ciencia de los Materiales y del Transporte, Escuela Politécnica Superior, Universidad de Sevilla, 41011 Seville, Spain; ptrueba@us.es (P.T.); jarortiz@us.es (J.A.R.-O.); erndelpuj@alum.us.es (E.J.D.-P.); lralbelo@us.es (L.M.R.-A.); ytorres@us.es (Y.T.)
[2] Departamento de Ingeniería Mecánica y Fabricación, Escuela Técnica Superior de Ingeniería, Universidad de Sevilla, 41092 Seville, Spain; cnp@us.es
[3] Departamento de Citología e Histología Normal y Patológica, Universidad de Sevilla, 41009 Sevilla, Spain
[4] Departamento de Medicina, Universidad de Sevilla, 41009 Sevilla, Spain; pmontoya@us.es
* Correspondence: mginer@us.es; Tel.: +34-(9)-54551796

Abstract: In this work, the fatigue and cellular performance of novel superficially treated porous titanium dental implants made up using conventional powder metallurgy and space-holder techniques (30 vol.% and 50 vol.%, both with a spacer size range of 100–200 μm) are evaluated. Before the sintering stage, a specific stage of CNC milling of the screw thread of the implant is used. After the consolidation processing, different surface modifications are performed: chemical etching and bioactive coatings (BG 45S5 and BG 1393). The results are discussed in terms of the effect of the porosity, as well as the surface roughness, chemical composition, and adherence of the coatings on the fatigue resistance and the osteoblast cells' behavior for the proposed implants. Macro-pores are preferential sites of the nucleation of cracks and bone cell adhesion, and they increase the cellular activity of the implants, but decrease the fatigue life. In conclusion, SH 30 vol.% dental implant chemical etching presents the best bio-functional (in vitro osseointegration) and bio-mechanical (stiffness, yield strength and fatigue life) balance, which could ensure the required characteristics of cortical bone tissue.

Keywords: porous dental implant; fatigue resistance; cellular behavior; surface roughness; chemical etching; bioglass coating

1. Introduction

Commercially pure titanium (c.p. Ti) and its alloys are among the most widely employed metallic biomaterials for the convenient replacement of damaged cortical bone tissues [1,2], considering their appropriate biocompatibility, corrosion resistance and mechanical strength. However, these materials have two main drawbacks which compromise the clinical success of implants. On one hand, their stiffness (100–110 GPa) is much greater than that of cortical bone (20–25 GPa). This difference can cause the loss of bone and raise the threat of the rupture of the nearby bone [1,3]. Therefore, different methods are proposed to solve these problems. An approach to the reduce the stress-shielding phenomenon and to encourage bone-in-growth into the implant is the introduction of porosity into the implants [3,4]. Biomechanical properties such as the hardness, resistance to corrosion, fracture, fatigue, and wear of implants rely on the amount, volume, and form of the pores, besides the size and geometry of the dental implant. Another way to decrease the stress-shielding phenomenon is the use of β-titanium alloys, as their Young's modulus (60–80 GPa) is closer to cortical bone´s, and additionally, their biomechanical performances are also improved [2,5–9].

Other drawbacks are the poor osseointegration of the implants, which is inherent in the inert biological behavior of titanium surfaces [10], and the potential infections [11] (proliferation and growth of bacteria) that occur during surgery or the scarring period, which inhibit the formation of new bone. This fact increases the loading times of the dental implant and can cause its uselessness at the medium and long term. In order to address these problems, the literature proposes the manipulation and optimization of the topography (roughness and texture) and the chemistry (bioactive surfaces) of implants, including the immobilization of proteins on the surface [1,11–13]. The macro-topography is determined by the geometric design [12] (the presence of threads and conical shapes, etc.), while the surface micro-topography and nano-topography cause effects at the cellular and protein levels, respectively [14].

One of the routes used for this purpose is chemical etching [15], considering its attractive cost, versatility and repeatability. The nature and concentration of the chemical reagent used are particularly important, as are the etching times implemented. On the other hand, among the ways to facilitate the chemical interaction between the implant and cortical bone tissue are the use of bioactive coatings [16,17], thermochemical treatments [18] or the bio-functionalization of the surface [19].

Furthermore, the failure in service of dental implants can also occur due to an overload and/or a poor resistance to fatigue [10,20–22]. In this scenario, it is widely known that the accomplishment of successful dental implants depends not only on the quantity/quality of the patient's bone tissue [23], material used, size, and implant design [24] but also on the surface treatments [25]. In the literature, we found some works in which different routes of manufacture and superficial modification of porous implants have been used. There are also studies that propose computational models and simulations to investigate the stress distribution at the threads, section changes, and pores, etc., or to estimate and understand the static and cyclical behavior in service [26–28]. However, few investigations focus on the fatigue behavior of porous dental implants, or the influence of surface modification treatments on their response.

Unfortunately, although enormous advances in this field have been accomplished, dental implants as perfect medical prosthesis devices still remain as an enormous clinical challenge. In this study, the influence of porosity (content, size, and morphology), surface modification treatments (chemical etching and bioactive coating), and fatigue resistance are studied for porous dental implants that were previously obtained via conventional powder metallurgy and space-holder techniques. Finally, exploratory studies of the cellular characterization (the attachment and proliferation of osteoblasts) of new manufactured dental implants are also addressed. The final objective is to propose an implant with a better balance of bio-mechanical (stiffness, yield strength and fatigue limit) and bio-functional (osseointegration and bone ingrowth) performances.

2. Materials and Methods

In this work, the conventional powder metallurgical route (PM) and space-holder technique (SH: ammonium bicarbonate as a spacer–NH_4HCO_3) were used for the manufacture of the porous titanium dental implants. The titanium grade IV powder and the spacer particles (30 vol.% and 50 vol.%, both with size range between 100 μm and 200 μm) were mixed in a Turbula® T2C Shaker-Mixer for 40 min to achieve good homogenization. Next, the powder mix was uniaxially compressed in a cylindrical die (8 mm in diameter) at 300 MPa using an Instron 5505 universal testing machine. Later, the green samples were micro-milled using a CNC machine (Roland, Model MDX-40, Shinmiyakoda, Japan) to obtain the thread of the dental implant. Before the sintering step (1250 °C for 2 h, and 10^{-5} mbar), the spacer particles were removed in a conventional oven (60 °C and 110 °C, both for 12 h and 10^{-2} mbar). Furthermore, the surfaces of the dental implants were chemically etched or coated with two bioglasses (BG 45S5 and BG 1393). All of the details of the protocols related to the fabrication and surface modification treatments implemented in this study were used by the authors using first porous titanium disks [15,29] and then similar porous dental implants [30]. In these investigations, the details of the porosity measurements (Archimedes' method and image analysis) and surface roughness (scanning electron microscopy and confocal laser) of the dental implants used in this investigation can also be consulted. Three of the superficially modified porous implants are shown in Figure 1.

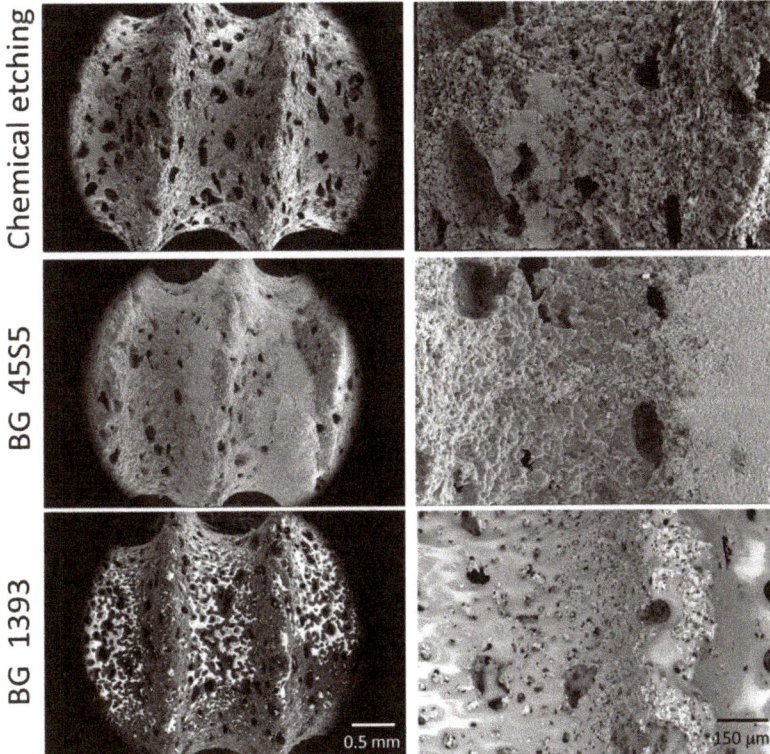

Figure 1. SEM image of the superficially treated porous dental implants.

2.1. Characterization of the Fatigue Behavior of the Porous Dental Implants Studied

In this section, firstly, the study of the mechanical behavior (static and cyclical) of the virgin porous dental implants (without surface modification) is presented, in order to obtain the influence of the pores (percentage, size, and irregularity). Finally, a preliminary study of the fatigue behavior of superficially treated porous dental implants is shown.

The mechanical characterization was performed following the test setup proposed in the ISO 14801 standard (see Figure 2) [31]. In these investigations, the load must be applied with an angle of 30° from the axis of the implant, and the piece applying the load was allowed to rotate about the semispherical part of the dental implant. For its part, the fixing plane was placed at a distance of $l = 11$ mm from the center of this semi-sphere, and 3 mm below the plane where the bone level would be, as in a real application. In this work, the first static tests were conducted at a load rate of 10 N/s in order to estimate the loads in the corresponding fatigue tests. On the other hand, the fatigue resistance testing of porous dental implants obtained by PM and SH (30 vol.% and 50 vol.%) routes was carried out at a load ratio of $R = 0.1$, at a frequency of 15 Hz, and until the complete fracture of the implant. An adequate parameter to compare the mechanical behavior of dental implants is nominal stress, σ, as the sizes of implants are included in this calculation, such that implants with different dimensions could be compared. In order to calculate this stress, the applied force,

F, can be decomposed into a component generating a compression in the dental implant ($F \cdot \cos 30°$) and a component generating a bending moment at the fixing plane of the implant ($F \cdot \sin 30°$). The bending moment was obtained by multiplying this force by the distance to the fixing plane, l. Only the bending moment will be used to calculate the nominal stress because it was the one generating the tensile stress, which produces the fatigue damage. Assuming, in this case, an implant with a solid circular section with diameter $d = 3.45$ mm, the nominal stress can be calculated using the well-known expression for a circular beam subjected to a bending moment, M:

$$\sigma = \frac{32 \times M}{\pi \times d^3} = \frac{32 \times F \times \sin 30 \times l}{\pi \times d^3} \qquad (1)$$

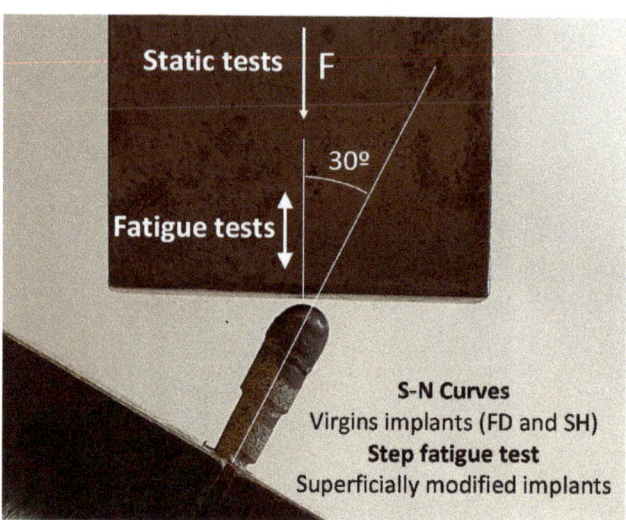

Figure 2. Setup and parameters of the static and fatigue test of the porous dental implants, following the standard ISO 14801.

In this work, seven virgin porous dental implants were tested in order to assess their fatigue resistance. However, considering the additional economic cost of the implemented surface modification treatments (a chemically etched surface or being coated with BG 45S5 or BG 1393), a different fatigue test procedure was performed, to be compared to the conventional tests (S-N curves). In this other test protocol (step fatigue test), the configuration parameters and the load ratio are the same as those used in the conventional test. For the new fatigue test proposed, instead of applying a constant amplitude load, the test starts with a cycle with a maximum load of 40 N, to be later increased by 10% every 50,000 cycles until failure. The fatigue resistance evaluated with the described protocol depends on the potential accumulation of damage in the different previous steps. Then, the results should be rationalized considering this fact. In this context, the maximum stress before failure in this last step is not appropriate to compare the fatigue life with the results of the conventional fatigue tests, because it does not reflect the damage accumulated in the previous steps with lower loads. Therefore, an equivalent constant amplitude stress level that produces the same fatigue damage as in the step test was defined. This equivalent stress can be directly compared to the conventional fatigue tests. In order to achieve this purpose, the concept of fatigue linear damage accumulation was used [32,33]. The fatigue damage was obtained by adding the damage in each load step, with the damage in each step being the ratio between the applied number of cycles, n_i, and the number of cycles to failure, if the corresponding load was the only one applied, N_i. Mathematically, it is possible to calculate the equivalent load that would have to be applied in a constant amplitude load test in order to produce the same damage, D, to the implant in the same total number of cycles, n (this is the sum of all of the n_i). This is shown in Equation (2):

$$D = \sum \frac{n_i}{N_i} = \frac{n}{N_{eq}} \qquad (2)$$

where N_i and N_{eq} can be calculated if the fatigue curve of the material is known, $\sigma \cdot N^b = C$. In Equation (2), N_{eq} is the number of cycles to failure in a fatigue test in which only the equivalent load is applied. Equation (2) was then transformed into Equations (3) and (4):

$$\sum \frac{n_i}{C^{1/b}} \times \sigma_i^{1/b} = \frac{n}{C^{1/b}} \times \sigma_{eq}^{1/b} \qquad (3)$$

$$\sigma_{eq} = \left(\frac{1}{n} \times \sum n_i \times \sigma_i^{1/b} \right)^b \qquad (4)$$

where n_i and σ_i are the number of cycles and the nominal stress in each segment, respectively, and n is the total number of cycles in the test.

Finally, a study of the possible fracture surfaces associated with the monotonic and cyclic tests was carried out by scanning electron microscopy, SEM (Teneo, FEI, Eindhoven, The Netherlands), in order to figure out the origin and the responsible mechanisms of the fracture of the different porous dental implants (with and without surface modification).

2.2. Cellular Characterization of Superficially Modified Porous Dental Implants

In this section, the effect of the porosity and surface treatment on the cell behavior of the dental implants studied is addressed.

2.2.1. In Vitro Cell Culture

The MC3T3-E1 mouse pre-osteoblast cell line was grown (CRL-2593 from the American Type Culture Collection (ATCC), Manassas, VA, USA). The implants were sterilized in an autoclave (121 °C, 1.05 kg·cm^{-2}, 20 min). We seeded 30,000 cells/cm^2 above each implant. In order to calculate the number of cells to be seeded, the area of the implant was considered [30]. The cells were grown in Minimum Essential Medium (αMEM) plus 10% fetal bovine serum (FBS) and antibiotics (100 U/mL penicillin and 100 mg/mL streptomycin sulphate) (Invitrogen, Carlsbad, CA, USA), at 37 °C and 5% CO$_2$. At 48 h, the medium was changed to osteogenic induction with α-MEM medium, 10% FBS, 10 mM ascorbic acid (Merck, Darmstadt, Germany), and 50 μg/mL β-glycerophosphate (StemCell Technologies, Vancouver, BC, Canada). The medium was changed every 2 days. The in-vitro cell experiments were carried out at 21 days.

2.2.2. Cell Differentiation by Alkaline Phosphatase (ALP) Evaluation

The MC3T3 differentiation levels by alkaline phosphatase (ALP) activity were conveniently evaluated using the Alkaline Phosphatase Assay kit (Colorimetric) (Abcam, Cambridge, UK). All of the determinations were performed in triplicate in order to measure the absorbance at 405 nm of 4-nitrophenol. The data were expressed as U/mL of p-nitrophenyl Phosphate (PNPP).

2.2.3. Cell Morphology

After 21 days, the cells were fixed in 10% formalin, followed by a dehydration step using ethanolic solutions (in 30%, 50%, 60%, 70%, 80% and 90% ethanol for 10 min each); then, they were gold-coated using a sputter coater (Pelco 91000, Ted Pella, Redding, CA, USA). The culture was analysed using scanning electron microscopy (SEM) (Zeiss EVO LS 15 scanning electron microscope (Zeiss, Oberkochen, Germany)) with an acceleration voltage of 10 kV.

2.2.4. Statistical Analysis

All of the results are expressed as the mean and standard deviation. The statistical test used was a two-way ANOVA and Tukey's post-test (SPSS v.22.0 for Windows, IBM Corp., Ar-monk, NY, USA). All of the determinations were analysed in triplicate. $p < 0.05$ was considered statistical difference.

3. Results and Discussion

Table 1 shows results of the experimental static behavior (fracture load and nominal stress) of the studied virgin dental implants, together with the collected values of the Young's (E_N) and dynamic Young's modulus (E_d, by Nielsen approximation), and the yield strength (σ_y) of the virgin implants, using equations previously described in the literature [34–36], which establish the relationships between microstructural parameters (the porosity and morphology of the pores) and their mechanical behavior. Although porous implants have reliable yield-strength values, 200 and 135 MPa for SH 30 vol.% and 50 vol.%, respectively, which are close to the values for cortical bone tissue (150–180 MPa [37,38]), the stiffness (90 GPa) and yield strength (638 MPa) of the conventional PM implants were not satisfactory to find a solution to the stress-shielding phenomenon (20–25 GPa, [37,39]). In this case, it would be necessary to manufacture implants with greater porosity, with the intact structural integrity of the implant during the micro-milling stage and/or under service conditions. In light of this, the decrease in compaction pressure, temperature, and/or sintering time, as well as the use of spacers (included in this study) could arise as possible solutions to cope this problem. Furthermore, although it will be discussed below, it is worthwhile to point out that the gradient of porosity of the SH implants could influence both the stiffness and fatigue behavior between the core and the threads (in contact with the bone) of the implants, to create a gradient of the corresponding Young's modulus [29]. Finally, focusing on the implant rigidity, the influence of the size and geometry of the implant should also be considered [23,24].

Table 1. Static behavior of the virgin porous dental implants.

		Fracture Load (N)	Nominal Stress (MPa)	Estimated Mechanical Behavior [34–36]		
				E_N (GPa)	E_d (GPa)	σ_y (MPa)
PM		140 ± 3	191 ± 1	90.9 ± 0.5	86.2 ± 0.6	638 ± 5
SH	30 vol.%	70 ± 4	95.5 ± 1.5	44.6 ± 0.9	45.8 ± 1.0	200 ± 8
	50 vol.%	52 ± 6	71 ± 2	30.3 ± 1.1	35.6 ± 1.0	135 ± 14

Note: The static behavior of superficially modified dental implants is similar to the corresponding virgin implant.

Figure 3 shows the results of the conventional fatigue tests for PM, SH 30 vol.% and SH 50 vol.%, where the load applied in each test is rationalized with the static strength shown in Table 1. This shows that the PM and SH 30 vol.% have a similar qualitative trend: a fatigue limit between 40% and 50% of the static strength, and a similar fatigue behavior in the rest of the curve. The SH 50 vol.% implant has a different behavior with a slightly lower relative fatigue limit, but a much lower fatigue strength compared to its static strength for the lower lives. It seems that, at this level of porosity, a different fatigue mechanism appears compared to the other two types of implants. The trend lines in Figure 3 represent the statistical fit to the experimental results using the typical fatigue curve mentioned earlier, $\sigma \cdot N^b = C$. The coefficients of regression, R^2, of these fitted curves are 0.79, 0.92 and 0.47 for the implants PM, SH 30 vol.% and SH 50 vol.%, respectively. A very high scatter appears in SH 50 vol.%, which explains the poor fit. Table 2 shows the nominal stresses and the number of cycles in the fatigue tests for the three porous implants studied. An inverse relationship is observed between the porosity of the dental implants (the pore content and size) and the fatigue resistance values obtained for 10^5 cycles: 100.3 MPa (PM), 44.9 MPa (SH 30 vol.%), and 29.0 MPa (SH 50 vol.%). The fatigue strength at 10^7 cycles has values of 27–35 MPa and 200–430 MPa for the cortical bone and the commercially pure titanium implants (obtained by a forging process), respectively [40]. In this context, it could be indicated that superficially modified dental implants potentially guarantee the mechanical requirements of the bone tissue to be replaced. An increase in fatigue resistance could even be expected, once the bone in-growth and osseointegration of the implant have occurred.

Table 2. Nominal stresses and the number of cycles in the fatigue tests.

PM		SH 30 vol.%		SH 50 vol.%	
Cycles	MPa	Cycles	MPa	CYCLES	MPa
300	129.3	50	64.1	150	34.1
34,820	113.0	5900	54.6	200	30.0
8324	106.4	42,855	45.0	52,403	27.3
5900	99.6	-	-	315	28.7
310,056 *	86.6 *	-	-	5×10^6	28.7
5×10^6	89.7	-	-		
2.6×10^6	87.2	-	-		

Note: * Step fatigue test for the PM implant.

Furthermore, in Figure 3 a data point called "PM step test" is drawn. This point comes from a step fatigue test, as described in Section 2.1, in which the total number of cycles was 310,056 and the load in the last step was 70.9 N. As was also explained in the experimental section, this equivalent stress can be directly compared to the conventional fatigue tests. In this case, the parameters of the fatigue curves are obtained from the conventional fatigue tests already shown, although, as seen in Equation (4), only the slope of the curve, b, is needed. The values are $b = 0.035$ for PM and $b = 0.049$ for SH. This particular point matches the data trend obtained in the conventional fatigue tests, assuming the typical scatter in fatigue. Therefore, we can conclude that it is perfectly valid to use the technique of the step fatigue test together with the equivalent stress to analyze and compare different implants with different surface treatments using only one fatigue test. However, the ideal would be a complete fatigue curve; this procedure gives the opportunity to achieve a first discrimination when there is little material available. In this sense, in Figure 4, the maximum loads vs. the number of cycles in step fatigue tests of superficially modified implants are presented. The fatigue behavior showed dependency on the accumulated damage over a certain number of cycles with different load levels. Another semi-quantitative comparison could be made using the nominal stress instead of the applied load. As mentioned earlier, this would be useful in the future in order to compare the results with an implant of distinct size. In this context, the results (equivalent stress, σ_{eq}, vs. N) are shown in Figure 5. This equivalent stress is calculated using Equation (4), where the slope of the fatigue curve is assumed to be the same as in the fatigue tests of the virgin implants. Furthermore, the fatigue curves for these treated implants could be estimated using the points in Figure 5 and the already mentioned slope of the fatigue curve, b (0.035 for PM and 0.049 for SH). Table 3 shows the fatigue data of the tests, including the estimated fatigue strength for a fatigue life of 10^5 cycles, assuming the same fatigue slope as in the virgin implants.

Table 3. Step fatigue tests of porous dental implants (PM and SH 30 vol.%).

Porous Dental Implants		Maximum Fatigue Load (N)	Nominal Stress (MPa)	Equivalent Maximum Stress (MPa) (See Figure 5)	Number of Total Cycles	Estimated Fatigue Strength at 10^5 Cycles (MPa)
Virgin	PM	70.9	96.7	86.6	310,056	90.1
Chemical Etching	PM	114.1	155.7	150.5	611,850	160.4
	SH	64.4	87.9	81	298,754	85.5
BG 45S5	PM	58.6	79.9	75	235,260	77.3
	SH	64.4	87.9	80.8	294,670	85.2
BG 1393	PM	94.3	128.7	115.3	466,920	121.7
	SH	103.7	141.5	120.3	510,240	130.3

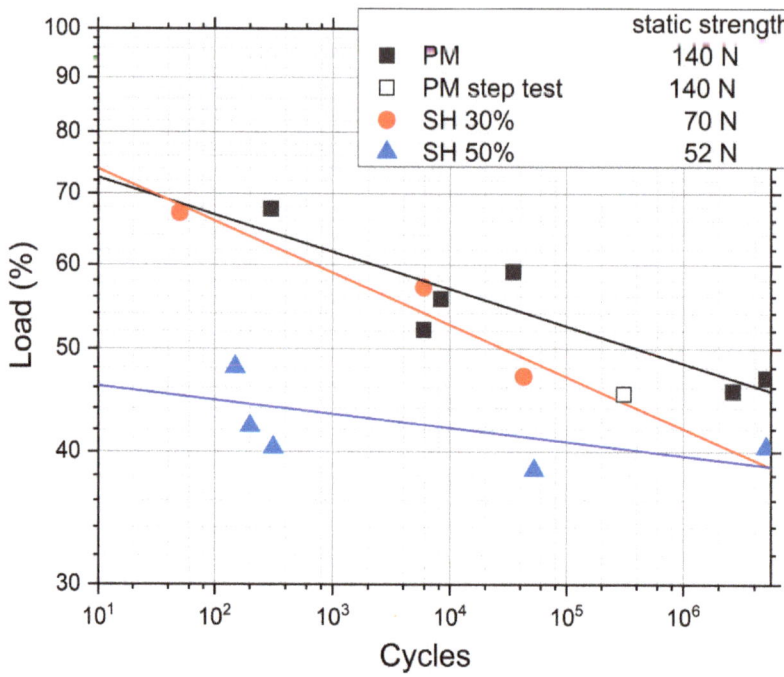

Figure 3. Results of the conventional fatigue tests for implants with different porosities, and the result of one test with the procedure of the step fatigue test. The load percentage is with respect to the static strength.

The general analysis of these fatigue results allowed (Figures 4–6) us to indicate the following aspects:

(i) Modified PM implants: The chemically etched implant and the bioactive glass BG 1393 coated implant presented a higher fatigue resistance than the virgin PM implant, while this was less for the implant coated with BG 45S5. On the other hand, the improvement in the fatigue life of the chemically etched implant may be associated with the formation of a more stable oxide layer on the surface of the implant, usually rutile. This oxide hardened the surface, and thus hindered the movement of dislocations and/or nucleation of micro-cracks under cyclic loads [41]. Comparable results were already reported by Apachitei et al. [42]. They studied in detail the effect of plasma electrolytic oxidation coatings on the fatigue properties of Ti6Al4V and Ti6Al7Nb alloys under physiological conditions (Hank's solution at 37 °C) in order to describe the fact that oxidized Ti6Al7Nb alloys exhibit an improved fatigue behavior if compared to oxidized Ti6Al4V alloys, independently from the coating thickness. Furthermore, the best fatigue behavior of the implant coated with BG 1393 could be explained by its better adhesion with the Ti implant [43,44] compared to BG 45S5. This fact could be associated with the best compatibility between its thermal expansion coefficients [45]. Furthermore, the temperature used during the coating treatment (exceeding the melting temperature of BG 1393) allowed its infiltration into the macro-pores (see Figure 6).

(ii) Modified SH implants: As previously described, the fatigue behavior of SH virgin implants was conditioned by the role of macro-pores (associated with the use of spacer particles). However, the resistance under cyclical loads of the modified implants clearly depended on what happened on their surface and how it took place. In this context, after chemical etching, the macro-pores were larger and more irregular, justifying the sudden drop in mechanical strength (see Figure 6). Furthermore, the intrinsic micro-porosity of the BG 45S5 coating and its poor adherence (see the red arrow in Figure 7) compromised their use for this type of solicitation. Finally, despite the good infiltration and adherence of BG 1393, the presence of pre-existing microcracks—originating in the macro-pores after the thermal treatment of this coating—could explain its resistance to fatigue (see Figure 6).

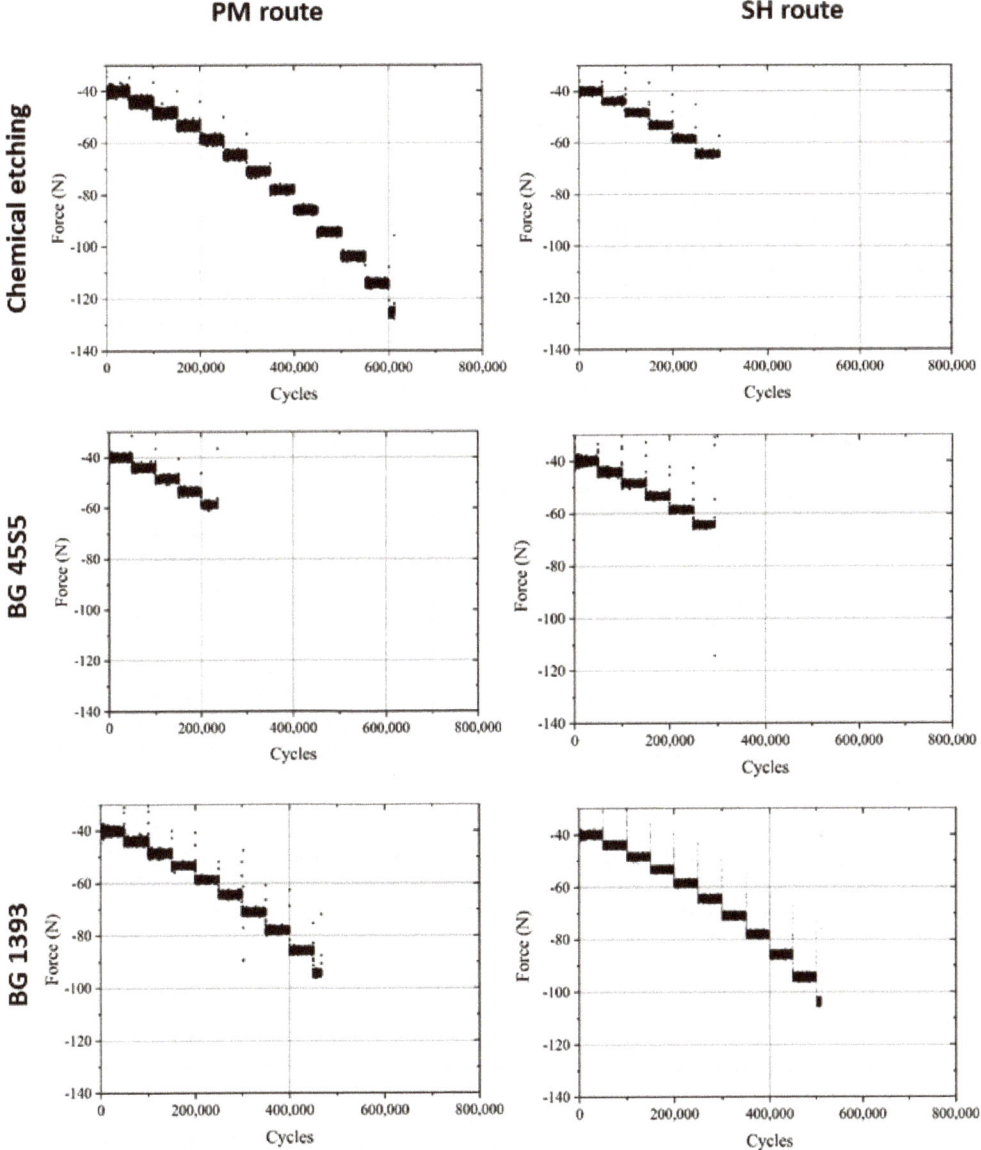

Figure 4. Fatigue behavior of PM and SH (30 vol.%) superficially modified implants (chemical etching, coated with BG 1393 or BG 45S5).

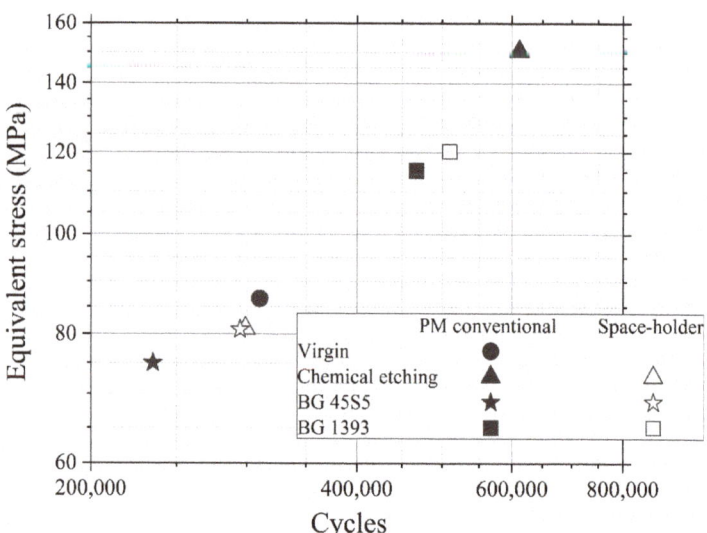

Figure 5. Equivalent stress vs number of cycles. Note: Superficially modified dental implants could guarantee the mechanical requirements (the cortical bone tissue presents a fatigue strength at 10^7 cycles of 27–35 MPa [40]).

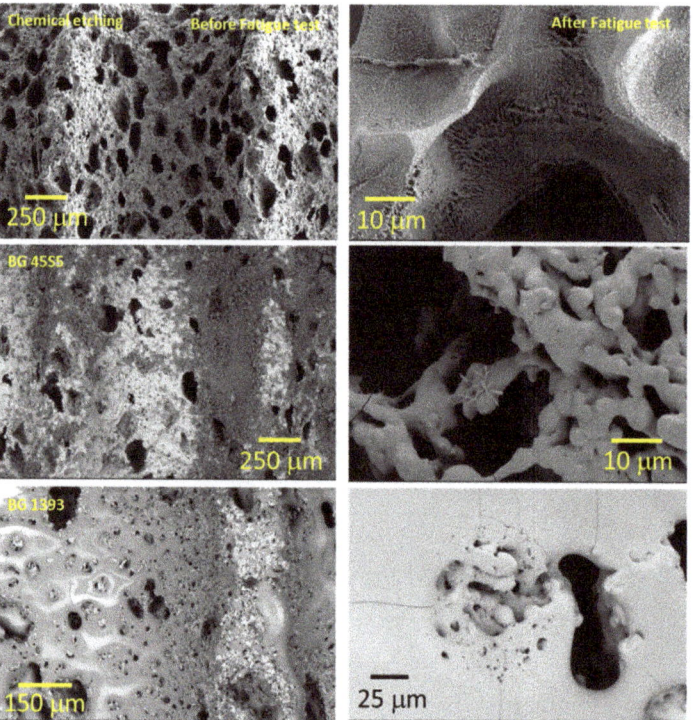

Figure 6. Coated implant surface before and after the fatigue tests (SH: 30 vol.%). Note: Observe the more irregular macro-pores when using chemical etching, and the nucleation of micro cracks in the pores under cyclic loading.

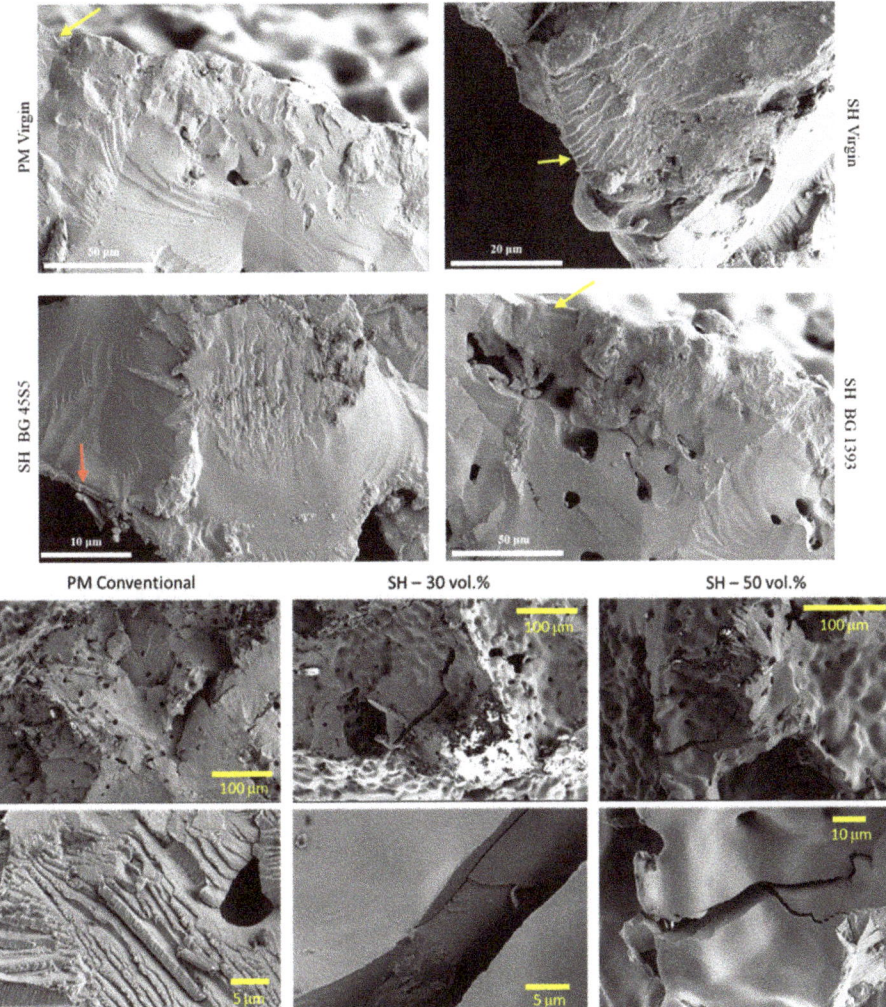

Figure 7. Fracture surfaces under cyclic loads of PM and SH dental implants (virgin and superficially treated), comparing virgin and modified surfaces. Note: Observe the presence of small areas of fatigue striations (see the yellow arrows), and the poor adherence of the BG 45S5 coating (see the red arrow).

It should be noted that the number of cycles that the tested implants resisted in the last step depended on the history of previous fatigue to which each implant was subjected; that is, it depended on the accumulation of damage that was generated on the surface of the material. Previous cyclic solicitations (minor variations in the applied cyclic stresses) may have caused two types of effects:

(1) surface hardening (virgin c.p. Ti implants);
(2) the nucleation and accumulation of damage to the treated surface, chemically or in the interlayer of the coating-implant joint; and
(3) the subcritical growth of pre-existing micro-cracks in the coating.

On the other hand, the fracture surfaces of the studied dental implants are shown in Figure 7. The presence of cleavage was observed, which is the mechanism responsible for the brittle fracture under static mechanical conditions. In this context, it was difficult to identify and measure the size of the defect that caused the fracture under these two types of mechanical solicitation (monotonic and cyclical). However, small areas of subcritical growth (close to the implant surface) could be elucidated, with the presence of fatigue pseudo-striation.

Finally, the ALP activity is used to assess the enhanced in-vitro osseointegration capacity as markers of the early differentiation of osteoblast-like cells. In Figure 8, a general trend of ALP activity is shown, being similar in all the samples; additionally, osteoblast cells have presented an activity of around 2.5 U/L. However, the BG 45S5 implants show the highest cellular activity, although no significant differences were found among the other implants. The higher activity of ALP indicates that they are cultures with more differentiated cells [46], such that the osteoblasts grown on the BG 45S5 implants present greater differentiation and—it is expected—more hydroxyapatite deposits, as was observed on the SEM images (Figure 9). SEM was used to observe the spreading of the MC3T3-E1 cells on each sample. In Figure 9, the presence of osteoblasts cells could be observed on the surface (marked with yellow arrows). Osteoblasts have a triangular shape, which is typical of differentiated cells, with pseudopod protrusion and microfilament extension. Its plasma membrane presents a surface full of sector vesicles, which is indicative of good enzymatic functional activity. In addition, small precipitates with a morphology like hydroxyapatite (marked with red asterisks) are appreciated. However, a future study of XRD should be carried out in order to corroborate this fact. An in vivo test will be performed in the future to validate the reliability of the results. However, their behavior is promising, considering previous results reported in the literature for this type of surface [46].

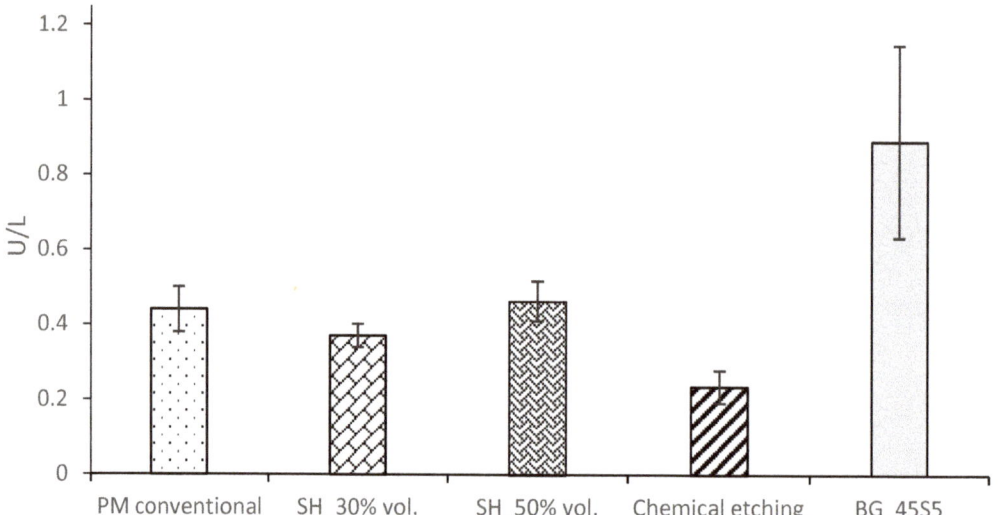

Figure 8. Alkaline phosphatase enzymes (U/mL) in the culture of MC3T3 in different conditions.

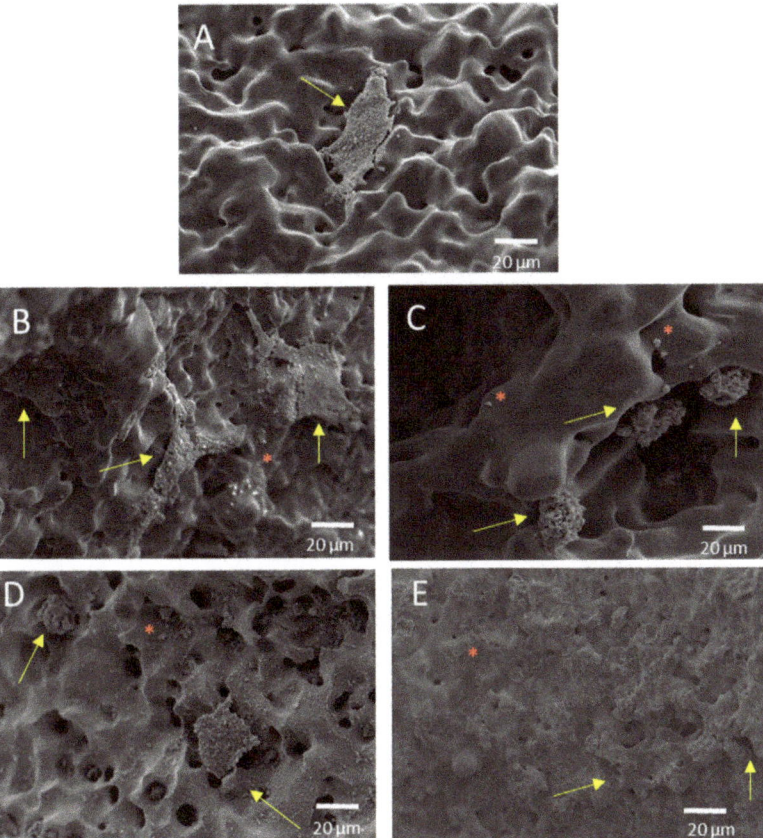

Figure 9. Representative SEM images of the cell adherence in different surfaces of porous dental implants. Virgins: (**A**) PM conventional, (**B**) SH 30 vol.%, (**C**) SH 50 vol.%, and superficially modified. (**D**) Chemical etching and (**E**) BG 45S5, both on conventional PM implants. Note: The cells are indicated in the images (yellow arrow), along with calcium phosphate (red asterisk).

4. Conclusions

The influences of the porosity and surface treatments of dental implants manufactured using PM and SH routes on their fatigue and cellular behavior have provided the following conclusions:

(1) The virgin SH dental implants have a lower fatigue resistance than those obtained by the conventional PM route. The macro-pores control the crack nucleation process, although they can also hinder the propagation of cracks (stop-hole mechanism—the tip of the crack is blunted). On the other hand, the roughness of the walls of these implants favors the adhesion of osteoblasts. Furthermore, an increase in the behavior (ALP activity or cells differentiation) of the in vitro cell cultures is observed after the surface modifications, and the differences between the treatments used are not statistically significant.

(2) The high micro-porosity of the BG 45S5 coating compromised the fatigue behavior of the implant, being 17% less than the value corresponding to PM dental implants without surface treatment. In the case of SH 30 vol.% implants, it also decreased by 65% compared to the virgin implant. On the other hand, the fatigue resistance of conventional PM implants coated with BG 1393 improves by 25%. This increase may

be related to the improved infiltration and/or better thermal compatibility (coefficients of expansion) between Ti and the BG 1393. Finally, the increase of the fatigue resistance of the superficially chemically etched porous dental implant (38% vs. PM virgin) is related to the formation of a hard layer of titanium oxide formed during the chemical treatment of the surface.

In summary, SH 30 vol.% the chemical etching of dental implants presents the best bio-functional (in vitro osseointegration) and bio-mechanical (stiffness, yield strength and fatigue life) balance, which could guarantee the requirements of cortical bone tissue (E = 20–25 GPa, $\hat{O}_y$ = 150–180 MPa, and K_{Ic} = 3.5 MNm$^{-3/2}$).

Author Contributions: Investigation, formal analysis, validation, P.T., C.N., M.G., E.J.D.-P. and L.M.R.-A.; writing—original draft preparation, writing—review and editing, C.N., M.G., P.T., J.A.R.-O., L.M.R.-A., E.J.D.-P. and Y.T.; conceptualization, methodology, project administration, and funding acquisition, C.N., J.A.R.-O., M.J.M.-G. and Y.T. All authors have read and agreed to the published version of the manuscript.

Funding: This work was supported by the Ministry of Science and Innovation of Spain [grant number PID2019-109371GB-I00] and by the Junta de Andalucía (Spain) through the Project PAIDI [grant Ref. P20_00671].

Institutional Review Board Statement: Not applicable.

Informed Consent Statement: Not applicable.

Data Availability Statement: Not applicable.

Acknowledgments: The authors also thank the investigation support technician J.M. Manzano for the contribution to the implant fabrication.

Conflicts of Interest: The authors declare no conflict of interest.

References

1. Kaur, M.; Singh, K. Review on titanium and titanium based alloys as biomaterials for orthopaedic applications. *Mater. Sci. Eng. C* **2019**, *102*, 844–862. [CrossRef] [PubMed]
2. Niinomi, M.; Nakai, M.; Hieda, J. Development of new metallic alloys for biomedical applications. *Acta Biomater.* **2012**, *8*, 3888–3903. [CrossRef] [PubMed]
3. Pałka, K.; Pokrowiecki, R. Porous titanium implants: A review. *Adv. Eng. Mater.* **2018**, *20*, 1700648. [CrossRef]
4. Ghouse, S.; Reznikov, N.; Boughton, O.R.; Babu, S.; Ng, K.G.; Blunn, G.; Cobb, J.P.; Stevens, M.M.; Jeffers, J.R. The design and in vivo testing of a locally stiffness-matched porous scaffold. *Appl. Mater. Today* **2019**, *15*, 377–388. [CrossRef]
5. Mohammed, M.T.; Khan, Z.A.; Siddiquee, A.N. Beta titanium alloys: The lowest elastic modulus for biomedical applications: A review. *Int. J. Chem. Mol. Nucl. Mater. Metall. Eng.* **2014**, *8*, 726.
6. Biesiekierski, A.; Lin, J.; Li, Y.; Ping, D.; Yamabe-Mitarai, Y.; Wen, C. Investigations into Ti–(Nb, Ta)–Fe alloys for biomedical applications. *Acta Biomater.* **2016**, *32*, 336–347. [CrossRef]
7. Gudkov, S.V.; Simakin, A.V.; Sevostyanov, M.A.; Konushkin, S.V.; Losertová, M.; Ivannikov, A.Y.; Kolmakov, A.G.; Izmailov, A.Y. Manufacturing and study of mechanical properties, structure and compatibility with biological objects of plates and wire from new Ti-25Nb-13Ta-5Zr alloy. *Metals* **2020**, *10*, 1584. [CrossRef]
8. Sevostyanov, M.A.; Kolmakov, A.G.; Sergiyenko, K.V.; Kaplan, M.A.; Baikin, A.S.; Gudkov, S.V. Mechanical, physical–chemical and biological properties of the new Ti–30Nb–13Ta–5Zr alloy. *J. Mater. Sci.* **2020**, *55*, 14516–14529. [CrossRef]
9. Giner, M.; Chicardi, E.; Costa, A.d.F.; Santana, L.; Vázquez-Gámez, M.Á.; García-Garrido, C.; Colmenero, M.A.; Olmo-Montes, F.J.; Torres, Y.; Montoya-García, M.J. Biocompatibility and cellular behavior of tinbta alloy with adapted rigidity for the replacement of bone tissue. *Metals* **2021**, *11*, 130. [CrossRef]
10. Goharian, A.; Abdullah, M.R. 7—Bioinert Metals (Stainless Steel, Titanium, Cobalt Chromium). In *Trauma Plating Systems*; Goharian, A., Ed.; Elsevier: Amsterdam, The Netherlands, 2017; pp. 115–142.
11. Olmedo, M.M.; Godino, F.I.; Liétor, P.F.; Iglesias, F.C. Corrosion and fracture analysis in screws of dental implants prostheses. New coatings. *Eng. Fail. Anal.* **2017**, *82*, 657–665. [CrossRef]
12. Le Guehennec, L.; Soueidan, A.; Layrolle, P.; Amouriq, Y. Surface treatments of titanium dental implants for rapid osseointegration. *Dent. Mater.* **2007**, *23*, 844–854. [CrossRef] [PubMed]
13. Wang, M.; Tang, T. Surface treatment strategies to combat implant-related infection from the beginning. *J. Orthop. Transl.* **2019**, *17*, 42–54. [CrossRef] [PubMed]
14. Smeets, R.; Stadlinger, B.; Schwarz, F.; Beck-Broichsitter, B.; Jung, O.; Precht, C.; Kloss, F.; Gröbe, A.; Heiland, M.; Ebker, T. Impact of dental implant surface modifications on osseointegration. *BioMed Res. Int.* **2016**, *2016*, 6285620. [CrossRef]

15. Civantos, A.; Domínguez, C.; Pino, R.J.; Setti, G.; Pavón, J.J.; Martínez-Campos, E.; Garcia, F.J.G.; Rodríguez, J.A.; Allain, J.P.; Torres, Y. Designing bioactive porous titanium interfaces to balance mechanical properties and in vitro cells behavior towards increased osseointegration. *Surf. Coat. Technol.* **2019**, *368*, 162–174. [CrossRef]
16. Lobato, J.; Hussain, N.S.; Botelho, C.; Maurício, A.; Lobato, J.; Lopes, M.; Afonso, A.; Ali, N.; Santos, J. Titanium dental implants coated with Bonelike®: Clinical case report. *Thin Solid Film.* **2006**, *515*, 279–284. [CrossRef]
17. Jones, J.R. Reprint of: Review of bioactive glass: From Hench to hybrids. *Acta Biomater.* **2015**, *23*, S53–S82. [CrossRef]
18. Torres, Y.; Sarria, P.; Gotor, F.J.; Gutiérrez, E.; Peon, E.; Beltrán, A.M.; González, J.E. Surface modification of Ti-6Al-4V alloys manufactured by selective laser melting: Microstructural and tribo-mechanical characterization. *Surf. Coat. Technol.* **2018**, *348*, 31–40. [CrossRef]
19. Sarma, J.; Kumar, R.; Sahoo, A.K.; Panda, A. Enhancement of material properties of titanium alloys through heat treatment process: A brief review. *Mater. Today Proc.* **2020**, *23*, 561–564. [CrossRef]
20. Gherde, C.; Dhatrak, P.; Nimbalkar, S.; Joshi, S. A comprehensive review of factors affecting fatigue life of dental implants. *Mater. Today Proc.* **2021**, *43*, 1117–1123. [CrossRef]
21. Shemtov-Yona, K.; Rittel, D. Identification of failure mechanisms in retrieved fractured dental implants. *Eng. Fail. Anal.* **2014**, *38*, 58–65. [CrossRef]
22. Chen, Q.; Thouas, G.A. Metallic implant biomaterials. *Mater. Sci. Eng. R Rep.* **2015**, *87*, 1–57. [CrossRef]
23. Li, J.; Jansen, J.A.; Walboomers, X.F.; van den Beucken, J.J.J.P. Mechanical aspects of dental implants and osseointegration: A narrative review. *J. Mech. Behav. Biomed. Mater.* **2020**, *103*, 103574. [CrossRef] [PubMed]
24. Rojo, R.; Prados-Privado, M.; Reinoso, A.J.; Prados-Frutos, J.C. Evaluation of Fatigue Behavior in Dental Implants from In Vitro Clinical Tests: A Systematic Review. *Metals* **2018**, *8*, 313. [CrossRef]
25. Rokaya, D.; Srimaneepong, V.; Wisitrasameewon, W.; Humagain, M.; Thunyakitpisal, P. Peri-implantitis update: Risk indicators, diagnosis, and treatment. *Eur. J. Dent.* **2020**, *14*, 672–682. [CrossRef] [PubMed]
26. Velmurugan, D.; Alphin, M.S. Influence of geometric design variable and bone quality on stress distribution for zirconia dental implants-A 3D finite element analysis. *Comput. Modeling Eng. Sci.* **2018**, *117*, 125–141. [CrossRef]
27. Hedayati, R.; Hosseini-Toudeshky, H.; Sadighi, M.; Mohammadi-Aghdam, M.; Zadpoor, A. Computational prediction of the fatigue behavior of additively manufactured porous metallic biomaterials. *Int. J. Fatigue* **2016**, *84*, 67–79. [CrossRef]
28. Zhang, J.; Fatemi, A. Surface roughness effect on multiaxial fatigue behavior of additive manufactured metals and its modeling. *Theor. Appl. Fract. Mech.* **2019**, *103*, 102260. [CrossRef]
29. Domínguez-Trujillo, C.; Ternero, F.; Rodríguez-Ortiz, J.A.; Pavón, J.J.; Montealegre-Meléndez, I.; Arévalo, C.; García-Moreno, F.; Torres, Y. Improvement of the balance between a reduced stress shielding and bone ingrowth by bioactive coatings onto porous titanium substrates. *Surf. Coat. Technol.* **2018**, *338*, 32–37. [CrossRef]
30. Trueba, P.; Navarro, C.; Rodríguez-Ortiz, J.A.; Beltrán, A.M.; García-García, F.J.; Torres, Y. Fabrication and characterization of superficially modified porous dental implants. *Surf. Coat. Technol.* **2021**, *408*, 126796. [CrossRef]
31. *AENOR UNE-EN ISO 14801*; Dentistry—Implants—Dynamic Fatigue Test for Endosseous Dental Implants. ISO: Geneva, Switzerland, 2007.
32. Miner, M.A. Cumulative Damage in Fatigue. *J. Appl. Mech.* **1945**, *12*, A159–A164. [CrossRef]
33. Topoliński, T.; Cichański, A.; Mazurkiewicz, A.; Nowicki, K. Applying a stepwise load for calculation of the SN curve for trabecular bone based on the linear hypothesis for fatigue damage accumulation. *Mater. Sci. Forum* **2012**, *726*, 39–42. [CrossRef]
34. Nielsen, L.F. On strength of porous material: Simple systems and densified systems. *Mater. Struct.* **1998**, *31*, 651–661. [CrossRef]
35. Lascano, S.; Arévalo, C.; Montealegre-Melendez, I.; Muñoz, S.; Rodriguez-Ortiz, J.A.; Trueba, P.; Torres, Y. Porous titanium for biomedical applications: Evaluation of the conventional powder metallurgy frontier and space-holder technique. *Appl. Sci.* **2019**, *9*, 982. [CrossRef]
36. Trueba, P.; Beltrán, A.M.; Bayo, J.M.; Rodríguez-Ortiz, J.A.; Larios, D.F.; Alonso, E.; Dunand, D.C.; Torres, Y. Porous titanium cylinders obtained by the freeze-casting technique: Influence of process parameters on porosity and mechanical behavior. *Metals* **2020**, *10*, 188. [CrossRef]
37. Black, J.; Hastings, G. *Handbook of Biomaterial Properties*; Springer Science & Business Media: Berlin/Heidelberg, Germany, 2013.
38. Collings, E. *The Physical Metallurgy of Titanium Alloys*; ASM Series in Metal Processin; ASM International: Novelty, OH, USA, 1984; Volume 3, ISBN-13: 978-0871701817.
39. Rho, J.Y.; Ashman, R.B.; Turner, C.H. Young's modulus of trabecular and cortical bone material: Ultrasonic and microtensile measurements. *J. Biomech.* **1993**, *26*, 111–119. [CrossRef]
40. Niinomi, M. Fatigue characteristics of metallic biomaterials. *Int. J. Fatigue* **2007**, *29*, 992–1000. [CrossRef]
41. Pramanik, A.; Islam, M.; Basak, A.; Littlefair, G. Machining and tool wear mechanisms during machining titanium alloys. *Adv. Mater. Res.* **2013**, *651*, 338–343. [CrossRef]
42. Apachitei, I.; Lonyuk, B.; Fratila-Apachitei, L.; Zhou, J.; Duszczyk, J. Fatigue response of porous coated titanium biomedical alloys. *Scr. Mater.* **2009**, *61*, 113–116. [CrossRef]
43. Domínguez-Trujillo, C.; Ternero, F.; Rodríguez-Ortiz, J.A.; Heise, S.; Boccaccini, A.R.; Lebrato, J.; Torres, Y. Bioactive coatings on porous titanium for biomedical applications. *Surf. Coat. Technol.* **2018**, *349*, 584–592. [CrossRef]
44. Beltrán, A.M.; Begines, B.; Alcudia, A.; Rodríguez-Ortiz, J.A.; Torres, Y. Biofunctional and tribomechanical behavior of porous titanium substrates coated with a bioactive glass bilayer (45S5–1393). *ACS Appl. Mater. Interfaces* **2020**, *12*, 30170–30180. [CrossRef]

45. Bellucci, D.; Cannillo, V.; Sola, A. Coefficient of thermal expansion of bioactive glasses: Available literature data and analytical equation estimates. *Ceram. Int.* **2011**, *37*, 2963–2972. [CrossRef]
46. Hisham Zainal Ariffin, S.; Manogaran, T.; Zarina Zainol Abidin, I.; Megat Abdul Wahab, R.; Senafi, S. A perspective on stem cells as biological systems that produce differentiated osteoblasts and odontoblasts. *Curr. Stem Cell Res. Ther.* **2017**, *12*, 247–259. [CrossRef] [PubMed]

Article

Influence of Femtosecond Laser Modification on Biomechanical and Biofunctional Behavior of Porous Titanium Substrates

Ana M. Beltrán [1], Mercè Giner [2], Ángel Rodríguez [3,*], Paloma Trueba [1], Luisa M. Rodríguez-Albelo [1,*], Maria Angeles Vázquez-Gámez [4], Vanda Godinho [1], Ana Alcudia [5], José M. Amado [3], Carmen López-Santos [6,7] and Yadir Torres [1]

[1] Departamento de Ingeniería y Ciencia de los Materiales y del Transporte, Escuela Politécnica Superior, Universidad de Sevilla, 41011 Seville, Spain; abeltran3@us.es (A.M.B.); ptrueba@us.es (P.T.); vfortio@us.es (V.G.); ytorres@us.es (Y.T.)
[2] Departamento de Citología e Histología Normal y Patológica, Universidad de Sevilla, 41009 Seville, Spain; mginer@us.es
[3] Departamento Ingeniería Naval e Industrial, Escuela Politécnica Superior, Campus Industrial, Universidade da Coruña, 15403 Ferrol, Spain; jose.amado.paz@udc.es
[4] Departamento de Medicina, Universidad de Sevilla, 41009 Seville, Spain; mavazquez@us.es
[5] Departamento de Química Orgánica y Farmacéutica, Facultad de Farmacia, Universidad de Sevilla, 41005 Seville, Spain; aalcudia@us.es
[6] Departamento de Física Aplicada I, Escuela Politécnica Superior, Universidad de Sevilla, 41011 Seville, Spain; mlopez13@us.es
[7] Nanotecnología en Superficies y Plasma, Instituto de Ciencia de Materiales de Sevilla, 41092 Seville, Spain
* Correspondence: angel.rcarballo@udc.es (Á.R.); lralbelo@us.es (L.M.R.-A.)

Citation: Beltrán, A.M.; Giner, M.; Rodríguez, Á.; Trueba, P.; Rodríguez-Albelo, L.M.; Vázquez-Gámez, M.A.; Godinho, V.; Alcudia, A.; Amado, J.M.; López-Santos, C.; et al. Influence of Femtosecond Laser Modification on Biomechanical and Biofunctional Behavior of Porous Titanium Substrates. *Materials* 2022, *15*, 2969. https://doi.org/10.3390/ma15092969

Academic Editor: Jānis Andersons

Received: 19 March 2022
Accepted: 13 April 2022
Published: 19 April 2022

Publisher's Note: MDPI stays neutral with regard to jurisdictional claims in published maps and institutional affiliations.

Copyright: © 2022 by the authors. Licensee MDPI, Basel, Switzerland. This article is an open access article distributed under the terms and conditions of the Creative Commons Attribution (CC BY) license (https://creativecommons.org/licenses/by/4.0/).

Abstract: Bone resorption and inadequate osseointegration are considered the main problems of titanium implants. In this investigation, the texture and surface roughness of porous titanium samples obtained by the space holder technique were modified with a femtosecond Yb-doped fiber laser. Different percentages of porosity (30, 40, 50, and 60 vol.%) and particle range size (100–200 and 355–500 μm) were compared with fully-dense samples obtained by conventional powder metallurgy. After femtosecond laser treatment the formation of a rough surface with micro-columns and micro-holes occurred for all the studied substrates. The surface was covered by ripples over the micro-metric structures. This work evaluates both the influence of the macro-pores inherent to the spacer particles, as well as the micro-columns and the texture generated with the laser, on the wettability of the surface, the cell behavior (adhesion and proliferation of osteoblasts), micro-hardness (instrumented micro-indentation test, P–h curves) and scratch resistance. The titanium sample with 30 vol.% and a pore range size of 100–200 μm was the best candidate for the replacement of small damaged cortical bone tissues, based on its better biomechanical (stiffness and yield strength) and biofunctional balance (bone in-growth and in vitro osseointegration).

Keywords: porous titanium; femtosecond laser; surface modification; instrumented micro-indentation; scratch test; wettability; cellular behavior

1. Introduction

Today, the demand for implants obtained from natural and synthetic biomaterials for different parts of the human body is exponentially increasing. Within metallic biomaterials, titanium (Ti) and its alloys are considered one of the best choices for the manufacture of dental and bone implants due to their acceptable biomechanical behavior and corrosion resistance in biological surroundings [1,2]. However, there are still challenging problems to solve, such as bone resorption of tissues adjacent to the implant, related to the phenomenon of stress-shielding [3], as well as implant loosening caused by poor osseointegration and/or bacteria proliferation. In this context, the use of β-titanium alloys [4–7] and porous titanium implants [8–10] have been widely recognized as valid approaches for eliminating the effect of stress-shielding on titanium implants. The latter also allows proper vascularization through the interconnected pores, for desirable bone in-growth [11]. Furthermore, good implant osseointegration implies adequate adhesion, proliferation, and differentiation of bone tissue cells have been achieved on its surface [12,13].

Several methods were proposed to improve the osseointegration of prosthetics, mainly based on surface, chemical, or physical modifications. On the one hand, chemical techniques alter the composition of implant surfaces by coating, impregnation, immersion, or deposition of bioactive glasses, ceramics, polymers, or peptides [1,14–16]. On the other hand, among the techniques to modify the texture and roughness of the implant surface, it is worth highlighting sand- and grit-blasting, acid-etching, ultraviolet treatment, electrochemical anodizing, spark anodizing, direct irradiation synthesis methods, and laser surface modification [17–24]. The main goal of physical modifications of the surface of Ti and Ti-alloy implants is the creation of micro- and nano-structures to stimulate osseointegration [25] by increasing the porosity for cell adhesion and proliferation or adapting roughness to better wettability, protein adsorption, and bactericidal response. Furthermore, the high roughness in terms of patterned surfaces is also suitable for preventing bacterial colonization [26–28].

Recently, laser surface modifications were exponentially employed due to their superior advantages over other physical techniques such as more accurate control of specific topological designed features on the surface, high efficiency, and low material consumption [29]. In particular, femtosecond laser (FSL) ablation of Ti surfaces has been widely investigated in the past two decades as this technique allows for high precision and control of desired patterns on the surfaces, as well as it being low cost and having a reliable process [24,30]. A wide variety of micro- and nano-structures [29,31] can be designed on Ti surfaces depending on the FSL beam parameters [31–35]. Different authors, [36–38], have validated the improvement of biocompatibility of Ti and Ti-alloy surfaces using the FSL technique, generating bioinspired micro- and nano-features such as laser-induced periodic surface structure (LIPSS), ripples, columns, pits, or spikes among others [39,40]. In particular, Liang et al. [41] demonstrated better osseointegration and cell proliferation of pure Ti implants using FSL ablation combined with Ca/P deposition. The formation of a micro-pattern on the Ti surface helped in accelerating the cellular integration compared with those of pure Ti and sand-blasted Ti implants. Furthermore, Wang et al. [42] have demonstrated the capabilities of micro-grooved Ti6Al4V implants by FSL and chemical assembly of graphite (G) and graphene oxide (GO), which could improve cell adhesion, proliferation, and osteogenic differentiation and also induce surface wettability and bone-like apatite formation.

However, an appropriate roughness of metallic implant surfaces could allow for the control of wettability [43] and therefore enhance the hydrophilicity or hydrophobicity [44,45] of these surfaces. This aspect is very interesting to prevent bacterial adhesion and biofilm formation [9]. In this regard, Cunha et al. [46] tested the adhesion and biofilm formation of *Staphylococcus aureus* on surfaces of FSL-patterned titanium alloys compared to polished ones. The nano topography size of single features and the distance between them induce a significantly reduced contact area interface between the individual bacterium and the metal to make bacteria agglomeration difficult for ulterior biofilm formation.

Few works have reported on the study of surface modification of porous titanium implants. In particular, the authors of this work have performed preliminary studies in which hierarchical micro- and nano-structures such as micro-holes, micro-columns, and laser-induced periodic surface structure (LIPSS) could be created on porous titanium surfaces. Despite these promising results, understanding the phenomena that occur on femtosecond laser-modified surfaces remains a challenge [47]. Therefore, this current work studies how the macro-porosity (percentage and range size) of titanium substrates influences the final surface roughness generated by laser irradiation, as well as its relationship with the tribo-mechanical behavior, wettability of the surface, and the in vitro cellular response.

2. Materials and Methods

2.1. Substrates Preparation

Figure 1 shows a diagram of the procedure followed by substrate fabrication, superficial modification, as well as tribo-mechanical and in vitro cell characterization. Commercially pure Ti (C.p. Ti), (SEJONG Materials Co. Ltd. Seoul, Korea) with a chemical composition according to the standard [48] and a particle size distribution $d_{(50)}$ = 23.3 µm and $d_{(90)}$ = 48.8 µm [49] was employed to fabricate the fully-dense discs as reference samples, as well as porous substrates, using powder metallurgy technology (PM). Fully-dense samples were fabricated by pressing at 1300 MPa and then sintering at 1300 °C. Porous samples were manufactured using the space holder technique (SH) with ammonium bicarbonate (NH_4HCO_3) (Cymit Quimica S.L., Barcelona, Spain) as spacer particles in different volume percentages (30, 40, 50, and 60 vol.%) and range sizes (100–200 and 355–500 µm). C.p. Ti was mixed with the corresponding amount of spacer particles (percentage and range size), pressed at 800 MPa, and then sintered at 1250 °C for 2 h in a high vacuum atmosphere (~10^{-5} mbar). The spacer particles were removed before sintering in two stages (60 °C and 110 °C) for 12 h each. Before laser modification, the surface of the discs was carefully ground and polished to preserve the porosity fraction, size, and morphology of the pores.

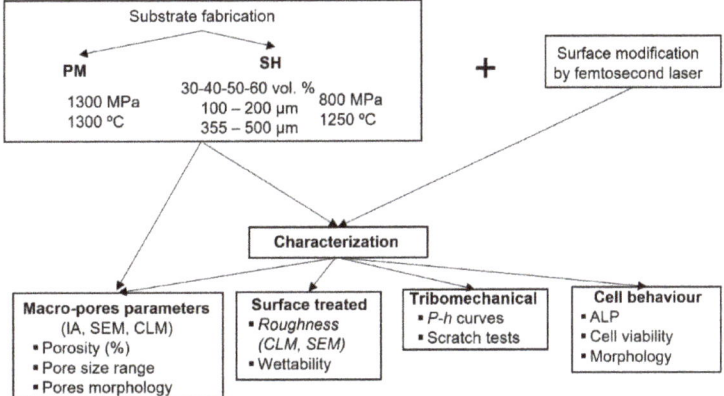

Figure 1. Scheme of the procedure followed by sample fabrication, modification, and characterization. Abbreviations: PM (Pulvimetallurgy), SH (Space Holder), IA (Image Analysis), SEM (Scanning Electron Microscopy), CLM (Confocal Laser Microscopy), ALP (Alkaline Phosphatase) and P-h (micro-indentation curve Power or Force applied vs. penetration deep).

Next, both types of titanium substrates were subjected to surface modification using a femtosecond Yb-doped fiber laser (Spirit 1040-4, Spectra-Physics, Santa Clara, CA, USA). The laser system generates 396 fs pulses with a maximum pulse energy of E_p = 49.7 µJ, at a repetition rate of f = 100 kHz. A computer-controlled galvanometric scanning system was used to direct the laser beam across the target surface. A flat field lens kept the laser focused on the surface to a spot with radius w0 = 12 µm. The irradiation was carried out along parallel lines at a constant speed of v = 960 mm/s in the scan direction and the laser paths were laterally overlapped with an overlap of s = 50% until the entire surface of the workpiece was processed. The surface was irradiated 20 times after this procedure, resulting in 100 accumulated laser pulses per spot and fluency of F = 21.98 J/cm^2. Experiments were performed in air using an Ar jet to reduce surface oxidation These parameters were selected after performing preliminary tests with the goal of obtaining a hierarchical surface structure consisting of both laser-induced micro-structures and laser-induced periodic surface structures (LIPSS) at the nano-metric level.

The modification of the surface due to the femtosecond laser treatment was evaluated by scanning electron microscopy (SEM), (FEI TENEO, Eindhoven, The Netherlands) and confocal laser microscopy (CLM), (Sensofat S Nexox; Barcelona, Spain). CLM allowed the acquisition of two-dimensional (2D) and three-dimensional (3D) images and parameters related to roughness such as the arithmetical mean deviation (S_a) and the root-mean-square height (S_q). Furthermore, the percentage of total porosity (P_T) and the equivalent diameter of the pores (D_{eq}) of the surface (before and after femtosecond treatment) were evaluated, using SEM images and Image-Pro Plus 6.2 software (Rockville, MD, USA).

Wettability was evaluated by static contact angle (CA) measurements obtained with an OCA 20 (Data Physics Instruments GmbH, Filderstadt, Germany) goniometer set up by depositing macroscopic droplets on the surface of the samples according to Young's method. Measurements were based on a minimum of 3 data points per sample, taking the average as the CA value. Totals of 2 μL bidistilled water (pH 7) and 5 μL bovine serum albumin (Merck Life Science S.L.U., Madrid, Spain) droplets were used.

2.2. Tribo-Mechanical Characterization of Modified Substrates

First, macro-mechanical behavior of the porous c.p. Ti substrates (yield strength, σ_y, and dynamic Young's modulus, E_d) were estimated from porosity data, using equations already reported in the literature [50]. For these equations, Young's modulus for bulk c.p. Ti grade IV was ~ 110 GPa [51] and the yield strength of the bulk c.p. Ti grade IV was ~650 GPa [52]. The micro-mechanical characterization and scratch-resistance of the surface of modified substrates were evaluated using instrumented micro-indentation (P–h curves) and scratch tests, respectively. Static loading–unloading tests were performed on a MICROTEST machine (Microtest Company, Madrid, Spain). A preload of 0.05 N was used to ensure contact between the Vickers indenter and the surface, 0.9 N being the maximum load, which was applied with a rate of 0.5 N/min and a dwell time of 40 s. The micro-hardness and Young's modulus were calculated from these data applying Oliver and Pharr method [53]. Additionally, scratch resistance was measured using the same commercial MICROTEST device with a constant applied load of 3 N at a rate of 0.5 mm/min for a scar of 3 mm, using a Rockwell diamond tip of 200 μm of diameter. Variation width was recorded with applied load of the in situ penetration depth and permanent plastic deformation depth. In addition, elastic recovery of the material and the damage inherent to the imposed tribo-mechanical stresses were also evaluated by SEM and CLM.

2.3. Cellular Behavior of Modified Surfaces by Femtosecond Laser Texture

Finally, cell behavior was evaluated in terms of growth, proliferation, and morphology using different techniques.

2.4. In Vitro Cell Culture Techniques

MC3T3-E1, a murine pre-osteoblast cell line (CRL-2593 from the American Type Culture Collection (ATCC), Manassas, VA, USA), was used to analyze the possible influence of surface modified with FSL on bone cells. All c.p. Ti substrates (fully-dense and porous with different percentages of porosity and pores range size) were tested for cell metabolism and viability during the cell adhesion, proliferation, and differentiation process.

2.5. Cell Culture

Routine cell line passaging was performed in 100 mm plates with Minimum Essential Medium (αMEM), containing 10% fetal bovine serum (FBS) plus antibiotics (100 U/mL penicillin and 100 mg/mL streptomycin sulfate) (Invitrogen, Carlsbad, CA, USA). The discs were autoclaved at 121 °C for 30 min and then placed on a 24-well plate. Osteoblast cells were seeded at a cellular density of 35,000 cells/cm². Plates were kept at 37 °C and 5% CO_2 atmosphere. Fully dense c.p. Ti discs were used as a reference.

At 48 h of osteoblast culture, they were induced to undergo differentiation using osteogenic induction medium consisting of α-MEM medium, 10% Fetal Calf Serum (FCS),

10 mM ascorbic acid (Merck, Darmstadt, Germany), and 50 µg/mL of β-glycerophosphate (StemCell Technologies, Vancouver, BC, Canada). The medium was replaced every 2 days. The in vitro cell experiments were carried out at 21 days.

2.6. Cell Viability and Proliferation Assay

Cell proliferation and viability tests were evaluated using AlamarBlue® reagent (Invitrogen, Carlsbad, CA, USA). According to the manufacturer's protocol. Subsequently, the absorbance at 570 nm (oxidized) and 600 nm (reduced) (TECAN, Infinity 200 Pro, Männedorf, Switzerland) was recorded.

2.7. Cell Differentiation by Alkaline Phosphatase (ALP) Evaluation

MC3T3 differentiation levels were evaluated by alkaline phosphatase (ALP) activity, using the Alkaline Phosphatase Assay Kit (Colorimetric) (Abcam, Cambridge, UK). The assay was performed in triplicate according to the manufacturer's protocol. The absorbance at 405 nm of 4-nitrophenol was measured in a 96-well microplate reader. Data were expressed as µmol/min/mL of *p*-Nitrophenyl Phosphate (*p*NPP).

2.8. Cell Morphology

Scanning electron microscopy images acquired with a Zeiss EVO LS 15 scanning electron microscope (SEM), (Zeiss, Oberkochen, Germany) with an acceleration voltage of 10 kV were used to evaluate cell behavior at 21 days. The samples were fixed in 10% formalin, followed by a dehydration step with ethanolic solutions, and then coated with gold-coating using a sputter coater (Pelco 91000, Ted Pella, Redding, CA, USA) before the SEM study.

2.9. Statistical Analysis

All experiments were carried out in triplicate to ensure reproducibility. The results were expressed in terms of mean and standard deviation to perform a two-way ANOVA followed by Tukey's post-test using SPSS v.22.0 for Windows (IBM Corp., Armonk, NY, USA). The significance level was considered at p-values of $p < 0.05$ (*) and $p < 0.01$ (**).

3. Results and Discussion

SEM (Figures 2a and 3) and CLM (Figures 2b and 4) images acquired after femtosecond laser treatment displayed the formation of a rough surface with micro-columns and micro-holes for all the studied substrates. The surface was covered with a pattern formed by nano-metric ripples over the micro-metric structures. These laser-induced ripples were periodic surface structures that appeared when the surface was subjected to ultra-short laser pulses. Moreover, the periodic ripples were aligned perpendicularly to the polarization of the laser beam. In the case of porous substrates, the pores resulting from the spacer particles were also observed. In the case of the fully-dense samples, the generation of a roughness pattern on the surface of the substrates was clearly observed (Figure 2a) with the formation of micro-holes and micro-columns.

In the case of the porous c.p. Ti substrates (Figures 3 and 4) independently of the pore size and percentage of porosity, the results were also evident in terms of the formation of these micro-structures, micro-columns, micro-holes, and ripples, among the pores generated by the spacer particles. The structures generated by the laser appeared both in the flat area and inside the macro-pores. In fact, there was no appreciable difference between the textures in the flat areas and the inner surface of the macro-pore walls. From the SEM micrographs (Figure 3), macro- and micro-porosity measurements were performed and parameters relating to the macro-mechanical behavior were estimated based on the porosity data (Table 1). The analyses of these results are shown in Table 1 and allowed us to indicate that: (1) three types of pore populations were identified: micro-pores inherent to the sintering of the substrates, micro-columns resulting from the femtosecond treatment, and macro-pores associated with the spacer particles; (2) an inverse relationship between porosity (percentage, size, and degree of interconnectivity) and the macro-mechanical behavior (Young's modulus and yield strength) of the studied materials was corroborated; and (3) it was also observed that the number and diameter of the micro-columns depended on the size of the titanium matrix that was modified by femtosecond laser radiation. This tendency of the micro-columns might be related to the heat evacuation phenomena during the interaction between the laser radiation and the titanium matrix.

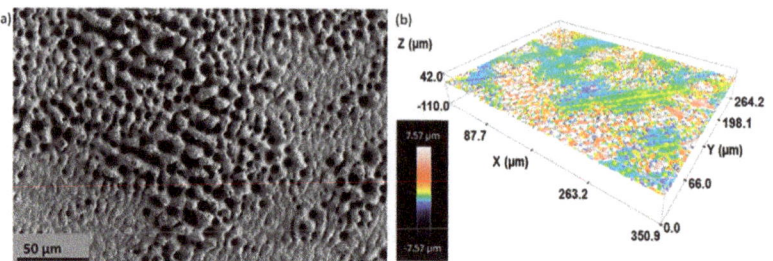

Figure 2. (a) SEM (Scanning Electron Microscopy) and (b) 3D-CLM (Confocal Laser Microscopy) images of the fully-dense c.p. Ti substrates after femtosecond laser treatment.

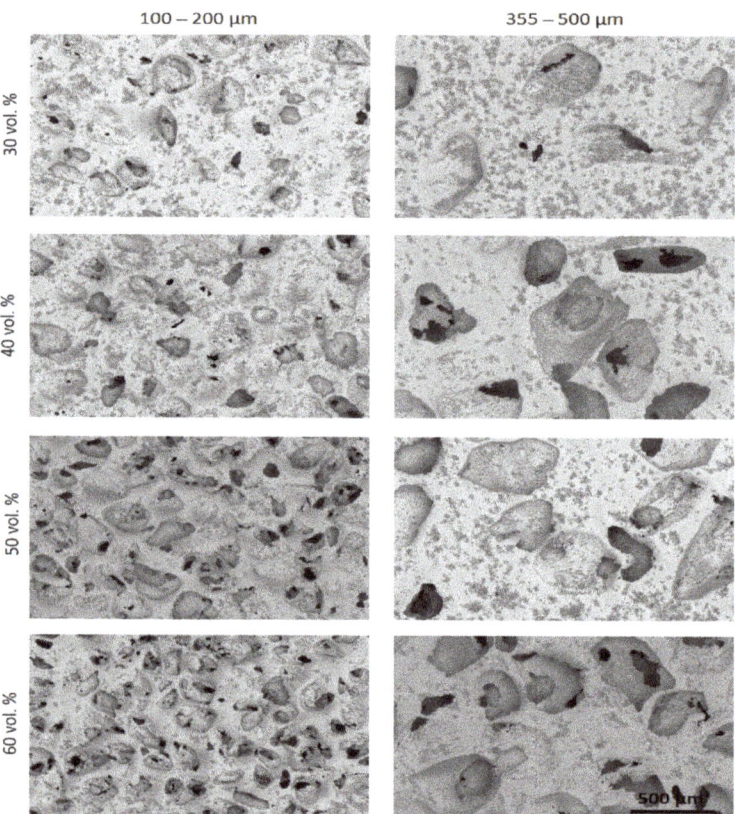

Figure 3. SEM images after the femtosecond laser treatment of the different porous c.p. Ti substrates. Images acquired using the topography view configuration gather both material and topographic contrast with the unique segmented in-lens backscattered electron detector (BSE). Common scale bar for all subfigures.

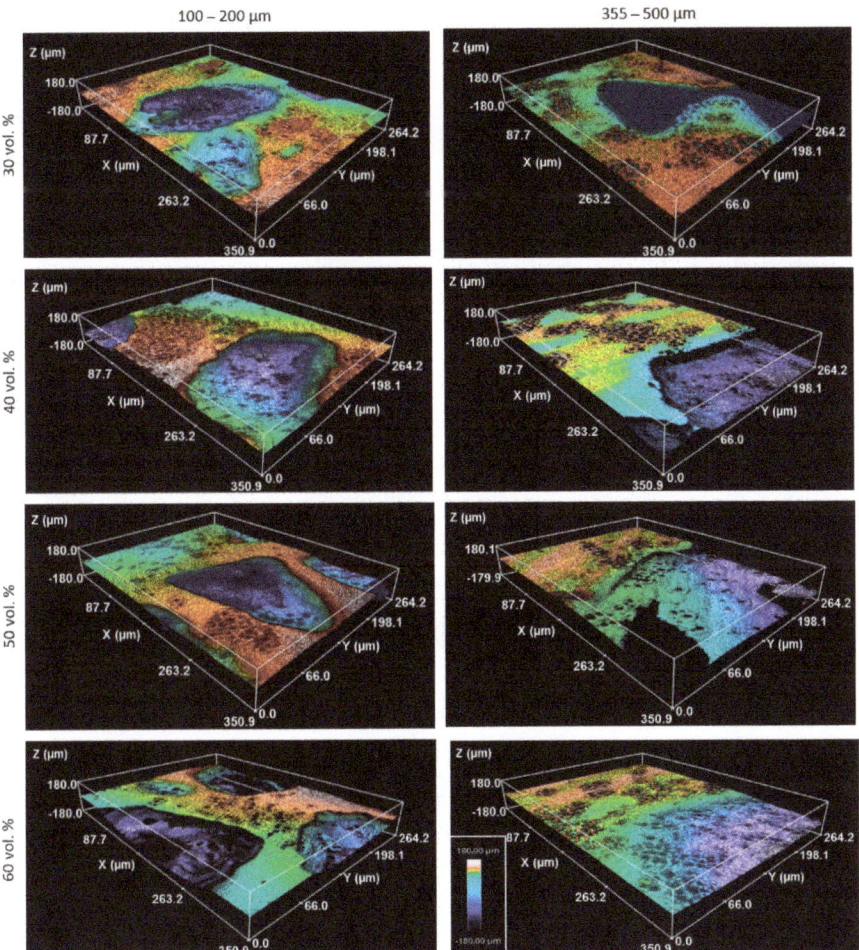

Figure 4. CLM images after the femtosecond laser treatment of the different porous c.p. Ti substrates. Common scale bar for all subfigures.

Furthermore, analysis of CLM images allowed us to measure the roughness parameters due to superficial modification of the substrates (Figure 5). For porous substrates, the roughness of the flat area among the pores was also measured. Before laser treatment, the fully-dense sample presented very low roughness, while the porous sample had slightly higher roughness due to the existence of pores (all substrates were previously mirror-polished). After laser irradiation, the roughness of all the samples increased. The changes in topography can be clearly seen in Figures 2 and 4. Laser radiation produced a surface with heterogeneous heights and a grainy texture. A relationship was also found between the porosity of the substrates and the final roughness, a higher increase in the surface roughness after laser irradiation was observed for highly porous substrates, as shown in Figures 3–5. It is evident that the Sa increases as the number of pores per unit area increases.

Table 1. Experimental porosity parameters and estimated macro-mechanical behavior of porous implants before and after FSL. Note: Flat surface (between macro-pores).

Discs Tested			Before FSL					After FSL		
			P_T (%)	D_{eq} (µm)	E_d (GPa)	σ_y (MPa) *		P_T (%)	D_{eq} (µm)	E_d (GPa) **
	Conventional PM		1.6 ± 0.3	3.9 ± 0.2	103.2 ± 1.3	739 ± 6		25.6 ± 0.9	6 ± 1.2	62.6 ± 1.2
Space-holder technique	30 vol.%	100–200 µm	30.2 ± 0.2	192 ± 117	56.8 ± 0.7	354 ± 26	P_F *** P_{SH}	9.8 ± 0.7 27.2 ± 0.9	10 ± 3 111 ± 2	50.2 ± 1.5
		355–500 µm	30.1 ± 0.1	335 ± 40	57.0 ± 1.4	341 ± 37	P_F *** P_{SH}	9.4 ± 0.8 32.2 ± 0.9	11 ± 3 250 ± 5	44.6 ± 1.1
	40 vol.%	100–200 µm	40.2 ± 1.1	226 ± 178	45.9 ± 1.2	234 ± 28	P_F *** P_{SH}	6.7 ± 1.1 45.4 ± 0.6	11 ± 3 138 ± 3	35.7 ± 1.0
		355–500 µm	40.8 ± 1.3	359 ± 123	45.3 ± 1.1	206 ± 26	P_F *** P_{SH}	6.4 ± 0.9 40.3 ± 1.2	12 ± 3 322 ± 4	39.7 ± 1.2
	50 vol.%	100–200 µm	52.3 ± 1.2	164 ± 28	35.4 ± 1.9	95 ± 30	P_F *** P_{SH}	5.8 ± 1.5 56.3 ± 1.1	12 ± 4 197 ± 3	30.0 ± 1.0
		355–500 µm	50.1 ± 1.0	365 ± 34	37.1 ± 1.6	118 ± 22	P_F *** P_{SH}	3.8 ± 1.3 46.5 ± 1.4	11 ± 4 340 ± 5	35.4 ± 1.4
	60 vol.%	100–200 µm	56.4 ± 0.5	189 ± 105	32.3 ± 1.5	91 ± 27	P_F *** P_{SH}	2.7 ± 1.0 63.6 ± 0.9	9 ± 2 205 ± 4	26.5 ± 1.4
		355–500 µm	57.8 ± 0.6	395 ± 131	31.3 ± 1.6	84 ± 31	P_F *** P_{SH}	2.5 ± 1.3 64.5 ± 1.5	8 ± 2 325 ± 2	25.4 ± 1.6

* The values of the yield stress estimated before and after femtosecond laser treatment were similar because the static mechanical behavior does not depend on the additional micro-porosity generated on the surface of the samples by the laser. ** The estimated E_d values after FSL corresponded to the surface of the samples (influences the additional porosity due to surface treatment). *** Micro-porosity generated with the femtosecond laser relative to the effective area of the titanium matrix between the pores.

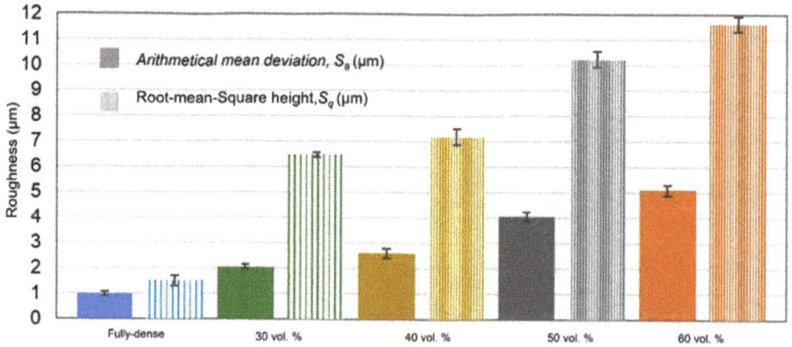

Figure 5. S_a and S_q values measured from the CLM images for all the studied c.p. Ti substrates (fully-dense and 100–200 µm for porous substrates).

The fully-dense surface showed a partial hydrophilic behaviour with water contact angles around 65° (Figure 6). The substrates prepared with smaller spacer particles tended to have a hydrophobic character, while larger pores induced a notable decrease in the water contact angle that agreed with the Wenzel model [54]. Thus, as the surface roughness and pore size increased, the surfaces with the highest porosity reached a completely hydrophilic character.

The first interaction of these porous c.p. Ti surfaces with a simulated physiological fluid was tested through the study of wettability with bovine serum albumin. In that case, larger pores were demonstrated to improve the wettability of the surfaces when compared to a fully-dense surface and even to those with smaller pore sizes. In general, after FSL treatment, the comparison of the water contact angle and the values of S_a, previously described in Figure 5, showed that for a roughness acquired with 30 vol.% and larger pores samples, hydrophobicity was promoted. This repulsive behaviour was also manifested when bovine serum droplets were deposited on the porous surfaces, reaching contact angle values greater than 100°. Special mention was deserved for the substrates with

60 vol.%, which went from being completely wet to maintaining a contact angle of over 70°. These results pointed to a remarkably stable and protective response when exposed to the action of the biological environment, as well as the expected antibacterial behavior of the surface, which prevented the generation of spores and bacterial contamination on the surface of implants [55,56]. Cell adhesion was not only affected by wettability but roughness and surface chemistry also play competitive roles, plus differences in the sizes and elastic responses of bacterial cells compared to bone tissue cells were added factors [57] that would allow the hydrophobic porous c.p. Ti surfaces to selectively induce a reduction in the bacteria attachment in favor of a greater living cell proliferation [58,59].

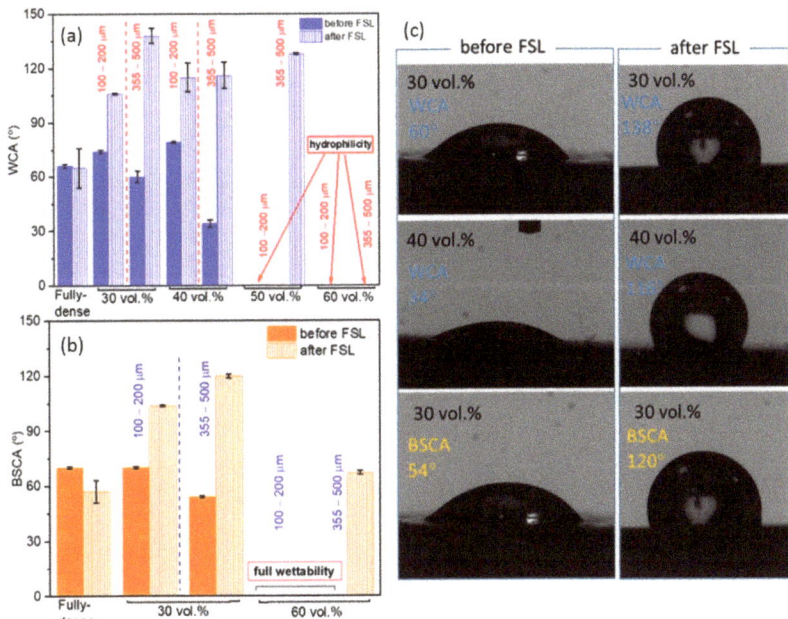

Figure 6. Wetting analysis of the porous c.p. Ti substrates: (**a**) Water Contact Angle (WCA) values of surfaces with different porosity percentages and pores range size before and after the FSL (Femto Second Laser) treatment; (**b**) bovine serum contact angle (BSCA) values of porous surfaces before and after the FSL treatment; (**c**) images representing the wetting of water and bovine serum droplets deposited on c.p. Ti porous surfaces (355–500 μm pores range size) before and after FSL treatment.

Figures 7 and S1 display the most relevant results inherent to the instrumented microindentation tests (*P–h* curves, static behavior). On the other hand, in Figures 8 and S2 the results of the scratch tests (dynamic characterization) are showne. Both figures compare the fully-dense substrate to the 40 vol.% and 60 vol.% porous substrates for both ranges of analyzed pore sizes. A higher penetration depth was observed for higher porosity (60 vol.%), almost independently of the pore size. Furthermore, mechanical properties such as the micro-hardness and Young's modulus were estimated from the resulting *P–h* loading and unload curves using the Oliver and Pharr method described above. In general, the micro-hardness, Young´s modulus, and scratch resistance decreased as the pore size and porosity increased. Regardless of the type of test, static (*P–h*) and dynamic (scratch test), the elastic recovery is proportional to the porosity of the titanium substrate. Figure 9 shows SEM and CLM images (2D and 3D) of the grooves generated by the scratch tests. As expected, the width and depth of the scar were inversely proportional to the resistance to scratching of the surfaces. An additional widening inherent to the presence of the macropores was observed, as well as a collapse (plastic deformation) of the micro-columns.

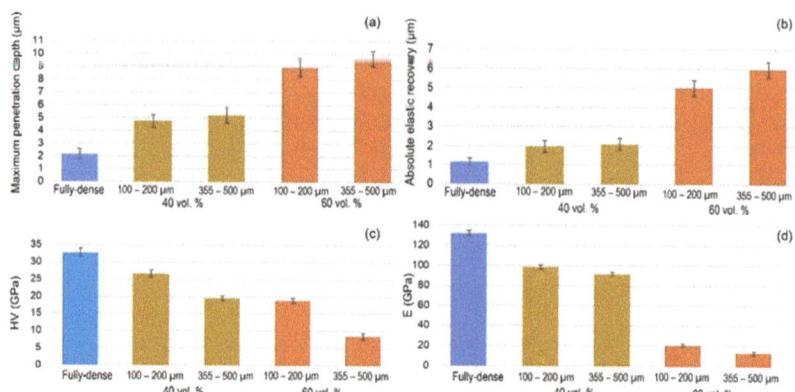

Figure 7. Static loading–unloading test (*P–h* curves) parameters of the different FSL modified substrates studied: (**a**) maximum penetration depth; (**b**) absolute elastic recovery; (**c**) microhardness (*HV*); and (**d**) Young's modulus (*E*). Note: c and d are calculated following the Oliver and Pharr method.

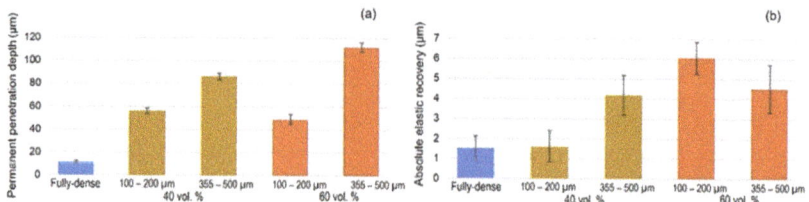

Figure 8. Scratch tests parameters on FSL-modified titanium substrates: (**a**) permanent penetration depth; and (**b**) absolute elastic recovery (difference between scratch in situ tests and permanent penetration depth).

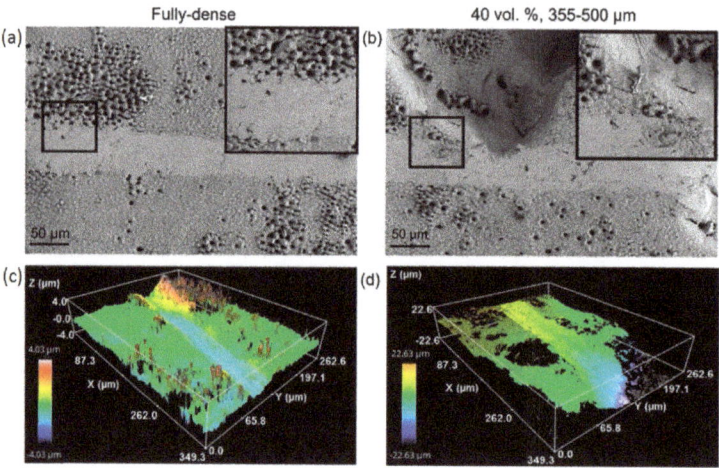

Figure 9. (**a**,**b**) SEM and (**c**,**d**) CLM images of scar (due to the scratch test) on fully-dense and porous c.p. Ti substrates. Inset SEM: zoomed details of area marked in subfigures.

3.1. Cell Viability and Proliferation Study

The study of the degree of proliferation and viability on modified laser surfaces reached at 21 days revealed the best results for 30 vol.% substrates, where porosity increased and cell proliferation decreased slightly (see results in Figure 10). When comparing cell growth according to pore size, it was found that the smallest pore size range was the most favorable condition for osteoblast cell growth, while growth over substrates with larger pores showed worse proliferation. That is, on porous substrates with pores of 100–200 µm, cells proliferated better than in the case of larger pore sizes. Only in discs with 30 vol.% porosity and 100–200 µm pore size, cell growth was greater than 80%. The other porous substrates presented values similar to those of the fully-dense c.p. Ti substrates. Regarding the degree of porosity, the 30 and 40 vol.% samples showed better cell growth, although the differences were not significant (Figure 10). These results confirmed the affinity of osteoblast cells for growth on a surface modified by FSL [45,47,60] and, as porosity increased, the contact surface and growth also increased [61]. In addition, as already reported in the literature [62], in this work it was also observed that 100–200 µm pore sizes that are more similar to cell area size were more favorable for monolayer growth of osteoblasts.

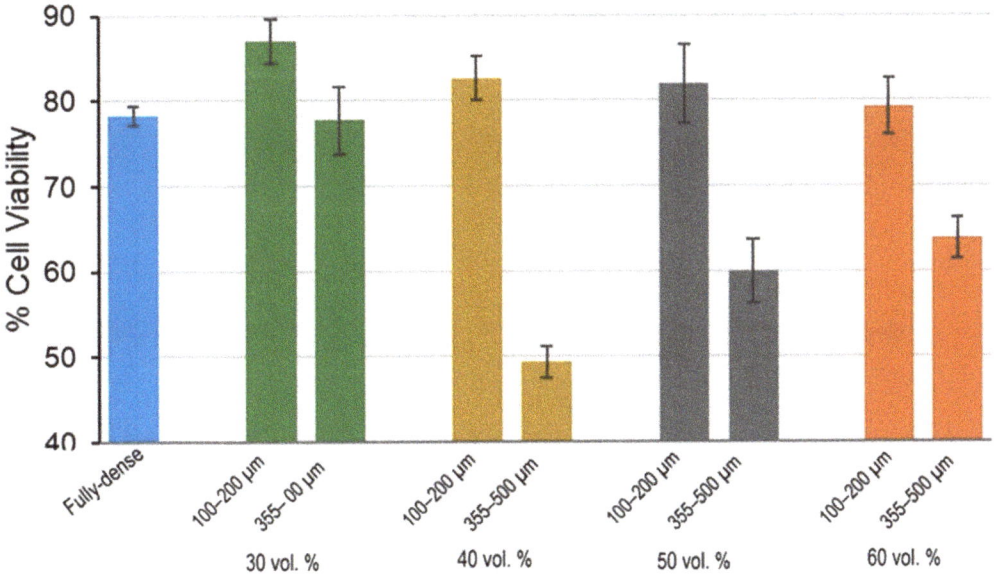

Figure 10. Effects of the percentage of porosity and pores range size on cell viability in MC3T3–E1 cells. Results are represented as % cell viability.

3.2. Cell Functional Activity

ALP activity was used to assess improved osseointegration capacity as markers of early differentiation of osteoblast-like cells [63]. The activity of ALP was observed to be similar in all cases (Figure 11), except for samples with 30 vol.% porosity and a smaller pore range size. Under this condition, osteoblasts showed activity more than double that of the other substrates. At the metabolic level, osteoblasts grown in 30 vol.% and 100–200 µm samples showed greater cellular functionality compared to other conditions, therefore there were more differentiated cells. No differences in the rest of the conditions were found, either with porosity percentage and/or with pore size range. Other authors also showed higher ALP values in MC3T3 cell cultures on FSL-treated surfaces [64,65]. In view of these results, it could be suggested that the better proliferation was due to the size of the osteoblast cells, which allowed them to create better cell–cell interactions on this type of surface and consequently, greater stimulation in cell growth and differentiation.

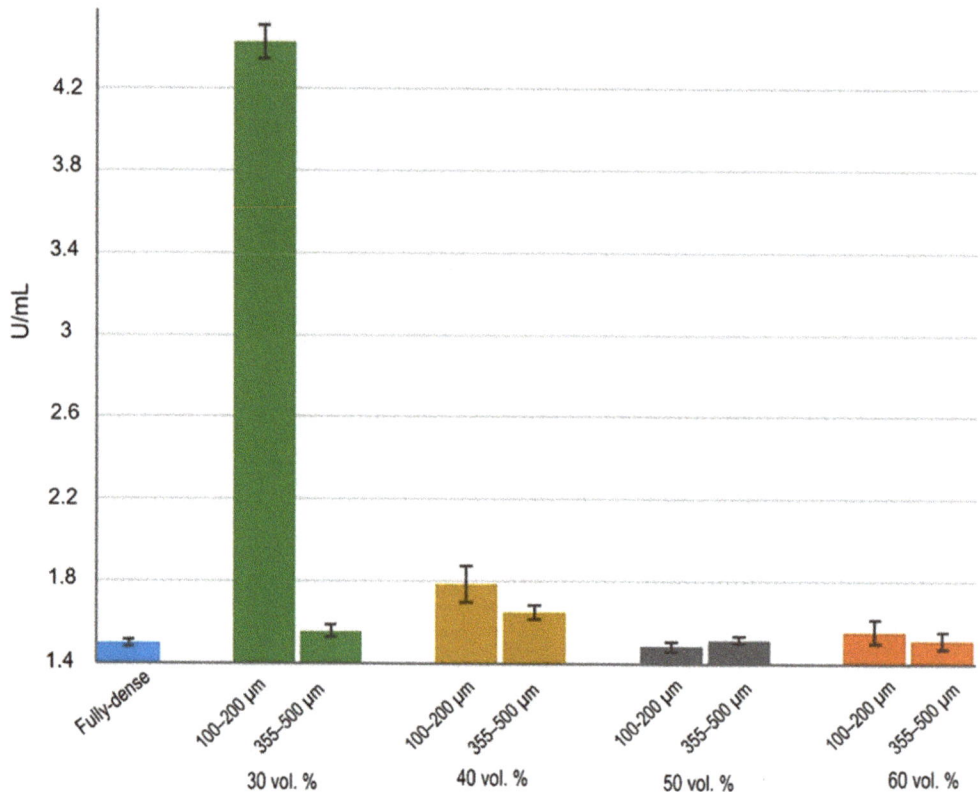

Figure 11. Cell differentiation of osteoblasts, in vitro evaluation of alkaline phosphatase enzyme (ALP) activity measured as U/mL. Statistical differences are indicated at * $p < 0.05$.

3.3. Morphological Study by SEM

The spread of MC3T3-E1 cells was observed in each studied condition by SEM images. As shown in Figure 12, osteoblasts showed a characteristic growth direction, although growth was lower for substrates with the largest pores, in which it was observed that cells could not grow in a monolayer throughout the surface. In samples with smaller pores, cell growth was observed both outside and inside the pores, there were even filopodium-like union structures among cells located on the periphery of the pores [60,66]. In general, the cells presented an elongated shape with slightly pronounced outstretched filopodium. Specifically, for samples with 30 and 40 vol.% porosity and pore sizes in the range of 100–200 µm, a more fibroblastic cell shape was observed, indicating a higher degree of differentiation, which was consistent with the measured ALP values.

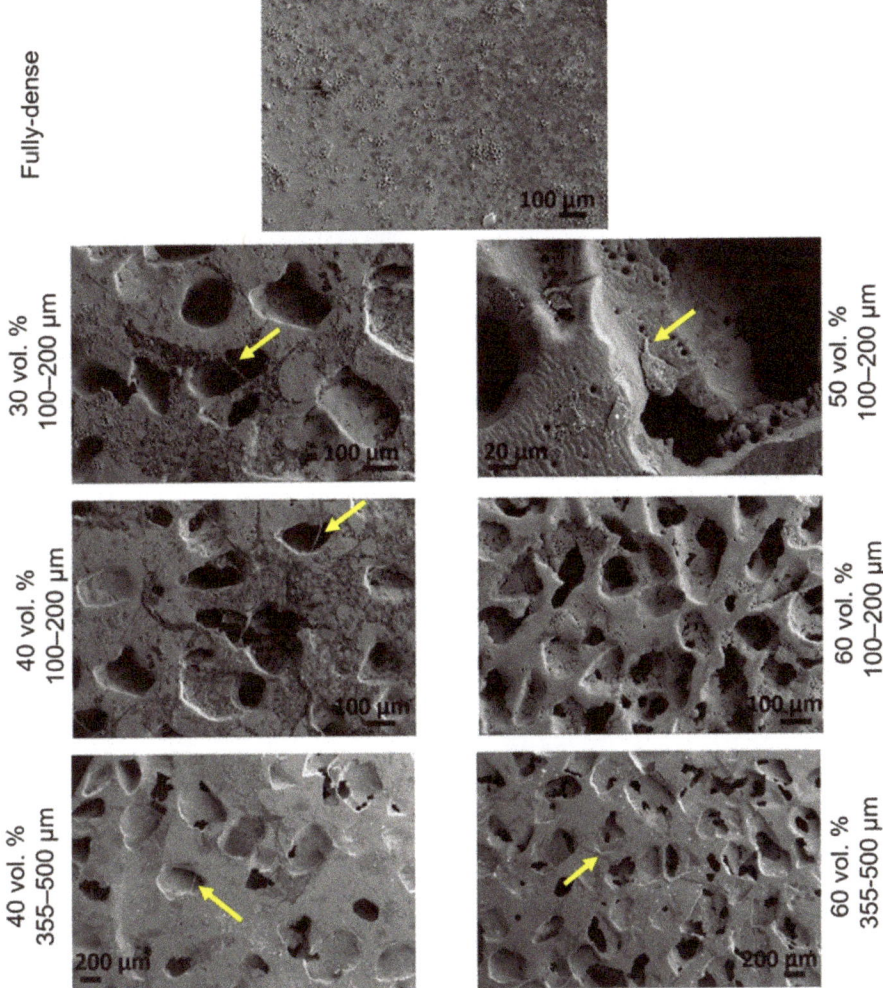

Figure 12. SEM images of the cell adherence of the fully-dense and porous substrates after femtosecond laser treatment. Cell–cell interactions and filopodium (yellow arrow) are indicated in the images.

4. Conclusions

In summary, in this work, micro-structural, tribo-mechanical characterization, and in vitro cellular behavior evaluations were performed on fully-dense and porous c.p. Ti samples that were superficially FSL-modified. The use of femtosecond laser radiation increased the surface roughness for all studied samples, generating characteristic textures and micro-columns. Treatment with FSL generated notably hydrophobic porous surfaces, with an increase in the water and bovine serum contact angle values as the pore size increased, as shown in the cases of 30, 40, and 50 vol.%, indicating a potential protective and antiseptic behavior on the surface of the implants working in a biological environment, preventing bacterial proliferation and dissemination without any detrimental effect on biocompatibility. In tribo-mechanical terms, Young´s modulus and scratch-resistance were found to decrease as the pore range size and the percentage of porosity increased, while the elastic recovery was directly proportional to the porosity of the titanium substrate. Surface modification with femtosecond laser treatment improved cell viability by 14%, mainly in the case in which osteoblasts were grown with 30 vol.% and a 100–200 μm pore size substrate, which showed better cell differentiation potential, proliferation, and viability performance.

Supplementary Materials: The following supporting information can be downloaded at: https://www.mdpi.com/article/10.3390/ma15092969/s1, Figure S1: *P–h* curves of the study of substrates modified by femtosecond and calculated mechanical properties based on Oliver and Pharr method; Figure S2: Resistance to penetration evaluated by scratch tests, for the pore range size 100–200 μm and two porosities; Figure S3: Scanning electron microscopy image of the scar on a substrate of 60% vol% and pore range size 355–500 μm.

Author Contributions: Conceptualization and project administration J.M.A., A.A. and Y.T.; supervision and methodology, M.G., M.A.V.-G., A.A. and Y.T.; investigation, formal analysis, validation, A.M.B., Á.R., L.M.R.-A., V.G., C.L.-S. and P.T.; discussion and writing—original draft preparation, all the authors. All authors have read and agreed to the published version of the manuscript.

Funding: This research was funded by the Ministerio de Ciencia e Innovación del Gobierno de España (PID2019-109371GB-100), Junta de Andalucía (Spain), through the Project PAIDI P20-00671.

Institutional Review Board Statement: Not applicable.

Informed Consent Statement: Not applicable.

Data Availability Statement: All the data is available within the manuscript.

Acknowledgments: The authors also thank the technician, J. Pinto for his support of the tribo-mechanical tests. C.L.S. acknowledges the "Ramon y Cajal" program from the Ministerio de Ciencia e Innovación.

Conflicts of Interest: The authors declare no conflict of interest.

References

1. Kaur, M.; Singh, K. Review on titanium and titanium based alloys as biomaterials for orthopaedic applications. *Mater. Sci. Eng. C* **2019**, *102*, 844–862. [CrossRef] [PubMed]
2. Khorasani, A.M.; Goldberg, M.; Doeven, E.H.; Littlefair, G. Titanium in Biomedical Applications—Properties and Fabrication: A Review. *J. Biomater. Tissue Eng.* **2015**, *5*, 593–619. [CrossRef]
3. Niinomi, M.; Nakai, M. Titanium-Based Biomaterials for Preventing Stress Shielding between Implant Devices and Bone. *Int. J. Biomater.* **2011**, *2011*, 836587. [CrossRef]
4. Niinomi, M.; Liu, Y.; Nakai, M.; Liu, H.; Li, H. Biomedical titanium alloys with Young's moduli close to that of cortical bone. *Regen. Biomater.* **2016**, *3*, 173–185. [CrossRef] [PubMed]
5. Chen, L.-Y.; Cui, Y.-W.; Zhang, L.-C. Recent Development in Beta Titanium Alloys for Biomedical Applications. *Metals* **2020**, *10*, 1139. [CrossRef]
6. Niinomi, M.; Nakai, M.; Hieda, J. Development of new metallic alloys for biomedical applications. *Acta Biomater.* **2012**, *8*, 3888–3903. [CrossRef] [PubMed]
7. Sidhu, S.S.; Singh, H.; Gepreel, M.A.-H. A review on alloy design, biological response, and strengthening of β-titanium alloys as biomaterials. *Mater. Sci. Eng. C* **2021**, *121*, 111661. [CrossRef]
8. Zhao, B.; Gain, A.K.; Ding, W.; Zhang, L.; Li, X.; Fu, Y. A review on metallic porous materials: Pore formation, mechanical properties, and their applications. *Int. J. Adv. Manuf. Technol.* **2018**, *95*, 2641–2659. [CrossRef]
9. Domínguez-Trujillo, C.; Peón, E.; Chicardi, E.; Pérez, H.; Rodríguez-Ortiz, J.A.; Pavón, J.; García-Couce, J.; Galvan, J.C.; García-Moreno, F.; Torres, Y. Sol-gel deposition of hydroxyapatite coatings on porous titanium for biomedical applications. *Surf. Coat. Technol.* **2018**, *333*, 158–162. [CrossRef]
10. Zhang, L.-C.; Chen, L.-Y. A Review on Biomedical Titanium Alloys: Recent Progress and Prospect. *Adv. Eng. Mater.* **2019**, *21*, 1801215. [CrossRef]
11. Murr, L.E. Strategies for creating living, additively manufactured, open-cellular metal and alloy implants by promoting osseointegration, osteoinduction and vascularization: An overview. *J. Mater. Sci. Technol.* **2018**, *35*, 231–241. [CrossRef]
12. Mavrogenis, A.; Dimitriou, R.; Parvizi, J.; Babis, G.C. Biology of implant osseointegration. *J. Musculoskelet. Neuronal. Interact.* **2009**, *9*, 61–71. [PubMed]
13. de Vasconcellos, L.M.R.; Carvalho, Y.R.; do Prado, R.F.; de Vasconcellos, L.G.O.; de Alencastro Graça, M.L.; Cairo, C.A.A. Porous titanium by powder metallurgy for biomedical application: Characterization, cell citotoxity and in vivo tests of osseointegration. In *Biomedical Engineering: Technical Applications in Medicine*; Intech: London, UK, 2012; pp. 47–74.

14. Devgan, S.; Sidhu, S.S. Evolution of surface modification trends in bone related biomaterials: A review. *Mater. Chem. Phys.* **2019**, *233*, 68–78. [CrossRef]
15. Ibrahim, M.Z.; Sarhan, A.A.; Yusuf, F.; Hamdi, M. Biomedical materials and techniques to improve the tribological, mechanical and biomedical properties of orthopedic implants—A review article. *J. Alloys Compd.* **2017**, *714*, 636–667. [CrossRef]
16. Accioni, F.; Vázquez, J.; Merinero, M.; Begines, B.; Alcudia, A. Latest Trends in Surface Modification for Dental Implantology: Innovative Developments and Analytical Applications. *Pharmaceutics* **2022**, *14*, 455. [CrossRef]
17. Pavón, J.; Galvis, O.; Echeverría, F.; Castaño, J.; Echeverry, M.; Robledo, S.; Jiménez-Piqué, E.; Mestra, A.; Anglada, M. Anodic oxidation of titanium for implants and prosthesis: Processing, characterization and potential improvement of osteointegration. In Proceedings of the V Latin American Congress on Biomedical Engineering CLAIB 2011, Havana, Cuba, 16–21 May 2011; pp. 176–179.
18. Saji, V.S. Superhydrophobic surfaces and coatings by electrochemical anodic oxidation and plasma electrolytic oxidation. *Adv. Colloid Interface Sci.* **2020**, *283*, 102245. [CrossRef] [PubMed]
19. Echeverry-Rendón, M.; Galvis, O.; Giraldo, D.A.Q.; Pavón-Palacio, J.-J.; López-Lacomba, J.L.; Jimenez-Pique, E.; Anglada, M.; Robledo, S.M.; Castaño, J.G.; Echeverria, F. Osseointegration improvement by plasma electrolytic oxidation of modified titanium alloys surfaces. *J. Mater. Sci. Mater. Electron.* **2015**, *26*, 1–18. [CrossRef]
20. Tsai, D.-S.; Chou, C.-C. Review of the Soft Sparking Issues in Plasma Electrolytic Oxidation. *Metals* **2018**, *8*, 105. [CrossRef]
21. Aditya, T.; Allain, J.P.; Jaramillo, C.; Restrepo, A.M. Surface Modification of Bacterial Cellulose for Biomedical Applications. *Int. J. Mol. Sci.* **2022**, *23*, 610. [CrossRef]
22. Echeverry-Rendon, M.; Allain, J.P.; Robledo, S.M.; Echeverria, F.; Harmsen, M.C. Coatings for biodegradable magnesium-based supports for therapy of vascular disease: A general view. *Mater. Sci. Eng. C* **2019**, *102*, 150–163. [CrossRef]
23. Mahajan, A.; Sidhu, S.S. Surface modification of metallic biomaterials for enhanced functionality: A review. *Mater. Technol.* **2018**, *33*, 93–105. [CrossRef]
24. Liu, W.; Liu, S.; Wang, L. Surface Modification of Biomedical Titanium Alloy: Micromorphology, Microstructure Evolution and Biomedical Applications. *Coatings* **2019**, *9*, 249. [CrossRef]
25. Wennerberg, A.; Albrektsson, T. Effects of titanium surface topography on bone integration: A systematic review. *Clin. Oral Implant. Res.* **2009**, *20*, 172–184. [CrossRef] [PubMed]
26. Valle, J.; Burgui, S.; Langheinrich, D.; Gil, C.; Solano, C.; Toledo-Arana, A.; Helbig, R.; Lasagni, A.; Lasa, I. Evaluation of surface microtopography engineered by direct laser interference for bacterial anti-biofouling. *Macromol. Biosci.* **2015**, *15*, 1060–1069. [CrossRef] [PubMed]
27. Perera-Costa, D.; Bruque, J.M.; González-Martín, M.L.; Gómez-García, A.C.; Vadillo-Rodríguez, V. Studying the Influence of Surface Topography on Bacterial Adhesion using Spatially Organized Microtopographic Surface Patterns. *Langmuir* **2014**, *30*, 4633–4641. [CrossRef]
28. Li, X.; Chen, T. Enhancement and suppression effects of a nanopatterned surface on bacterial adhesion. *Phys. Rev. E* **2016**, *93*, 052419. [CrossRef]
29. Cunha, A.; Serro, A.P.; Oliveira, V.; Almeida, A.; Vilar, R.; Durrieu, M.-C. Wetting behaviour of femtosecond laser textured Ti–6Al–4V surfaces. *Appl. Surf. Sci.* **2013**, *265*, 688–696. [CrossRef]
30. Vorobyev, A.Y.; Guo, C. Femtosecond laser structuring of titanium implants. *Appl. Surf. Sci.* **2007**, *253*, 7272–7280. [CrossRef]
31. Oliveira, V.; Ausset, S.; Vilar, R. Surface micro/nanostructuring of titanium under stationary and non-stationary femtosecond laser irradiation. *Appl. Surf. Sci.* **2009**, *255*, 7556–7560. [CrossRef]
32. Bonse, J.; Höhm, S.; Koter, R.; Hartelt, M.; Spaltmann, D.; Pentzien, S.; Rosenfeld, A.; Krüger, J. Tribological performance of sub-100-nm femtosecond laser-induced periodic surface structures on titanium. *Appl. Surf. Sci.* **2016**, *374*, 190–196. [CrossRef]
33. Shinonaga, T.; Tsukamoto, M.; Kawa, T.; Chen, P.; Nagai, A.; Hanawa, T. Formation of periodic nanostructures using a femtosecond laser to control cell spreading on titanium. *Appl. Phys. A* **2015**, *119*, 493–496. [CrossRef]
34. Schnell, G.; Duenow, U.; Seitz, H. Effect of Laser Pulse Overlap and Scanning Line Overlap on Femtosecond Laser-Structured Ti6Al4V Surfaces. *Materials* **2020**, *13*, 969. [CrossRef] [PubMed]
35. Menci, G.; Demir, A.G.; Waugh, D.; Lawrence, J.; Previtali, B. Laser surface texturing of β-Ti alloy for orthopaedics: Effect of different wavelengths and pulse durations. *Appl. Surf. Sci.* **2019**, *489*, 175–186. [CrossRef]
36. Schweitzer, L.; Cunha, A.; Pereira, T.; Mika, K.; Rego, A.M.B.D.; Ferraria, A.M.; Kieburg, H.; Geissler, S.; Uhlmann, E.; Schoon, J. Preclinical In Vitro Assessment of Submicron-Scale Laser Surface Texturing on Ti6Al4V. *Materials* **2020**, *13*, 5342. [CrossRef]
37. Klos, A.; Sedao, X.; Itina, T.E.; Helfenstein-Didier, C.; Donnet, C.; Peyroche, S.; Vico, L.; Guignandon, A.; Dumas, V. Ultrafast Laser Processing of Nanostructured Patterns for the Control of Cell Adhesion and Migration on Titanium Alloy. *Nanomaterials* **2020**, *10*, 864. [CrossRef] [PubMed]
38. Martínez-Calderon, M.; Martín-Palma, R.J.; Rodríguez, A.; Gómez-Aranzadi, M.; García-Ruiz, J.P.; Olaizola, S.M.; Manso-Silván, M. Biomimetic hierarchical micro/nano texturing of TiAlV alloys by femtosecond laser processing for the control of cell adhesion and migration. *Phys. Rev. Mater.* **2020**, *4*, 056008. [CrossRef]
39. Li, C.; Yang, Y.; Yang, L.; Shi, Z.; Yang, P.; Cheng, G. In Vitro Bioactivity and Biocompatibility of Bio-Inspired Ti-6Al-4V Alloy Surfaces Modified by Combined Laser Micro/Nano Structuring. *Molecules* **2020**, *25*, 1494. [CrossRef]

40. Dumas, V.; Guignandon, A.; Vico, L.; Mauclair, C.; Zapata, X.; Linossier, M.T.; Bouleftour, W.; Granier, J.; Peyroche, S.; Dumas, J.-C.; et al. Femtosecond laser nano/micro patterning of titanium influences mesenchymal stem cell adhesion and commitment. *Biomed. Mater.* **2015**, *10*, 055002. [CrossRef]
41. Liang, C.; Wang, H.; Yang, J.; Cai, Y.; Hu, X.; Yang, Y.; Li, B.; Li, H.; Li, H.; Li, C.; et al. Femtosecond Laser-Induced Micropattern and Ca/P Deposition on Ti Implant Surface and Its Acceleration on Early Osseointegration. *ACS Appl. Mater. Interfaces* **2013**, *5*, 8179–8186. [CrossRef]
42. Wang, C.; Hu, H.; Li, Z.; Shen, Y.; Xu, Y.; Zhang, G.; Zeng, X.; Deng, J.; Zhao, S.; Ren, T.; et al. Enhanced Osseointegration of Titanium Alloy Implants with Laser Microgrooved Surfaces and Graphene Oxide Coating. *ACS Appl. Mater. Interfaces* **2019**, *11*, 39470–39483. [CrossRef]
43. Webb, H.; Crawford, R.; Ivanova, E.P. Wettability of natural superhydrophobic surfaces. *Adv. Colloid Interface Sci.* **2014**, *210*, 58–64. [CrossRef] [PubMed]
44. Raimbault, O.; Benayoun, S.; Anselme, K.; Mauclair, C.; Bourgade, T.; Kietzig, A.-M.; Girard-Lauriault, P.-L.; Valette, S.; Donnet, C. The effects of femtosecond laser-textured Ti-6Al-4V on wettability and cell response. *Mater. Sci. Eng. C* **2016**, *69*, 311–320. [CrossRef] [PubMed]
45. Liu, Y.; Rui, Z.; Cheng, W.; Song, L.; Xu, Y.; Li, R.; Zhang, X. Characterization and evaluation of a femtosecond laser-induced osseointegration and an anti-inflammatory structure generated on a titanium alloy. *Regen. Biomater.* **2021**, *8*, rbab006. [CrossRef] [PubMed]
46. Cunha, A.; Elie, A.-M.; Plawinski, L.; Serro, A.; Rego, A.M.B.D.; Almeida, A.; Urdaci, M.C.; Durrieu, M.-C.; Vilar, R. Femtosecond laser surface texturing of titanium as a method to reduce the adhesion of Staphylococcus aureus and biofilm formation. *Appl. Surf. Sci.* **2016**, *360*, 485–493. [CrossRef]
47. Rodríguez, Á.; Trueba, P.; Amado, J.M.; Tobar, M.J.; Giner, M.; Amigó, V.; Torres, Y. Surface Modification of Porous Titanium Discs Using Femtosecond Laser Structuring. *Metals* **2020**, *10*, 748. [CrossRef]
48. *ASTM F67-00, Standard Specification for Unalloyed Titanium for Surgical Implant Applications (UNS R50250, UNS R50400, UNS R50550, UNS R50700)*; ASTM: West Conshohocken, PA, USA, 2002.
49. Torres, Y.; Trueba, P.; Pavón-Palacio, J.-J.; Montealegre-Melendez, I.; Rodríguez-Ortiz, J.A. Designing, processing and characterisation of titanium cylinders with graded porosity: An alternative to stress-shielding solutions. *Mater. Des.* **2014**, *63*, 316–324. [CrossRef]
50. Lascano, S.; Arévalo, C.; Montealegre-Melendez, I.; Muñoz, S.; Rodriguez-Ortiz, J.A.; Trueba, P.; Torres, Y. Porous Titanium for Biomedical Applications: Evaluation of the Conventional Powder Metallurgy Frontier and Space-Holder Technique. *Appl. Sci.* **2019**, *9*, 982. [CrossRef]
51. Shen, H.; Brinson, L. Finite element modeling of porous titanium. *Int. J. Solids Struct.* **2007**, *44*, 320–335. [CrossRef]
52. Imwinkelried, T. Mechanical properties of open-pore titanium foam. *J. Biomed. Mater. Res. Part A* **2007**, *81*, 964–970. [CrossRef]
53. Oliver, W.C.; Pharr, G.M. Measurement of hardness and elastic modulus by instrumented indentation: Advances in understanding and refinements to methodology. *J. Mater. Res.* **2004**, *19*, 3–20. [CrossRef]
54. Wenzel, R.N. Resistance of solid surfaces to wetting by water. *Ind. Eng. Chem.* **1936**, *28*, 988–994. [CrossRef]
55. Hasan, J.; Crawford, R.; Ivanova, E.P. Antibacterial surfaces: The quest for a new generation of biomaterials. *Trends Biotechnol.* **2013**, *31*, 295–304. [CrossRef] [PubMed]
56. Perez-Gavilan, A.; de Castro, J.V.; Arana, A.; Merino, S.; Retolaza, A.; Alves, S.A.; Francone, A.; Kehagias, N.; Sotomayor-Torres, C.M.; Cocina, D.; et al. Antibacterial activity testing methods for hydrophobic patterned surfaces. *Sci. Rep.* **2021**, *11*, 6675. [CrossRef] [PubMed]
57. Neoh, K.G.; Hu, X.; Zheng, D.; Kang, E.-T. Balancing osteoblast functions and bacterial adhesion on functionalized titanium surfaces. *Biomaterials* **2012**, *33*, 2813–2822. [CrossRef] [PubMed]
58. Mei, S.; Wang, H.; Wang, W.; Tong, L.; Pan, H.; Ruan, C.; Ma, Q.; Liu, M.; Yang, H.; Zhang, L.; et al. Antibacterial effects and biocompatibility of titanium surfaces with graded silver incorporation in titania nanotubes. *Biomaterials* **2014**, *35*, 4255–4265. [CrossRef] [PubMed]
59. Wang, J.; Li, J.; Qian, S.; Guo, G.; Wang, Q.; Tang, J.; Shen, H.; Liu, X.; Zhang, X.; Chu, P. Antibacterial Surface Design of Titanium-Based Biomaterials for Enhanced Bacteria-Killing and Cell-Assisting Functions Against Periprosthetic Joint Infection. *ACS Appl. Mater. Interfaces* **2016**, *8*, 11162–11178. [CrossRef]
60. Chen, P.; Miyake, M.; Tsukamoto, M.; Tsutsumi, Y.; Hanawa, T. Response of preosteoblasts to titanium with periodic micro/nanometer scale grooves produced by femtosecond laser irradiation. *J. Biomed. Mater. Res. Part A* **2017**, *105*, 3456–3464. [CrossRef]
61. Civantos, A.; Giner, M.; Trueba, P.; Lascano, S.; Montoya-García, M.-J.; Arévalo, C.; Vázquez, M.; Allain, J.P.; Torres, Y. In Vitro Bone Cell Behavior on Porous Titanium Samples: Influence of Porosity by Loose Sintering and Space Holder Techniques. *Metals* **2020**, *10*, 696. [CrossRef]
62. Wo, J.; Huang, S.-S.; Wu, D.-Y.; Zhu, J.; Li, Z.-Z.; Yuan, F. The integration of pore size and porosity distribution on Ti-6A1-4V scaffolds by 3D printing in the modulation of osteo-differentiation. *J. Appl. Biomater. Funct. Mater.* **2020**, *18*. [CrossRef]
63. Van den Beucken, J.J.J.P.; Walboomers, X.F.; Boerman, O.C.; Vos, M.R.J.; Sommerdijk, N.A.J.M.; Hayakawa, T.; Fukushima, T.; Okahata, Y.; Nolte, R.J.M.; Jansen, J.A. Functionalization of multilayered DNA-coatings with bone morphogenetic protein. *J. Control Release* **2006**, *113*, 63–72. [CrossRef]

64. Zheng, J.; Zhang, X.; Zhang, Y.; Yuan, F. Osteoblast differentiation of bone marrow stromal cells by femtosecond laser bone ablation. *Biomed. Opt. Express* **2020**, *11*, 885–894. [CrossRef] [PubMed]
65. Li, L.-J.; Kim, S.-N.; Cho, S.-A. Comparison of alkaline phosphatase activity of MC3T3-E1 cells cultured on different Ti surfaces: Modified sandblasted with large grit and acid-etched (MSLA), laser-treated, and laser and acid-treated Ti surfaces. *J. Adv. Prosthodont.* **2016**, *8*, 235–240. [CrossRef] [PubMed]
66. Alves, A.; Thibeaux, R.; Toptan, F.; Pinto, A.; Ponthiaux, P.; David, B. Influence of macroporosity on NIH/3T3 adhesion, proliferation, and osteogenic differentiation of MC3T3-E1 over bio-functionalized highly porous titanium implant material. *J. Biomed. Mater. Res. Part B Appl. Biomater.* **2019**, *107*, 73–85. [CrossRef] [PubMed]

Article

Electrical Impedance of Surface Modified Porous Titanium Implants with Femtosecond Laser

Paula Navarro [1,2], Alberto Olmo [1,3,*], Mercè Giner [4], Marleny Rodríguez-Albelo [2], Ángel Rodríguez [5] and Yadir Torres [2]

[1] Departamento de Tecnología Electrónica, Escuela Técnica Superior de Ingeniería Informática, Universidad de Sevilla, Av. Reina Mercedes s/n, 41012 Sevilla, Spain; paunavgon2296@gmail.com

[2] Departamento de Ingeniería y Ciencia de los Materiales y del Transporte, Escuela Politécnica Superior, Calle Virgen de África 7, 41011 Seville, Spain; lralbelo@us.es (M.R.-A.); ytorres@us.es (Y.T.)

[3] Instituto de Microelectrónica de Sevilla, IMSE-CNM (CSIC, Universidad de Sevilla), Av. Américo Vespucio s/n, 41092 Sevilla, Spain

[4] Departamento de Citología e Histología Normal y Patológica, Universidad de Sevilla, Av. Doctor Fedriani s/n, 41009 Sevilla, Spain; mginer@us.es

[5] Escuela Politécnica Superior, Universidad da Coruña, Calle Mendizábal s/n, 15403 Ferrol, Spain; angel.rcarballo@udc.es

* Correspondence: aolmo@dte.us.es; Tel.: +34-954556835

Abstract: The chemical composition and surface topography of titanium implants are essential to improve implant osseointegration. The present work studies a non-invasive alternative of electrical impedance spectroscopy for the characterization of the macroporosity inherent to the manufacturing process and the effect of the surface treatment with femtosecond laser of titanium discs. Osteoblasts cell culture growths on the titanium surfaces of the laser-treated discs were also studied with this method. The measurements obtained showed that the femtosecond laser treatment of the samples and cell culture produced a significant increase (around 50%) in the absolute value of the electrical impedance module, which could be characterized in a wide range of frequencies (being more relevant at 500 MHz). Results have revealed the potential of this measurement technique, in terms of advantages, in comparison to tiresome and expensive techniques, allowing semi-quantitatively relating impedance measurements to porosity content, as well as detecting the effect of surface modification, generated by laser treatment and cell culture.

Keywords: cell culture; electrical impedance; femtosecond laser; osseointegration; porous titanium

1. Introduction

In recent decades, due to the aging of the population and change in lifestyles, millions of people have been affected by orthopedic, oral, and maxillofacial diseases [1]. Bone tissues are also exposed to damage due to degenerative or traumatic diseases that can cause serious disabilities and, therefore, carry high economic and social costs [2]. Biomaterials is one promising solution to solve such problems, as it can be used to manufacture medical devices for replacement of human tissues, such as teeth, bones, and cartilages. In addition, the demand for biomaterials is dramatically growing due to increasing maturity of materials manufacturing technologies [1,2].

Titanium and its alloys are considered one of the best choices for modern metallic implants, owing to their excellent biomechanical compatibility, long-term stability, and corrosion resistance in biological surroundings [3,4]. However, vital issues, such as bone resorption of tissues adjacent to the implant, related to the stress shielding phenomenon [5] and poor osseointegration, caused by bacteria proliferation or implant loosening, are still challenging problems to solve.

Two main approaches have been established to lessen or eliminate the stress shielding phenomenon on titanium implants. The first potential way is the use of β-titanium alloys with elements of low toxicity (Nb, Ta, Mo, and Zr) [6]. However, the use of porous titanium could be a more economical route for manufacturing titanium implants, with a stiffness and yield strength close to cortical bone [7–13].

On the other hand, to achieve a good osseointegration, implant surface should promote adhesion, proliferation, and differentiation of bone tissue cells. At the same time, it is highly desirable to avoid adhesion and growth of bacteria at bone-implant interface, since this can cause infections and subsequent implant failure [14]. In order to improve osseointegration, many approaches have been studied, including implant chemical and physical surface modifications, the most usual procedure. Chemical techniques comprise the introduction of natural or artificial chemical compounds, with elements that favor the interaction between implant and bone tissue cells, promoting bone ingrowth on the implant's surface. In general, chemical techniques include coating, impregnation, immersion, or deposition of bioactive materials, such as hydroxyapatite, bioglasses, ceramics, polymers, or peptides [15–17], onto the surface. Physical methods consist of techniques focused on modifying the surface topography, altering its porosity, roughness, or smoothness. Some examples of physical modifications are sand- and grit-blasting, acid-etching, plasma-spraying, laser surface modification [14], ultraviolet treatment, electrochemical, and oxidation (anodization) methods [1]. Physical modifications of titanium and alloy implant surfaces could allow the creation of micro- and nanostructures to stimulate osseointegration [18,19] by increasing: porosity for cell adhesion and proliferation, as well as roughness to enhance wettability for protein adsorption, or smoothing surface for repelling bacterial infection.

Among the latest physical techniques, femtosecond laser ablation stands out as very advantageous due to its accurate control of designed features on the surface, its high efficiency, and low material consumption [1]. In Vorobyev and Guo's pioneering work [18] on titanium substrates, the use of femtosecond laser ablation allowed the creation of nano- (pores, spherical protrusions, and multiple grooved surface patterns) and micro-structures (such as varied roughness configurations and smooth surface with micro-inhomogeneities) with appropriate adjustment of laser parameters. Recently, Rodríguez et al. [14] studied the influence of femtosecond laser modifications performed in porous titanium discs, producing a hierarchical arrangement composed of micro-holes, micro-columns, and a periodic surface nanometric structure, both on the flat surface and inside the pores. These modifications boosted superficial porosity and roughness, without any significantly affected mechanical properties of the titanium samples. Moreover, other authors have performed in vitro experiments on porous titanium substrates treated with femtosecond laser ablation, showing improved cell viability, as well as better differentiation morphology and cell adhesions, with acceptable biological response [20].

Femtosecond laser surface modifications on titanium and alloys substrates have been proved as a feasible tool to improve cells adhesion, differentiation, proliferation, and all together, more effectively boost osseointegration of the implant. In general, this physical modification technique allows: (1) custom design of nano- and micro-structures, such as laser induced periodically surface structure (LIPSS), ripples, columns, pits, and spikes [21–23], with an appropriate selection of laser beam parameters and conditions [24–27]; (2) formation of roughness with enhancement of wettability [28,29] or hydrophilicity-hydrophobicity of treated surfaces [29,30]; (3) inducing protein adsorption and following localized adhesion formation and cell shape-based mechanical restraints that promote osteogenic differentiation and hence, superior osseointegration of implants [31,32]; (4) prevention of bacterial adhesion and biofilm formation [33,34]; (5) variation of chemical composition of laser modified surfaces, for instance, bone-like apatite precipitation [35,36] and formation of nano- or micro-layers of oxides [37] such as, for example, protective TiO_2 on titanium substrates.

The increasing demand of titanium and its alloys, as medical implants, requires a practical technique to control evolution in time of cells adhesion, proliferation, and differentiation, meaning the osseointegration process. Furthermore, it is also required to check implant surface features in the exposed biological surrounding, such as corrosion resistance, ions migration, durability, etc. It is, therefore, that researchers have thoroughly used Electric Impedance Spectroscopy (EIS) as an electrochemical tool for both purposes. In this sense, several corrosion studies have been carried out on titanium or titanium alloys surfaces, using electrochemical impedance spectroscopy [38–40] to follow the formation of passive TiO_2 layer. Moreover, other surface features could be evaluated using impedance spectroscopy, such as porosity or pores sizes, with a clear advantage over other techniques such as Image Analysis or Archimedes Test, since it is a non-invasive technique and enables measurements to be carried out in situ. For example, Olmo et al. [41] used EIS for the characterization of porous titanium substrates, obtaining superior differences of total porosities, higher frequencies measured at electrical impedances, being 355–500 μm range of pore size and more sensitive to slight variations in impedance. Similarly, Chen et al. showed the negative effect of pores in corrosion resistance and higher corrosive rates in the presence of flowing electrolyte [42].

In recent years, Electrochemical Impedance Spectroscopy has become a leading topic for monitoring the evolution of cells adhesion, differentiation, and proliferation. Many studies have been led by researchers in this field, so it could be highlighted that Giner et al. work [43] performed in-situ evaluation of osteoblast cells growth on porous titanium substrates, studying the biological response of MC3T3E1, a murine pre-osteoblast cell line, by analysis of viability, morphology, differentiation, and alkaline phosphatase activity. Huang et al. showed that Electrochemical Impedance Spectroscopy could be used for in vivo measurement of U-2 OS osteoblast-like cell adhesion, spreading, and proliferation stage, on titanium and Ti-6A-4V implants, proposing equivalent circuits for each system [44]. An outstanding study was done by Nodberg et al., showing the suitability of electrical impedance spectroscopy to monitor, in real time, osteogenic differentiation of human Adipose Stem Cells (hASCs) of age-grouped donors, resulting in distinctive complex impedance patterns for each age group of cells [45]. Besides, Hamal et al. have summarized a wide range of Electrical Impedance measurements in cellular assays and its usefulness in regenerative medicine [46].

The present work is focused on the assessment of femtosecond laser modified porous titanium substrates using Electrical Impedance Spectroscopy. The objective is to analyze the impedance response due to diverse types of surface topographies, as different pores and pore sizes, total porosities, and oxide layers generated by femtosecond laser treatment (FT). Furthermore, osteoblast cells adhesion, differentiation, and proliferation will be monitored by electrical impedance measurements in previously modified titanium substrates. The aim of this study is to validate the utility and high sensitivity of the Electrical Impedance Spectroscopy technique to detect and differentiate subtle surface changes and its direct influence on osteoblast cells responses.

2. Materials and Methods

2.1. Manufacturing of Surface Modified Porous Titanium Discs Using Femtosecond Laser Surface Treatment

All samples were manufactured according to a methodology previously published [14,20,41,43]. Fully-dense commercially pure titanium (c.p. Ti–Grade IV, SE-JONG Materials Co., Ltd., Incheon, Korea) discs were prepared using conventional Powder Metallurgy Technology, by pressing and sintering at 1300 MPa and 1250 °C, respectively. Meanwhile, porous titanium samples were manufactured by space holder technique, with a particle size range of 100–200 μm. Ammonium hydrogen carbonate (NH_4HCO_3) from (Cymit Química SL, Barcelona, Spain) was used as a space holder with different content (30, 40, 50, and 60 vol. %). Subsequently, the mixture of titanium powder and spacer particles were pressed at 800 MPa and then, spacer was removed using a low vacuum furnace (Heraeus, Hanau, Germany) (10^{-2} mbar) in two steps (60 °C and 110 °C) during 10 h each, and sintered at 1250 °C in a molybdenum chamber furnace (Termolab-Fornos Eléctricos, Lda., Agueda, Portugal) under high vacuum atmosphere (~10^{-5} mbar) for 2 h. The surface of the discs, with 12 mm in diameter and approximately 5 mm high, were polished with magnesium oxide (MgO) and hydrogen peroxide (H_2O_2) prior any surface treatments using femtosecond.

Femtosecond laser irradiation was performed following the methodology presented by Rodriguez [14] and Trueba [20] and collaborators, using a Yb-doped fiber laser (Spirit 1040-4, Spectra-Physics, Santa Clara, CA, USA) with a wavelength of 1040 nm and pulses of 396 fs, at a repetition rate of f = 100 kHz. A pulse energy of Ep = 49.7 µJ (100% of nominal power) and a scanning speed of v = 960 mm/s were chosen. After deflection by a galvanometer scanner, the laser beam was focused through an F-Theta lens (f = 160 mm) to a beam radius of approximately w0 = 12 µm on the working surface. The resulting laser fluency on the surface was F = 21.98 J/cm^2. The surface of the samples was scanned line by line with the moving laser beam, and the laser paths were separated from each other according to an overlap of s = 50%. The surface was processed multiple times (Nr = 20), to increase the energy deposited on the surface. Under these conditions, the resulting number of pulses per point (PPS) at the surface was PPS = 100. The experiments were performed in air, and Argon was used as shielding gas in order to reduce any undesirable oxidation on the surface of the workpiece.

Macrostructure (high resolution Nikon camera) and microstructure (by scanning electron microscopy, using a Zeiss EVO LS 15 scanning electron microscope (Zeiss, Oberkochen, Germany) with an acceleration voltage of 10 kV) of two types of discs fabricated: fully dense (FD) and 30% porosity volume, taken before and after the treatment with femtosecond laser, is shown in Figure 1. The laser surface treatment was similar to the one performed in [14], but it was applied over a greater range of total porosity percentages. The resulting surface morphology is mostly independent of the volume of porosity of the samples and, therefore, the results of the laser treatment are similar to those presented in the previously published work. The surface, on the one hand, consists of macro pores generated by the spacers, the size and quantity of which depends on the total volume of the spacer. The laser treatment, on the other hand, generates, on the surface, a multiscale hierarchical texture. This texture is based on a mixture of clusters of micropores and micro-pillars, with characteristic lengths less than 10 µm, as well as laser-induced periodic surface structures (LIPSS), which are self-organized periodic nanostructures that cover the entire surface. These nanoripples are aligned perpendicularly to the polarization of the laser beam, and the spatial period of the structure is close to the wavelength of the laser (1040 nm).

Before Femtosecond   **After Femtosecond**

Fully-dense

Figure 1. *Cont.*

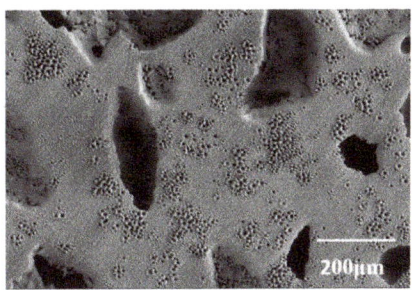

Figure 1. Optical and SEM images of some of the samples studied, before and after FS. All optical images are on the same scale, 3:1 (mm). FD: fully dense sample; 30 vol. %: sample with 30% porosity volume.

2.2. Electrical Impedance Characterization

As previously presented, the role of porosity to solve the stress-shielding phenomenon (mismatch of young modulus between the implant-cortical bone) guarantees the bone-ingrowth as well as allows infiltration and adhesion of the coatings, with it being widely recognized in the scientific literature. The improvement of osseointegration capacity is associated with the surface roughness patterns obtained with a femtosecond laser radiation [31–33,35,36]. In previous works, the authors have used the Archimedean method, image analysis, Micro-CT, and scanning electron microscopy to evaluate the macro and microporosity of porous titanium samples, with and without surface modification [12,17,19,20]. However, the experimental protocols, commonly used to characterize porosity and evaluate cell activity (presence of osteoblastic cells and mineralization), are relatively long, expensive, and destructive. In this work, the use of electrical impedance is proposed, not only as an interesting route to semi-quantitatively evaluate porosity but also as a potential changes inherent to surface modification treatments. They can be used to improve osseointegration as well as to detect, in real time, the changes that may occur in the implant/bone interface, during this process.

Hewlett–Packard 4395A (Agilent Technologies, Santa Clara, CA, USA), a network, spectrum, and impedance analyzer, available at IMSE-CNM-CSIC, was the equipment used to perform the electrical impedance measurements, as it is demonstrated in Figure 2a. Impedance measurements represent an affordable method to characterize, in a non-destructive way, different materials, while being especially useful to characterize surface modifications, as shown in different works [41]. To place the manufactured titanium samples on the impedance analyzer, the module HP 16092A was used, as Figure 2b indicates. Figure 2c also proves the implemented circuit by the impedance analyzer, where the sample is placed in the DUT (device under test).

Electrical impedance was measured in the frequency range from 150 MHz to 500 MHz. These measurements were performed three times for samples: before a femtosecond laser treatment, after a femtosecond laser treatment (FS), and with cell cultures (CC). Afterwards, the pore content, the effect of the femtosecond laser treatment [potential oxide layer, new microporosity (pillars), and new additional surface area (generated by the new texture of the roughness pattern)] were evaluated.

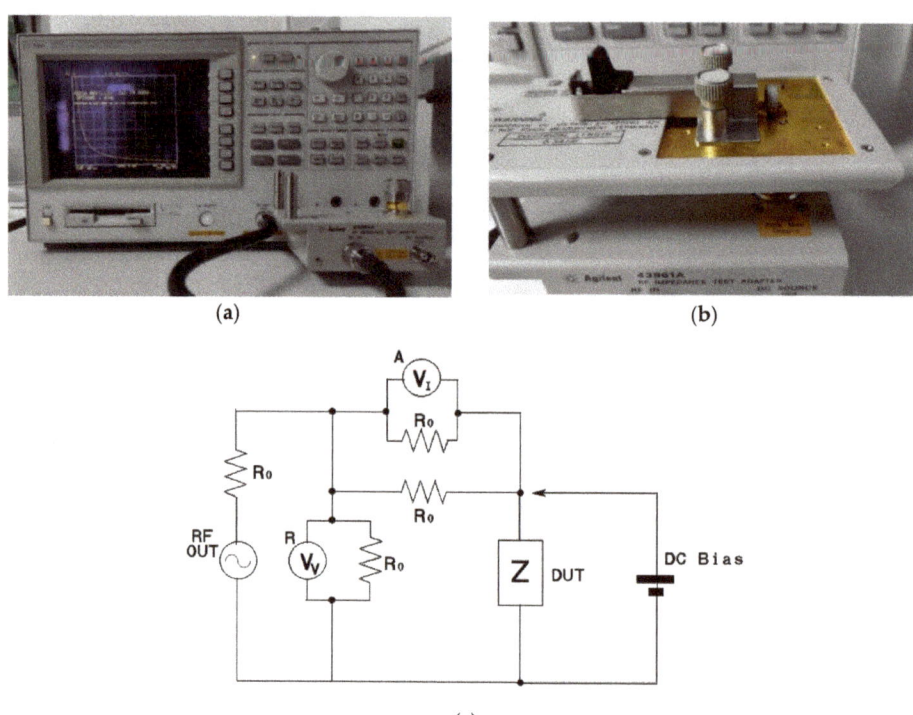

Figure 2. (**a**) impedance analyzer used (Hewlett–Packard 4395A). (**b**) placement of the sample in the HP 16,092 module. (**c**) measurement circuit used by the impedance analyzer, where the source signal is output from RF OUT port. Vv voltmeter is R port receiver that measures a voltage. Vi voltmeter is A port receiver that measures a voltage of Ro to obtain a current.

In-vitro cellular behavior (adhesion and proliferation of osteoblasts) in fully-dense and porous discs (before and after femtosecond treatment) was also evaluated. To get it, MC3T3E1, a murine pre-osteoblast cell line (CRL-2593, from ATCC), was used. Cell proliferation and viability tests were evaluated using AlamarBlue® reagent (Invitrogen, Carlsbad, CA, USA), in accordance with the manufacturer's protocol. The absorbance at 570 nm (oxidized) and 600 nm (reduced) (TECAN, Infinity 200 Pro, Männedorf, Switzerland) was subsequently recorded, and these experiments were performed in triplicate. The results were expressed in terms of mean and standard deviation to perform two-way ANOVA, followed by Tukey's post-test, using SPSS v.22.0 for Windows (IBM Corp., Armonk, NY, USA). The significance level was considered at p values of $p < 0.05$ (*).

Additionally, cell behavior at 21 days was evaluated by the acquisition of images with a scanning electron microscopy (SEM) (Zeiss EVO LS 15 scanning electron microscope) (Zeiss, Oberkochen, Germany). Once the osteoblast cells were grown along the surface of the discs, electrical impedance measurements were obtained again. In order to assess the effect of cell growth on the material and its electrical properties, this process uses the same equipment and configuration as the samples before femtosecond laser treatment and with femtosecond laser treatment (FS).

3. Results and Discussion

3.1. Electrical Characterization of Porous Discs and Femtosecond Laser Treatment

Figure 3 shows, as an example, the graphs obtained for an impedance value of samples, with 30% and 60% porosity volume, before and after femtosecond laser treatment. As can be appreciated in each image, the upper graph in yellow, corresponding to channel 1, represents the modulus |Z| in mΩ of the obtained impedance. The second graph in the color blue, corresponding to channel 2, represents the phase θz of the impedance in degrees. In this figure, it can be observed that the marker is at 250 MHz, and it shows the impedance and phase corresponding to that frequency. It allows one to see how the modulus impedance value increases with the percentage of porosity volume and with the femtosecond laser treatment.

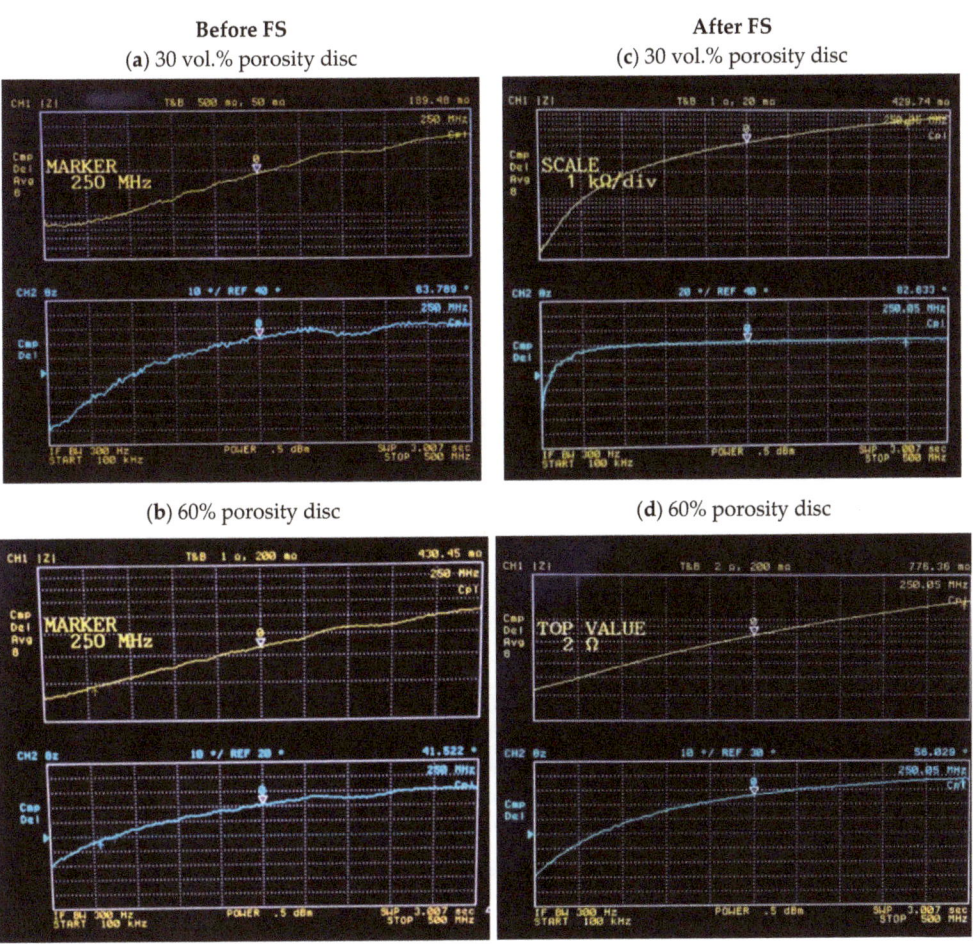

Figure 3. Graphs obtained with the modulus and phase impedance of titanium samples, with 30% and 60% porosity volume. (**a**) sample with 30% porosity volume before FS, (**b**) sample with 60% porosity volume before FS, (**c**) sample with 30% porosity volume after FS, and (**d**) sample with 60% porosity volume after FS.

The absolute value of impedance of the different samples, before and after the application of the laser, at the range of frequencies studied, is proven in Figure 4. It can be seen how the impedance values increase directly proportional to frequency, as expected, and in relation with other previous studies [41]. It is distinguished for a specific size of pore, by different samples with different porosity values. It is also observed that more sensitivity is obtained at higher frequencies, i.e., porosity volume can be better distinguished at higher frequencies, while still in relation with previous studies [41].

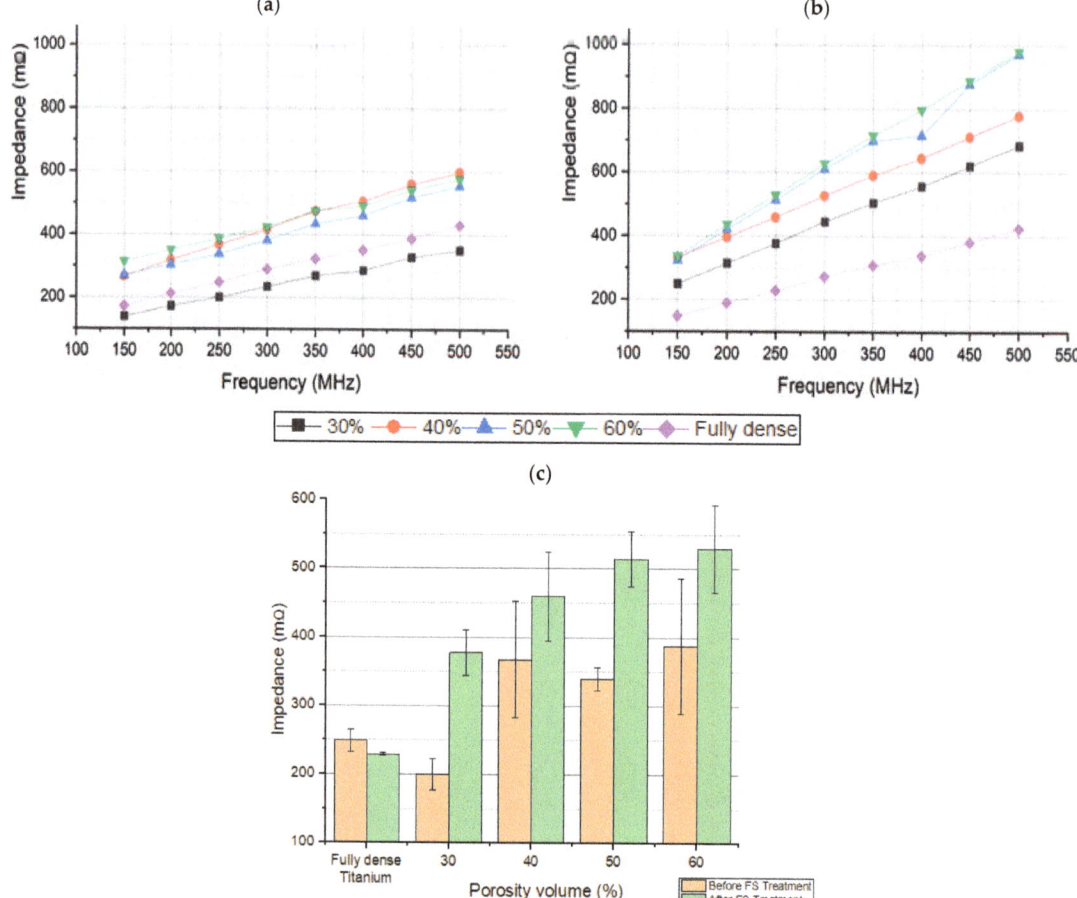

Figure 4. Impedance of the different samples with pore size 100–200 μm (**a**) before FS, And (**b**) after FS. The impedance values are directly proportional to the frequency, with higher sensitivities at higher frequencies. (**c**) Impedance values vs porosity volumes at 250 MHz.

Laser application produces an important increase in the absolute value of the impedance (Figure 4b). In Figure 4c, the dispersion of measurements at the specific frequency of 250 MHz is observed. It is also shown the importance of an increase in the absolute values of the impedance of the sample after the treatment with the femtosecond laser, in comparison with the samples without laser treatment. These results can be explained by the change on the surface, which was produced by the laser treatment.

The values of the electrical impedance modulus, together with the values of the phase (imaginary part of the impedance), are shown for three different frequencies in Table 1, for the samples before FS. With these measurements, the relationship between the increase in electrical impedance and the increase in porosity can be seen.

Table 1. Impedance modulus and phase values for different titanium samples with pore size 100–200 μm before FS.

Frequency	Fully-Dense Titanium		30 vol.%		40 vol.%		50 vol.%		60 vol.%	
	|Z| (mΩ)	θ	|Z| (mΩ)	θ	|Z| (mΩ)	θ	|Z| (mΩ)	θ	|Z| (mΩ)	θ
150 MHz	172.23	58.78°	139.32	52.88°	266.70	51.12°	271.96	48.83°	315.43	35.42°
250 MHz	249.02	66.85°	199.57	65.72°	367.48	56.82°	339.57	58.15°	387.5	44.93°
500 MHz	430.92	75.01°	351.56	72.24°	598.41	59.71°	555.33	65.44°	577.31	52.74°

Note: Impedance measurements have an error of ±0.1. |Z| is the impedance modulus measured in milliohms (mΩ), and θ is the phase measured in degrees.

Similarly, for the case of the samples after FS, data were collected at three different frequencies in Table 2. It shows the increase in the electrical impedance values with the increase in the porosity percentage. Moreover, these values have increased with respect to those obtained for the samples before FS.

Table 2. Impedance modulus and phase values for different titanium samples with pore size 100–200 μm after FS.

Frequency	Fully-Dense Titanium		30 vol.%		40 vol.%		50 vol.%		60 vol.%	
	|Z| (mΩ)	θ	|Z| (mΩ)	θ	|Z| (mΩ)	θ	|Z| (mΩ)	θ	|Z| (mΩ)	θ
150 MHz	148.81	63.37°	248.74	63.03°	333.19	53.33°	324.02	74.84°	337.07	72.66°
250 MHz	229.25	73.21°	377.33	69.02°	459.78	61.16°	514.23	78.37°	529.88	75.58°
500 MHz	426.06	79.17°	685.04	73.85°	779.71	70.5°	972.62	81.70°	981.51	77.34°

Note: Impedance measurements have an error of ±0.1. |Z| is the impedance modulus measured in milliohms (mΩ), and θ is the phase measured in degrees.

3.2. Electrical Characterization of Osteoblast Cell Cultures: Cell Proliferation and Viability Tests

Cell proliferation results (Figure 5) prove that there is better osteoblastic growth on the surface of samples with 30% porosity than on fully dense samples. It is observed that, in the 30% discs, the proliferation % is double compared to FD, making the increase statistically significant.

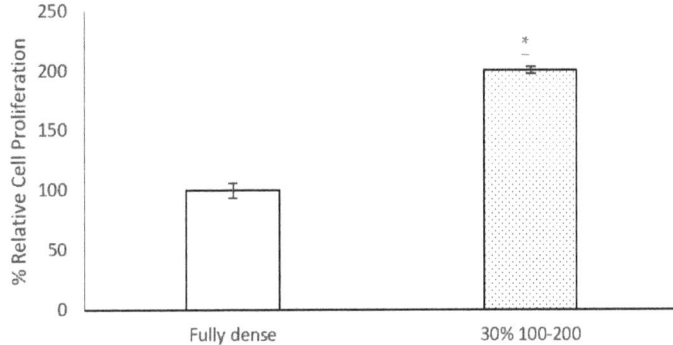

Figure 5. Percentage relative cell proliferation in MC3T3–E1 cultures after femtosecond laser treatment (by AlamarBlue). Results are represented vs fully dense growth. Significance level at p value < 0.05 (*).

Figure 6 also shows, as an example, a SEM image of adhered cells on a porous titanium surface modified with laser radiation, where the in-vitro osseointegration can be verified. These results correlate with the measured impedance of the samples, especially at higher frequencies, where higher impedance is measured for the 30% samples. On the other hand, Figure 7a,b show the absolute values of impedance measurements in samples with osteoblast cell cultures. An increase in the absolute value of the measured impedance is observed repeatedly, in line with previous studies [43]. A comparison between samples without treatment, with femtosecond laser treatment, and samples with cell cultures is presented in the whole range of frequencies. In Figure 7a, the absolute value of fully dense samples with osteoblast cell cultures is higher for all frequencies to the impedance measurements for samples with femtosecond laser treatment and without it. Absolute values of impedance measurements for 30% porosity volume samples is reproduced in Figure 7b. Again, the highest values of impedance are observed for samples with cell cultures.

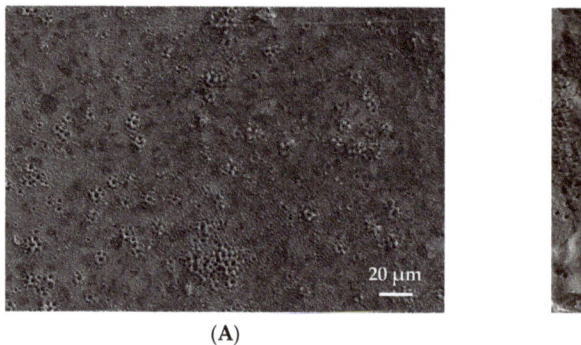
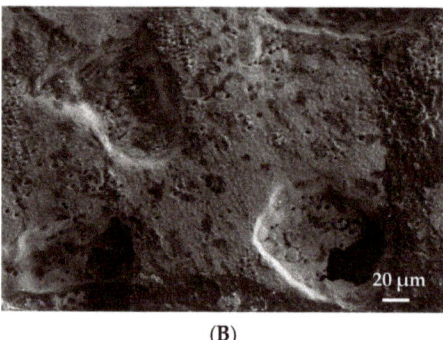

Figure 6. SEM Image of a samples of porous titanium after the femtosecond surface treatment in-vitro culture (observe the presence of adhered osteoblasts). (**A**) Fully-dense; (**B**) 30 vol. % (100–200 μm).

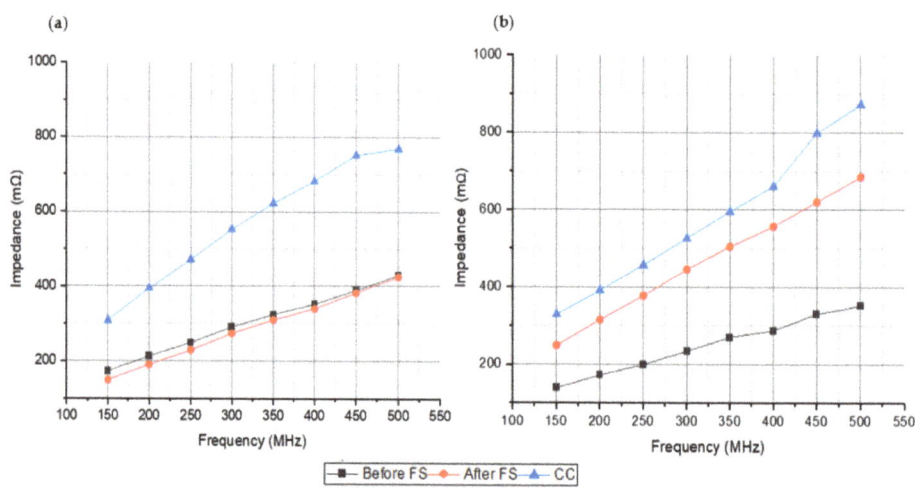

Figure 7. (**a**) Impedance values of discs with cell cultures (CC) vs previous impedance values before FS and after FS in fully dense sample. (**b**) impedance values in 30% porosity volume sample.

The increase in the absolute value of the impedance due to the osteoblast cell growth is, in all cases, higher than the increase due to the femtosecond laser treatment. This increase suggests that the capacitive layer, formed by the cell membranes on the disc's surface, has a more relevant effect on the overall impedance. Therefore, both cases can also be distinguished with electrical impedance in an affordable, non-destructive, and simple way (Table 3).

Table 3. Impedance modulus and phase values for different titanium samples with osteoblast cell cultures (CC).

Frequency	Fully-Dense Titanium		30 vol.%	
	|Z| (mΩ)	θ	|Z| (mΩ)	Θ
150 MHz	308.6	67.55°	328.967	50.62°
250 MHz	472.01	68.19°	457.557	63.71°
500 MHz	768.385	67.24°	873.023	74.54°

Note: Impedance measurements have an error of ±0.1.

The experimental protocols commonly used to characterize porosity and evaluate cell activity (presence of osteoblastic cells and mineralization) are relatively long, expensive, and destructive. To avoid the mentioned issues, an electrical impedance measurement protocol is proposed as an alternative. This simple protocol makes it possible to evaluate the pore content of a material and to detect physical and chemical changes that may occur on the surface of implants; resulting from surface modification treatments and the interaction of the implant with the surrounding tissue (osseointegration process). In this context, a clear direct relation is observed between electrical impedance and pore content, femtosecond laser treatment and cellular activity (adhesion and proliferation of osteoblasts).

Electrical impedance spectroscopy was used for the characterization of different porous titanium samples, modified with a femtosecond laser. Different volumes of porosities could be distinguished, in line with previous studies. The treatment of the samples with the femtosecond laser produced a significant increase in the absolute value of the electrical impedance, which can be perfectly characterized in a wide range of frequencies. It made them be more sensitive at higher frequencies. Furthermore, the in-vitro cellular behavior (adhesion and proliferation of osteoblasts) in porous discs was also evaluated, and an increase in the absolute values of the impedance was observed for all titanium samples where cells were cultivated, according to previous works. This increase in the impedance values was higher, in all cases, than the increase in the impedance produced by the treatment with femtosecond laser, for all the tested samples.

4. Conclusions

The method followed in this study has proven to be effective for the characterization of the treatment of the surface topography of titanium implants with femtosecond laser, through electrical impedance measurements. It has also demonstrated to be a valid tool in the study of subsequent osseointegration processes, with the characterization of the growth of osteoblast cell cultures in the same samples.

Significant differences can be appreciated in the impedance values obtained for the samples at different percentages of porosity volume and the fully dense sample. This occurs for the samples before FS treatment and after FS treatment, being higher frequencies (around 500 MHz) and the ones that show a better sensitivity to impedance changes. For the samples treated with femtosecond laser, a huge increase (higher than 50% in some cases) in the electrical impedance values is observed, compared to the values obtained for the untreated samples. This fact shows that the modification of the surfaces of the samples favors the increase in the electrical impedance.

In addition, it has also been observed that the cell culture (MC3T3-E1) influences the electrical impedance values obtained for the samples. The effect of the increase in electrical impedance is greater in the fully dense samples, where an increase higher than 70% is found at 500 MHz.

As a future work, it would be interesting to design a bioimpedance device that allowed real-time measurements while the cells are growing on the implant sample inside the incubator. This would enable one to obtain interesting data of the process of osseointegration, in the implant, in order to study possible future uses of the technique in medical applications.

Author Contributions: Conceptualization, project administration, supervision, methodology, A.O. and Y.T., investigation, formal analysis, validation, P.N., M.G., Á.R. and M.R.-A., discussion and writing—original draft preparation, all the authors. All authors have read and agreed to the published version of the manuscript.

Funding: This work was supported by the Ministry of Science and Innovation of Spain under the grant PID2019-109371GB-I00, by the Junta de Andalucía–FEDER (Spain) through the Project Ref. US-1259771 and by the Junta de Andalucía-Proyecto de Excelencia (Spain) P18-FR-2038.

Institutional Review Board Statement: Not applicable.

Informed Consent Statement: Not applicable.

Data Availability Statement: The data presented in this study are available on request from the corresponding author.

Conflicts of Interest: The authors declare no conflict of interest.

References

1. Liu, W.; Liu, S.; Wang, L. Surface Modification of Biomedical Titanium Alloy: Micromorphology, Microstructure Evolution and Biomedical Applications. *Coatings* **2019**, *9*, 249. [CrossRef]
2. Echeverry-Rendón, M.; Galvis, O.; Giraldo, D.A.Q.; Pavón-Palacio, J.-J.; López-Lacomba, J.L.; Jimenez-Pique, E.; Anglada, M.; Robledo, S.M.; Castaño, J.G.; Echeverria, F. Osseointegration improvement by plasma electrolytic oxidation of modified titanium alloys surfaces. *J. Mater. Sci. Mater. Med.* **2015**, *26*, 72. [CrossRef]
3. Kaur, M.; Singh, K. Review on titanium and titanium based alloys as biomaterials for orthopaedic applications. *Mater. Sci. Eng. C* **2019**, *102*, 844–862. [CrossRef]
4. Khorasani, A.M.; Goldberg, M.; Doeven, E.H.; Littlefair, G. Titanium in biomedical applications—properties and fabrication: A review. *J. Biomater. Tissue Eng.* **2015**, *5*, 593–619. [CrossRef]
5. Niinomi, M.; Nakai, M. Titanium-based biomaterials for preventing stress shielding between implant devices and bone. *Int. J. Biomater.* **2011**, *2011*. [CrossRef] [PubMed]
6. Niinomi, M.; Liu, Y.; Nakai, M.; Liu, H.; Li, H. Biomedical titanium alloys with Young's moduli close to that of cortical bone. *Regen. Biomater.* **2016**, *3*, 173–185. [CrossRef] [PubMed]
7. Pałka, K.; Pokrowiecki, R. Porous Titanium Implants: A Review. *Adv. Eng. Mater.* **2018**, *20*, 1700648. [CrossRef]
8. Naebe, M.; Shirvanimoghaddam, K. Functionally graded materials: A review of fabrication and properties. *Appl. Mater. Today.* **2016**, *5*, 223–245. [CrossRef]
9. Xu, J.; Weng, X.-J.; Wang, X.; Huang, J.-Z.; Zhang, C.; Muhammad, H.; Ma, X.; Liao, Q.-D. Potential use of porous titanium–niobium alloy in orthopedic implants: Preparation and experimental study of its biocompatibility in vitro. *PLoS ONE* **2013**, *8*, e79289. [CrossRef]
10. Teixeira, L.N.; Crippa, G.E.; Lefebvre, L.P.; de Oliveira, P.T.; Rosa, A.L.; Beloti, M.M. The influence of pore size on osteoblast phenotype expression in cultures grown on porous titanium. *Int. J. Oral Maxillofac. Surg.* **2012**, *41*, 1097–1101. [CrossRef] [PubMed]
11. Taniguchi, N.; Fujibayashi, S.; Takemoto, M.; Sasaki, K.; Otsuki, B.; Nakamura, T.; Matsushita, T.; Kokubo, T.; Matsuda, S. Effect of pore size on bone ingrowth into porous titanium implants fabricated by additive manufacturing: An in vivo experiment. *Mater. Sci. Eng. C* **2016**, *59*, 690–701. [CrossRef]
12. Civantos, A.; Domínguez, C.; Pino, R.J.; Setti, G.; Pavón, J.J.; Martínez-Campos, E.; Garcia Garcia, F.J.; Rodríguez, J.A.; Allain, J.P.; Torres, Y. Designing bioactive porous titanium interfaces to balance mechanical properties and in vitro cells behavior towards increased osseointegration. *Surf. Coat. Technol.* **2019**, *368*, 162–174. [CrossRef]
13. Muñoz, S.; Pavón, J.; Rodríguez-Ortiz, J.A.; Civantos, A.; Allain, J.P.; Torres, Y. On the influence of space holder in the development of porous titanium implants: Mechanical, computational and biological evaluation. *Mater. Charact.* **2015**, *108*, 68–78. [CrossRef]
14. Rodríguez, Á.; Trueba, P.; Amado, J.M.; Tobar, M.J.; Giner, M.; Amigó, V.; Torres, Y. Surface modification of porous titanium discs using femtosecond laser structuring. *Metals* **2020**, *10*, 748. [CrossRef]
15. Devgan, S.; Sidhu, S.S. Evolution of surface modification trends in bone related biomaterials: A review. *Mater. Chem. Phys.* **2019**, *233*, 68–78. [CrossRef]
16. Ibrahim, M.Z.; Sarhan, A.A.D.; Yusuf, F.; Hamdi, M. Biomedical materials and techniques to improve the tribological, mechanical and biomedical properties of orthopedic implants—A review article. *J. Alloys Compd.* **2017**, *714*, 636–667. [CrossRef]

17. Civantos, A.; Martínez-Campos, E.; Ramos, V.; Elvira, C.; Gallardo, A.; Abarrategi, A. Titanium Coatings and Surface Modifications: Toward Clinically Useful Bioactive Implants. *ACS Biomater. Sci. Eng.* **2017**, *3*, 1245–1261. [CrossRef]
18. Vorobyev, A.Y.; Chunlei, G. Femtosecond laser structuring of titanium implants. *Appl. Surf. Sci.* **2007**, *253*, 7272–7280. [CrossRef]
19. Wennerberg, A.; Albrektsson, T. Effects of titanium surface topography on bone integration: A systematic review. *Clin. Oral Implants Res.* **2009**, *20*, 172–184. [CrossRef] [PubMed]
20. Trueba, P.; Giner, M.; Rodríguez, Á.; Beltrán, A.M.; Amado, J.M.; Montoya-García, M.J.; Rodríguez-Albelo, L.M.; Torres, Y. Tribo-mechanical and cellular behavior of superficially modified porous titanium samples using femtosecond laser. *Surf. Coat. Technol.* **2021**, *422*, 127555. [CrossRef]
21. Schweitzer, L.; Cunha, A.; Pereira, T.; Mika, K.; Rego, A.M.B.D.; Ferraria, A.M.; Kieburg, H.; Geissler, S.; Uhlmann, E.; Schoon, J. Preclinical in vitro assessment of submicron-scale laser surface texturing on Ti6Al4V. *Materials* **2020**, *13*, 5342. [CrossRef]
22. Klos, A.; Sedao, X.; Itina, T.E.; Helfenstein-Didier, C.; Donnet, C.; Peyroche, S.; Vico, L.; Guignandon, A.; Dumas, V. Ultrafast laser processing of nanostructured patterns for the control of cell adhesion and migration on titanium alloy. *Nanomaterials* **2020**, *10*, 864. [CrossRef] [PubMed]
23. Martínez-Calderon, M.; Martín-Palma, R.J.; Gómez-Aranzadi, M.; García-Ruiz, J.P.; Olaizola, A.M.; Manso-Silva, M. Biomimetic hierarchical micro/nano texturing of TiAlV alloys by femtosecond laser processing for the control of cell adhesion and migration. *Phys. Rev. Mater.* **2020**, *4*, 056008. [CrossRef]
24. Oliveira, V.; Ausset, S.; Vilar, R. Surface micro/nanostructuring of titanium under stationary and non-stationary femtosecond laser irradiation. *Appl. Surf. Sci.* **2009**, *255*, 7556–7560. [CrossRef]
25. Bonse, J.; Höhm, S.; Koter, R.; Harlet, M.; Spaltmann, D.; Pentzinen, S.; Rosenfeld, A.; Krüger, J. Tribological performance of sub-100-nm femtosecond laser-induced periodic surface structures on titanium. *Appl. Surf. Sci.* **2016**, *374*, 190–196. [CrossRef]
26. Shinonaga, T.; Kinoshita, S.; Okamoto, Y.; Tsukamoto, M.; Okada, A. Formation of Periodic Nanostructures with Femtosecond Laser for Creation of New Functional Biomaterials. *Procedia CIRP* **2016**, *42*, 57–61. [CrossRef]
27. Schnell, G.; Duenow, U.; Seitz, H. Effect of Laser Pulse Overlap and Scanning Line Overlap on Femtosecond Laser-Structured Ti6Al4V Surfaces. *Materials* **2020**, *13*, 969. [CrossRef]
28. Rupp, F.; Gittens, R.A.; Scheideler, L.; Marmur, A.; Boyan, B.D.; Schwartz, Z.; Geis-Gerstorfer, J. A review on the wettability of dental implant surfaces I: Theoretical and experimental aspects. *Acta Biomater.* **2014**, *10*, 2894–2906. [CrossRef] [PubMed]
29. Raimbault, O.; Benayoun, S.; Anselme, K.; Mauclair, C.; Bourgade, T.; Kietzig, A.-M.; Girard-Lauriault, P.-L.; Valette, S.; Donnet, C. The effects of femtosecond laser-textured Ti-6Al-4V on wettability and cell response. *Mater. Sci. Eng. C* **2016**, *69*, 311–320. [CrossRef] [PubMed]
30. Liu, Y.; Rui, Z.; Cheng, W.; Song, L.; Xu, Y.; Li, R.; Zhang, X. Characterization and evaluation of a femtosecond laser-induced osseointegration and an anti-inflammatory structure generated on a titanium alloy. *Regen. Biomater.* **2021**, *8*, rbab006. [CrossRef]
31. Li, C.; Yang, Y.; Yang, L.; Shi, Z.; Yang, P.; Cheng, G. In vitro bioactivity and biocompatibility of bio-inspired Ti-6Al-4V alloy surfaces modified by combined laser micro/nano structuring. *Molecules* **2020**, *25*, 1494. [CrossRef] [PubMed]
32. Dumas, V.; Guignandon, A.; Vico, L.; Mauclair, C.; Zapata, X.; Linossier, M.T.; Bouleftour, W.; Granier, J.; Peyroche, S.; Dumas, J.-C.; et al. Femtosecond laser nano/micro patterning of titanium influences mesenchymal stem cell adhesion and commitment. *Biomed. Mater.* **2015**, *10*, 055002. [CrossRef]
33. Shaikh, S.; Singh, D.; Subramanian, M.; Kedia, S.; Singh, A.K.; Singh, K.; Gupta, N.; Sinha, S. Femtosecond laser induced surface modification for prevention of bacterial adhesion on 45S5 bioactive glass. *J. Non. Cryst. Solids* **2018**, *482*, 63–72. [CrossRef]
34. Cunha, A.; Elie, A.-M.; Plawinski, L.; Serro, A.; Rego, A.M.B.D.; Almeida, A.; Urdaci, M.C.; Durrieu, M.-C.; Vilar, R. Femtosecond laser surface texturing of titanium as a method to reduce the adhesion of Staphylococcus aureus and biofilm formation. *Appl. Surf. Sci.* **2016**, *360*, 485–493. [CrossRef]
35. Liang, C.; Wang, H.; Yang, J.; Cai, Y.; Hu, X.; Yang, Y.; Li, B.; Li, H.; Li, H.; Li, C.; et al. Femtosecond laser-induced micropattern and Ca/P deposition on Ti implant surface and its acceleration on early osseointegration. *ACS Appl. Mater. Interfaces* **2013**, *5*, 8179–8186. [CrossRef] [PubMed]
36. Wang, C.; Hu, H.; Li, Z.; Shen, Y.; Xu, Y.; Zhang, G.; Zeng, X.; Deng, J.; Zhao, S.; Ren, T.; et al. Enhanced Osseointegration of Titanium Alloy Implants with Laser Microgrooved Surfaces and Graphene Oxide Coating. *ACS Appl. Mater. Interfaces* **2019**, *11*, 39470–39483. [CrossRef]
37. Florian, C.; Wonneberger, R.; Undisz, A.; Kimer, V.S.; Wasmuth, K.; Spaltmann, D.; Krüger, J.; Bonse, J. Chemical effects during the formation of various types of femtosecond laser-generated surface structures on titanium alloy. *Appl. Phys. A Mater. Sci. Process* **2020**, *126*, 266. [CrossRef]
38. Pan, J.; Thierry, D.; Leygraf, C. Electrochemical impedance spectroscopy study of the passive oxide film on titanium for implant application. *Electrochim. Acta* **1996**, *41*, 1143–1153. [CrossRef]
39. Menini, R.; Dion, M.-J.; So, S.K.V.; Gauthier, M.; Lefebvre, L.-P. Surface and Corrosion Electrochemical Characterization of Titanium Foams for Implant Applications. *J. Electrochem. Soc.* **2006**, *153*, B13–B21. [CrossRef]
40. El Daym, D.A.A.; Gheith, M.E.; Abbas, N.A.; Rashed, L.A.; El Aziz, Z.A.A. Electrochemical assessment of laser-treated titanium alloy used for dental applications at acidic pH condition (in vitro study). *Dent. Res. J.* **2019**, *16*, 304–309. [CrossRef]
41. Olmo, A.; Hernández, M.; Chicardi, E.; Torres, Y. Characterization and monitoring of titanium bone implants with impedance spectroscopy. *Sensors* **2020**, *20*, 4358. [CrossRef] [PubMed]

42. Chen, X.; Fu, Q.; Jin, Y.; Li, M.; Yang, R.; Cui, X.; Gong, M. In vitro studying corrosion behavior of porous titanium coating in dynamic electrolyte. *Mater. Sci. Eng. C* **2017**, *70*, 1071–1075. [CrossRef]
43. Giner, M.; Olmo, A.; Hernández, M.; Trueba, P.; Chicardi, E.; Civantos, A.; Vázquez, M.Á.; Montoya-García, M.-J.; Torres, Y. Use of impedance spectroscopy for the characterization of in-vitro osteoblast cell response in porous titanium bone implants. *Metals* **2020**, *10*, 1077. [CrossRef]
44. Huang, H.-H. In situ surface electrochemical characterizations of Ti and Ti-6Al-4V alloy cultured with osteoblast-like cells. *Biochem. Biophys. Res. Commun.* **2004**, *314*, 787–792. [CrossRef]
45. Zhiyong, P.; Wong, C.L.; Guofeng, G.; Lin, M.N.; Chwee, T.L.; Jongyoon, H.; Krystyn, J.V.V. Bone Marrow Regeneration Promoted by Biophysically Sorted Osteoprogenitors From Mesenchymal Stromal Cells. *Stemcells Transl. Med.* **2015**, *4*, 56–65.
46. Gamal, W.; Wu, H.; Underwood, I.; Jia, J.; Smith, S.; Bagnaninchi, P.O. Impedance-based cellular assays for regenerative medicine. *Philos. Trans. R. Soc. B Biol. Sci.* **2018**, *373*, 1750. [CrossRef] [PubMed]

Article

Effect of the Processing Parameters on the Porosity and Mechanical Behavior of Titanium Samples with Bimodal Microstructure Produced via Hot Pressing

Ricardo Chávez-Vásconez [1], Sheila Lascano [1,*], Sergio Sauceda [1], Mauricio Reyes-Valenzuela [1], Christopher Salvo [2], Ramalinga Viswanathan Mangalaraja [3], Francisco José Gotor [4], Cristina Arévalo [5] and Yadir Torres [5]

[1] Departamento de Ingeniería Mecánica, Universidad Técnica Federico Santa María, Avenida Vicuña Mackenna 3939, Santiago 8940572, Chile; ricardo.chavezv@usm.cl (R.C.-V.); sergio.sauceda@usm.cl (S.S.); mauricio.reyes@usm.cl (M.R.-V.)

[2] Departamento de Ingeniería Mecánica, Facultad de Ingeniería, Universidad del Bío-Bío, Avda. Collao 1202, Casilla 5-C, Concepción 4081112, Chile; csalvo@ubiobio.cl

[3] Departamento de Ingeniería de Materiales, Universidad de Concepción, Edmundo Larenas 270, Concepción 4070409, Chile; mangal@udec.cl

[4] Instituto de Ciencia de Materiales de Sevilla (CSIC-US), Américo Vespucio 49, 41092 Sevilla, Spain; francisco.gotor@icmse.csic.es

[5] Departamento de Ingeniería y Ciencia de los Materiales y del Transporte, Escuela Politécnica Superior, Calle Virgen de África 7, 41011 Seville, Spain; carevalo@us.es (C.A.); ytorres@us.es (Y.T.)

* Correspondence: sheila.lascano@usm.cl

Citation: Chávez-Vásconez, R.; Lascano, S.; Sauceda, S.; Reyes-Valenzuela, M.; Salvo, C.; Mangalaraja, R.V.; Gotor, F.J.; Arévalo, C.; Torres, Y. Effect of the Processing Parameters on the Porosity and Mechanical Behavior of Titanium Samples with Bimodal Microstructure Produced via Hot Pressing. *Materials* **2022**, *15*, 136. https://doi.org/10.3390/ma15010136

Academic Editor: Javier Gil

Received: 5 December 2021
Accepted: 23 December 2021
Published: 25 December 2021

Publisher's Note: MDPI stays neutral with regard to jurisdictional claims in published maps and institutional affiliations.

Copyright: © 2021 by the authors. Licensee MDPI, Basel, Switzerland. This article is an open access article distributed under the terms and conditions of the Creative Commons Attribution (CC BY) license (https://creativecommons.org/licenses/by/4.0/).

Abstract: Commercially pure (c.p.) titanium grade IV with a bimodal microstructure is a promising material for biomedical implants. The influence of the processing parameters on the physical, microstructural, and mechanical properties was investigated. The bimodal microstructure was achieved from the blends of powder particles with different sizes, while the porous structure was obtained using the space-holder technique (50 vol.% of ammonium bicarbonate). Mechanically milled powders (10 and 20 h) were mixed in 50 wt.% or 75 wt.% with c.p. titanium. Four different mixtures of powders were precompacted via uniaxial cold pressing at 400 MPa. Then, the specimens were sintered at 750 °C via hot pressing in an argon gas atmosphere. The presence of a bimodal microstructure, comprised of small-grain regions separated by coarse-grain ones, was confirmed by optical and scanning electron microscopies. The samples with a bimodal microstructure exhibited an increase in the porosity compared with the commercially available pure Ti. In addition, the hardness was increased while the Young's modulus was decreased in the specimens with 75 wt.% of the milled powders (20 h).

Keywords: porous titanium; bimodal microstructure; hot-pressing; powder metallurgy; mechanical milling; mechanical behavior

1. Introduction

The research field related to biomedical materials has been grown in recent decades as a result of the demand for implants for bone replacement [1]. Titanium and its alloys are considered the most suitable option for biomedical applications due to their low density, high biocompatibility, specific mechanical strength, and corrosion resistance, as well as their in vitro and in vivo acceptable behavior [2,3]. However, there is mismatch between the stiffness of bone and metallic biomaterials. This incompatibility generates the stress-shielding phenomenon which promotes bone resorption at the implant–bone interface and can even induce implant failure [4,5].

It is well known that porous structures exhibit lower elastic modulus than their fully-dense counterparts [3,6]. It has also been identified that titanium components that possess optimal macro/micro porosities allow for the tuning of the elastic modulus in a considerably wide range, which also favors bone cell ingrowth and vascularization [3].

Among the manufacturing processes employed to obtain porous metallic materials are tied freeze casting [7,8], selective laser melting (SLM) [9,10], field assisted sintering (FAST) [11,12], and powder metallurgy [13,14]. Powder metallurgy, in combination with the space-holder technique, represents a cost-effective and flexible way to obtain components with a high-degree of porosity (35–80%) and a homogeneous distribution of pores throughout the volume [15–17]. Particles commonly used as space-holders include NH_4HCO_3 [5,18–22], NaCl [17,18,21,23,24], starch [25,26], Mg [27–29], PMMA [30,31], saccharose crystals [26,32], PVA [33], and carbamide [15,34,35], which can be eliminated at a relatively low temperature, or can be easily removed by a dissolution process, generally in water [3]. NH_4HCO_3 is one of the preferred spacer particles due to its moderate decomposition temperature, which makes it easily and completely removable, ensuring a low uptake of impurities such as oxygen, nitrogen, and carbon [17].

However, although this increase in porosity leads to a reduction in Young's modulus, it also reduces mechanical strength. Previous work [36] has shown that, to obtain Young's modulus close to that of human bone, porosity percentages greater than 45% are required. This leads to a drastic reduction in mechanical resistance, below what is required for bone replacement [17,36,37].

Concerning the mechanical performance, titanium components that exhibit a fine grain structure have gained attention due to the grain boundary strengthening effect [38,39], which results in an important increase in strength and hardness when compared to their coarse-grained counterparts. However, regardless of the processing methods, fine-grained metallic materials usually suffer from poor plastic deformation at room temperature [40]. An approach based on tailoring the microstructure by the development of bimodal/multimodal grain size distributions has been implemented, aiming to optimize the balance between ductility and strength [41,42]. Thus, fine grains provide a strength increase, while coarse grains allow ductility to be retained [43]. This method presents the potential to produce materials with a porous structure and suitable combination of elastic modulus, strength, and ductility, which allows an adequate balance between biological and mechanical behavior for biomedical applications.

In recent investigations, titanium samples with a bimodal microstructure synthesized by spark plasma sintering (SPS) have shown yielding stress and ultimate tensile strength values that exceeded twice the value of conventional α-titanium coarse-grained components [44] and hardness that exceeded the nominal value of commercially pure (c.p.) titanium by 3–4 times [45]. Hot consolidation techniques, such as hot pressing sintering (HP) [46], hot isostatic pressing (HIP) [47], and SPS [45], allow the fabrication of components with bimodal structure due to their characteristics of rapid heat/cooling rates and low sintering temperatures, which limits the excessive grain growth preserving the fine-grained microstructure when compared to the conventional sintering processes [48,49]. Besides, it has also been shown that HP increases the chemical homogeneity of the phases present in titanium alloys, and effectively controls grain growth in their microstructure [50,51]. However, the formation of porous structures, which is essential for Ti implants, by pressure-assisted techniques is rather complicated. To circumvent this problem, the use of the space-holder methodology is proposed in this work.

Therefore, the aim of this research is to obtain porous samples of titanium with bimodal microstructure via powder metallurgy from blends of powder particles with different sizes, using NH_4HCO_3 as a temporary spacer particle, and consolidated by hot pressing. The effect of the processing parameters on the bimodal microstructure, porosity, and microhardness of titanium samples is studied.

2. Materials and Methods

2.1. Starting Materials Preparation and Characterization

C.p. titanium grade IV was used as raw powder to produce the bimodal microstructure. According to the supplier's information (Alfa Aesar, Tewksbury, MA, USA), the mean particle size was less than 45 μm. In order to form the bimodal microstructure, the size and morphology of titanium powders were modified by mechanical milling in a planetary mill RETSCH® PM 400 (Retsch, Haan, Germany). Thus, 20 g of titanium powder, ZrO_2 (YSZ) ceramic balls with 5 and 10 mm of diameter (ball to powder weight ratio 10:1), and a 2 wt.% of stearic acid as processing control agent (PCA) were placed in a 250 cm³ ZrO_2 vial. One of the studied parameters was the influence of milling time over the obtained bimodal microstructure; thus, the milling was done using two effective milling times: 10 and 20 h. The milling procedures were carried out at 250 rpm, in cycles of milling and resting of 30 min, in argon atmosphere (ultrapure with <3 ppm O_2) to prevent excessive heating and the oxidation of the powder, respectively. From here, the resulting powders are named Ti_{10} and Ti_{20}.

Regarding the amount of fine powder, two different blends of powders were established: 50 wt.% and 75 wt.% of milled powder, where the remaining percentage corresponds to the as-received Ti powder. The processing parameters were chosen in order to detect improvements related to the capability of retaining porosities, although the high densification technique and the mechanical properties achieved this due to the hardening effect during milling stage. Table 1 summarizes the powder parameters used in order to produce each blend. Hence, the four blends are named Ti_{10-50}, Ti_{10-75}, Ti_{20-50}, and Ti_{20-75}.

Table 1. Powder parameters for blend processing.

Milling Time (h)	Nomenclature	Portion of Milled Powder (wt.%)	Nomenclature
10	Ti_{10}	50	Ti_{10-50}
		75	Ti_{10-75}
20	Ti_{20}	50	Ti_{20-50}
		75	Ti_{20-75}

Before the consolidation stage, morphological analysis of starting powders and NH_4HCO_3 particles was carried out by Scanning Electron Microscopy (SEM) with a JSM-6380LV SEM JEOL (JEOL Ltd., Tokyo, Japan) microscope equipped with an Energy Dispersive X-Ray Spectroscopy (EDS) device, according to the ASTM F1877 [52] standard. A particle size analysis of the milled and as-received Ti powders were performed using laser diffraction analysis in an Analysette 22 (Frisch GmbH, Idar-Oberstein, Germany) equipment, whereas the granulometric test of NH_4HCO_3 study was accomplished by sieving on an SS3 Gilson® (Gilson Incorporated, Global Headquarters, Middleton, WI, USA) device, according to ASTM E2651 standard [53]. Furthermore, X-ray diffraction (XRD) analysis was performed on the as-received and milled (Ti_{10} and Ti_{20}) powders. The XRD patterns were obtained with STOE STADI MP (STOE & Cie GmbH, Darmstadt, Germany) using CuKα1 radiation (λ = 0.15406 nm) and a step size of 0.12°; the scan was recorded in the 2θ range comprised from 20° to 120. Once the starting powders were characterized, the specimens were consolidated in order to study the effect of the processing parameters on the final properties of the sintered samples.

2.2. Green Specimens Preparation and Hot Consolidation

Before consolidation of specimens, powder blends (Table 1) and NH_4HCO_3 (50 vol.%) particles were mixed for 40 min in a TURBULA® T2F (WAB, Muttenz, Switzerland) to reach good homogenization. The amount of prepared mixture was stablished regarding the final dimensions of the specimens, having a cylindrical geometry with a diameter of 12.7 mm and height of 20 mm, according to ISO 13314 [54] and ASTM E9 [55] standards. Next, uniaxial cold compaction of mixtures was performed in a universal testing machine Zwick/Roell Z100 (Zwick/Roell, Ulm, Germany) in two stages. The first one was at a compaction pressure of 20 MPa (3 mm/min) and the second one at 500 MPa (5 mm/min), using 2 min dwell time and 15 min unloading time.

Subsequently, specimens were sintered in a hot press HP20-4560-20 (Thermal Technology LLC, Santa Rosa, CA, USA) in two stages: first, at 100 °C for 1 min in vacuum

(10^{-2} mbar), and then at 750 °C and 15 MPa pressure for 15 min in an argon atmosphere (heating rate in both stages was 10 K/min). A graphite die was used, previously coated by a boron nitride-sprayed film to avoid direct contact between the die and the green specimen. The first stage was performed to remove impurities and humidity, and to start the elimination of the space-holder particles, and the second one to continue removing remnants of the spacers and to sinter the green specimens. After sintering, the titanium samples were ground to remove the boron nitride residues. Then, the samples were characterized in order to study the microstructural aspects, porosity, and mechanical properties achieved after the sintering stage.

2.3. Microstructural and Mechanical Characterization of Sintered Samples

The specimens were prepared for microstructural analysis and microhardness measurements by conventional steps of metallographic preparation, according to ASTM E3 [28], with a final step consisting of a mechanical-chemical polishing using colloidal silica and hydrogen peroxide.

Size, morphology and porosity distribution, pores roughness, and bimodal microstructure were analyzed by image analysis (IA) of the micrographs obtained by optical microscopy (OM) in a Nikon Eclipse MA100N (Nikon Corporation, Tokyo, Japan) microscope, and by SEM in a Quanta FEG-250 SEM (Thermo Fisher Scientific, Waltham, MA, USA). IA was performed using Image Pro Plus software. The evaluated and studied porosity parameters were: (i) total porosity ($P(IA)$), (ii) equivalent diameter of pores (D_{eq}), and (iii) pore shape factor ($F_f = 4\pi A/(PE)^2$), where A is the pore area and PE is its perimeter. An F_f value close to 1 suggests a rounded pore, while a value close to 0 suggest a needle-shaped pore.

An analysis of variance (ANOVA) was carried out by a Tukey's test using Statgraphics® software, considering a significance level of $p < 5\%$.

The microhardness measurements were performed in an HMV-G (Shimadzu, Kyoto, Japan) tester, applying a 98.07 mN load for 10 s.

Young's modulus, E_p, was estimated by the Nielsen's equation [56], expressed as follows:

$$E_p = E_{Ti} \times \left[\frac{\left(1 - \frac{P(IA)}{100}\right)^2}{1 + \left(\frac{1}{F_f} - 1\right) \frac{P(IA)}{100}} \right] \quad (1)$$

where, E_{Ti} is the Young's modulus for c.p. Ti grade IV bulk (~110 GPa [57]), $P(IA)$ is the percentage of total porosity of the sample, and F_f is the shape factor calculated from the results of the image analysis.

In addition, the yield strength values were assessed from the correlation proposed by Jha et al. [58], expressed as follows:

$$\sigma_{y,f} = 0.74 \times \sigma_{y,b} \left(\frac{\rho_f}{\rho_b}\right)^{2.206} \quad (2)$$

where, σ_y is the yield strength, ρ is the material's density, and the subscripts f and b are for the porous and bulky material, respectively. The density values were estimated from the measured values of mass and volume of sintered specimens, and the yield strength and density of Ti used were the ones provided by the raw material supplier. This is a preliminary approach to estimate the yield strength due to the employed model and did not consider the effect of bimodal microstructure.

3. Results and Discussion

3.1. Characterization of Starting Materials

The results of the morphological analysis of the starting powders by SEM images and the particle size distribution, obtained by laser diffraction analysis, are depicted in Figure 1. The SEM images of the as-received titanium powders are shown in the Figure 1a, where an irregular shape is appreciated, typically from its processing via hydrogenation/dehydrogenation. The mean particle size is 50 µm in a range of 10–100 µm (Figure 1e). NH$_4$HCO$_3$ particles, Figure 1b, exhibit a polygonal and cubic morphology with a high dispersion of particle size. Particle size distribution of spacer present a normal distribution with a particle size between 50 µm and 400 µm (Figure 1); this distribution was the result of the sieving stage performed during preparation. Some investigations have shown the relevance of spacer particle size and morphology to generate a structure where the total porosity is controlled in an appropriate manner [17,36,59].

Morphologies and the particle size of milled titanium powders, after 10 h (Ti$_{10}$) and 20 h (Ti$_{20}$), are presented in Figure 1c,g and Figure 1d,h, respectively. In the case of milled Ti, a reduction in particle size and changes in the particle size distribution are obtained as a consequence of the mechanical milling. A mean particle size of 10 µm in the range between 3 and 70 µm was confirmed for Ti$_{10}$ and a bimodal distribution was evidenced for the Ti$_{20}$ with two mean sizes: 15 µm in the 3–30 µm range, and 0.6 µm in a range of 0.2–2 µm. Despite the increase in milling time, the mean particle size of the main distribution of Ti$_{20}$ powders was larger than that of the Ti$_{10}$ powders. This is due to the fact that, in the initial stage of milling, the dominant phenomenon is fracture, whereas, after increasing milling time, the active phenomena are fracture and cold welding [60], which results in an agglomeration of particles and the formation of two distributions for the Ti$_{20}$ powders. This phenomenon has been recognized as the final stage of the particle–particle interactions in dry-milling processes [61].

The XRD patterns of the as-received and milled Ti powders are shown in Figure 2, providing structural information. In the case of the as-received powder, only diffraction peaks that correspond to the Ti-hcp phase were detected. On the other hand, both milled powders (Ti$_{10}$ and Ti$_{20}$) exhibited the Ti-hcp peaks, accompanied by several peaks and a "hump" visualized as 2θ = 35°– 42°. The Ti-hcp peaks of both milled powders are broader in shape than the Ti-hcp peaks of the as-received powders, which suggest that, after milling, the crystallite size of this phase may decrease. This effect is frequently observed in high-energy milled powders due to the continuously welding and fracture processes occurring between them during the process [62]. Studying the pattern of both milled powders in detail, the diffraction peaks of Ti-γ were identified (in 2θ = 36.2°, 42.1°, and 61.2°), which is a metastable phase (FCC) with the Fm3m space group, which should not be confused with the Ti-β phase, which is also a cubic structure (BCC) but has an Im3m space group and should have diffraction peaks in 2θ = 38.5°, 55°, and 69°, peaks that are missing in the pattern. It has been reported that the formation of this metastable phase by ball-milling is driven by the accumulation of partial dislocations and stacking faults induced by high plastic deformation and nanocrystalline grain size [63–67]. In addition, the presence of diffraction peaks of YSZ is associated with the contamination that comes from the milling media.

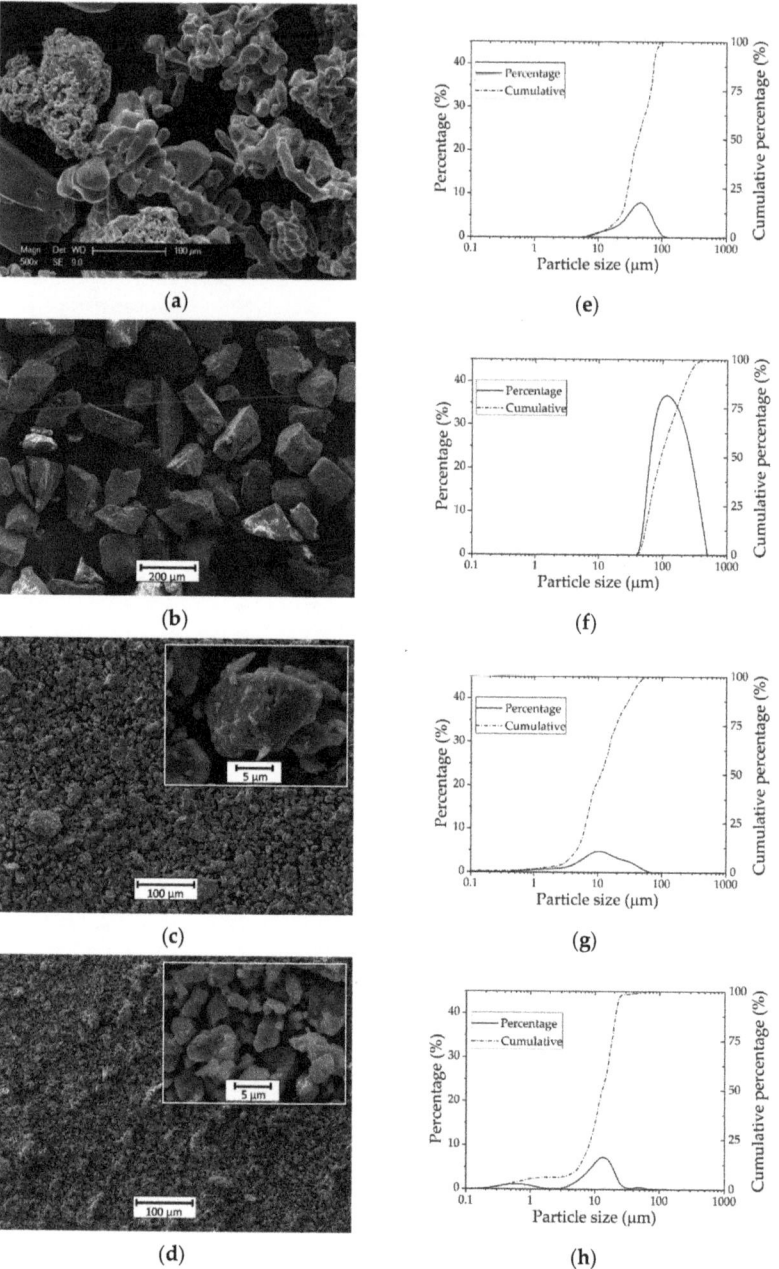

Figure 1. SEM images of: (**a**) as-received titanium powders; (**b**) NH$_4$HCO$_3$ particles; (**c**) Ti$_{10}$ milled powder; (**d**) Ti$_{20}$ milled powder. Particle size distribution of: (**e**) as-received titanium powder; (**f**) NH$_4$HCO$_3$ particles; (**g**) Ti$_{10}$ milled powder; (**h**) Ti$_{20}$ milled powder.

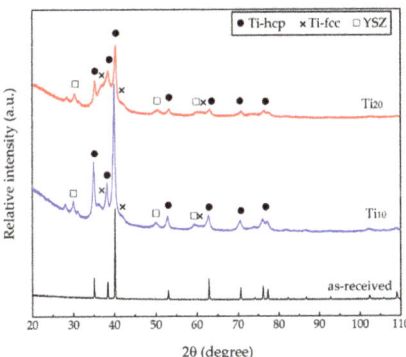

Figure 2. XRD patterns of titanium powders: as-received (black), Ti$_{10}$ (blue), and Ti$_{20}$ (red).

3.2. Microstructural Characterization of Sintered Samples

The results from the optical microscopy of the obtained samples are shown in Figure 3. Figure 3a represents an optical microscopy of the c.p. Ti specimen with 50 vol.%. In this sample, a titanium matrix with a very low porosity is observed, with isolated pores with sizes close to 20 microns. According to the results, the porosity is well below that obtained with the conventional powder metallurgy technique, where 50% spacer generates porosities between 40% and 45% [59]. However, as indicated in Table 2, the porosity obtained is less than 2%. This means that hot pressing has eliminated both the spacer and the porosity that it could have generated, so that it would not be named a "metal foam" with a final porosity of 45–60% pores. In addition, in Figure 3b–e, the micrographs of the Ti$_{10}$ and Ti$_{20}$ samples are shown, respectively. In general, a porous matrix with a bimodal microstructure, with higher porosity compared to the c.p Ti, is observed. The addition of spacer particles and milled powders promotes a porous structure that exhibits irregular pores with rounded borders. These pores reach a mean size of between approximately 40 and 250 µm, and they are surrounded by a fine-particle structure; pores with a lower size are homogenously and randomly distributed throughout the titanium matrix. It is observed that the porosity percentage goes up as the milling time and the percentage of milled powders increase. It can be stated that porosity depends equally on both factors. An increase in the pore size with the milling time is also noticed, where the maximum pore sizes were achieved for those samples that contain the Ti$_{20}$ powder. In addition, the porosity in the sample was affected by the amount of milled powder used to produce the bimodal microstructure, where the small pores go up when the milled powder percentage increases.

Table 2. Morphological parameters of pores.

Sample Name	P(IA) (%)	D_{eq} (µm)	F_f
Ti$_{10-50}$	1.5 ± 0.4	84 ± 8	0.70 ± 0.16
Ti$_{20-50}$	12.5 ± 3.8	121 ± 6	0.75 ± 0.11
Ti$_{10-75}$	13.2 ± 0.4	100 ± 4	0.75 ± 0.07
Ti$_{20-75}$	35.5 ± 1.5	141 ± 4	0.80 ± 0.09

The differences in pore size could be a consequence of the particle size distribution of the used powders. In this sense, it is possible that smaller particles were used to redistribute its mass to larger particles [68]. If a small particle forms a neck with large particles and redistributes its mass with them, and these large particles were restricted in movement together (as is the case when surrounded by denser regions), then the smaller particle would break away (de-sinter) from one of the larger particles and be absorbed by the other. In consequence, pore growth takes place to reduce the surface to volume ratio of a powder compact when the compact is restricted from shrinking, as is the case for the bimodal microstructure, where pore growth occurs in regions where densification is locally restricted by the denser zones (unmilled powders), although these partially dense zones are globally subjected to densification particles [68]. Furthermore, it is known that agglomerates have a strong influence on densification because they prevent the effective transfer of heat and pressure to the particles during sintering [69–72]. Another de-sintering cause in the powder compact could be the presence of inclusions [73]. This behavior could be caused by the contamination of the ground powders during the milling process, caused by the release of ZrO_2 from the milling medium. This lack of sintering is related to the fact that these contaminants are refractory materials [74]. Furthermore, the oxide content in the milled powders hinders the particle–boundary motion of coarse particles during sintering, hereby lowering the particle coalescence [75].

Backscattering Electron–SEM images (BSE-SEM) of sintered samples are shown in Figure 3f–j, where the microstructure of the titanium sample (Figure 3f) was compared with those compacts prepared from the mixed powders (Figure 3g–j). On one hand, in the sample of pure titanium with 50 vol.% NH_4HCO_3 (Figure 3f), a typical microstructure of c.p. Ti with equiaxial grains and some micropores in the matrix was observed. However, the spacer and the pores, which should have been generated from it, had disappeared due to the densification action of the HP process. On the other hand, the samples prepared from the mixed powders exhibited a bimodal particle microstructure consisting of coarse and fine grains, where the bimodal microstructure can be identified by the presence of clusters of microporosities, which are caused by the fine particles. The coarse particles that originate from the unmilled powders are randomly distributed and surrounded by porous regions with fine particle microstructure. The porosity of compact prepared from the mixed powders is higher than that observed in the c.p. Ti samples (Figure 3a,f). The porosity increases as the amount of milled powder and milling time increases. It should be noticed that only in Ti_{20-75} could the observed porosity values correspond to a porosity close to that added by the spacer. Relatively large pores in the order of 250 microns are observed in this sample (Figure 3e), but also small pores left by the spacer, as seen in Figure 3f. This figure presents and describes the surface of a pore caused by the spacer with a relatively small size, but it is also possible to distinguish the bimodal microstructure.

Interesting aspects about the bimodal microstructure and porosity in Ti_{20-75} samples are presented in Figure 4. The bimodal microstructure is evenly distributed throughout the entire specimen (Figure 4a), which indicates that a good homogenization was achieved during the mixing process. Three zones were identified in the samples (Figure 4b) as follows: (i) Zone A: fine particles zone, where the milled powders are predominant, surrounding a coarse particle zone (Zone B); (ii) Zone B: coarse particle zone, promoted by unmilled powders; (iii) Zone C: an intermediate or mixed zone with both fine and coarse particles. The formation of the three zones can possibly be attributed to assembly mechanisms between the powder particles, forming clusters through agglomeration. Although different mechanisms lead to the agglomeration of particles in a sample, in this case, the agglomeration could occur during mechanical milling and even blending, where the particles collide and can stick together as a result of completely random movement within the confined space such as the grinding vessel or mixing vessel [76]. The agglomeration of the particles by size was observed in the particle size distributions for the titanium ground at 10 h and 20 h (Figure 1g,h). Another mechanism that can carry out the agglomeration of the particles during these milling or mixing events is known as gravitational agglomeration, which depends on the size of the particles and their speed, where the particles that settle slower are trapped by those that settle faster [76]. The latter may explain the mixed zone (Zone C, Figure 4). In addition, the matrix composition in the different zones present in the bimodal microstructure was evaluated by EDS analysis. The presence of ZrO_2 is detected in Zones A and C, comprised by milled powders, which is the result of the contamination of the milling media.

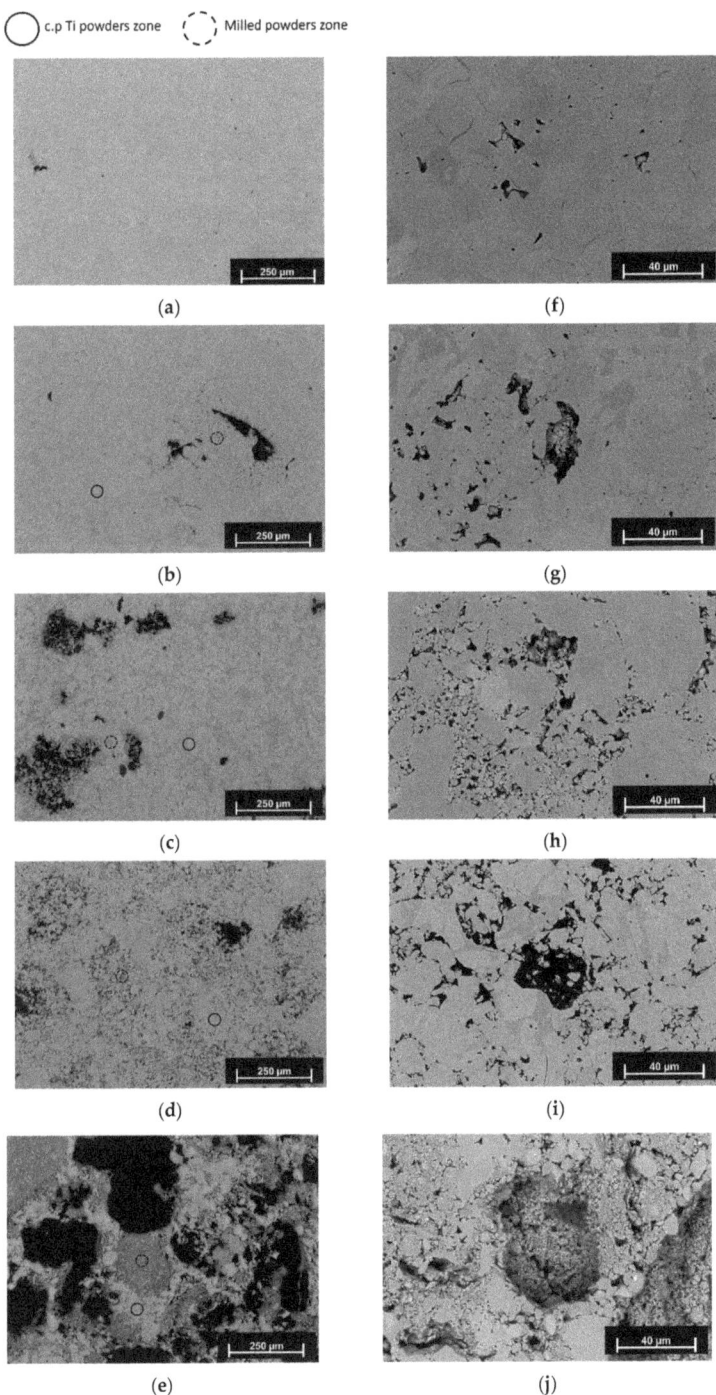

Figure 3. Optical micrographs of Ti samples: (**a**) c.p. Ti; (**b**) Ti$_{10-50}$; (**c**) Ti$_{20-50}$; (**d**) Ti$_{10-75}$; (**e**) Ti$_{20-75}$; BSE-SEM images of Ti foams: (**f**) c.p. Ti; (**g**) Ti$_{10-50}$; (**h**) Ti$_{20-50}$; (**i**) Ti$_{10-75}$; (**j**) Ti$_{20-75}$.

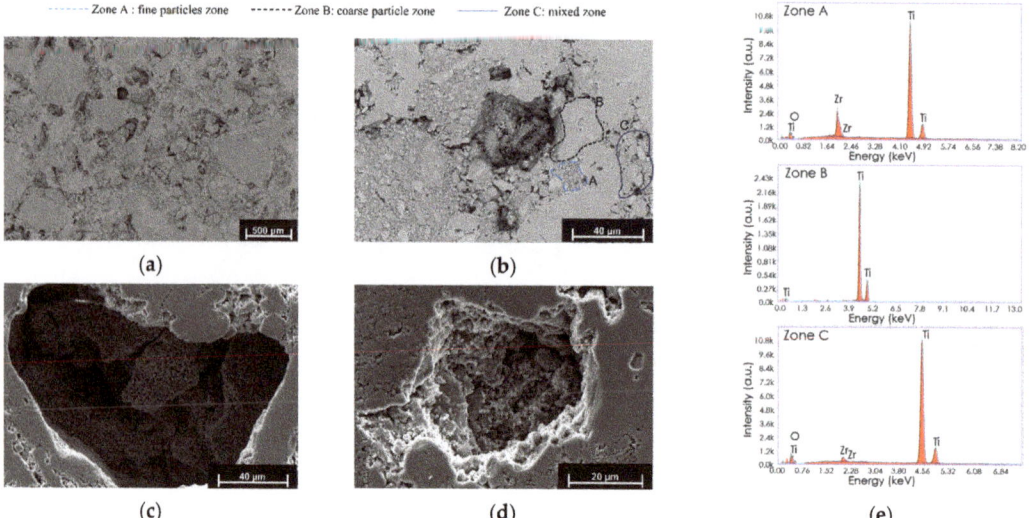

Figure 4. Microstructural aspects of synthesized Ti$_{20-75}$ specimen: BSE-SEM images (**a**) general view of bimodal microstructure; (**b**) characteristic microstructure of sample with three zones: Zone A—fine particles; Zone B—coarse particle; Zone C—mixed zone. SE-SEM images (**c**) Macro-pore surface; (**d**) Micro-pore surface; (**e**) EDS analysis for the different zones present in bimodal structure: Zone A, Zone B, and Zone C.

Two types of porosity were distinguished, as is observed in the Secondary Electron–SEM (SE-SEM) images shown in Figure 4c,d: the porosities were promoted by the spacer particles with a pore size of around 100 μm, a rough surface, rounded borders, and were interconnected (Figure 4c), and the porosities inherent to the powder metallurgy process with a pore size of around 10–50 μm and a rough surface were surrounded by the Zones A and B (Figure 4d).

The results of image analysis are summarized in Figure 5, where the stacked frequency distribution histograms of the pore size for each of the consolidated specimens are presented. Each bar of the histograms represents a quantity of pore counts with size in the comprised range. It is important to remember that all of the study samples in this graph have a 50% spacer added. The objective, then, is to evaluate the effect of the amount of ground powder and the grinding time on the pore size and on the porosity that could be obtained despite the use of the HP process. This figure shows the accumulated pore size corresponding to spacers with significant amounts of sizes between 100 and 150 microns, where it is possible to observe that specimens with higher amounts of fine powders (Ti$_{10-75}$ and Ti$_{20-75}$, Figure 5b) show higher frequencies of larger pores. The mentioned effect on porosity is observed in Figure 3d,e, where there are bigger pores in specimens using 75 wt.% of milled powders. A similar behavior has been documented by Dirras et al. [75], who produced Ni samples using different particle sizes and observed a "shielding effect" by introducing coarse particles onto a fine-particle matrix because the coalescence of coarse particles was hindered.

In addition, the pore size distribution shows that the samples Ti$_{10-50}$ and Ti$_{10-75}$ (10 h of milling time) exhibit porosities up to 200 μm, while the samples with the Ti$_{20}$ powder have 80% of its total porosity in this range, and the remaining 20% is between 200 and 400 μm.

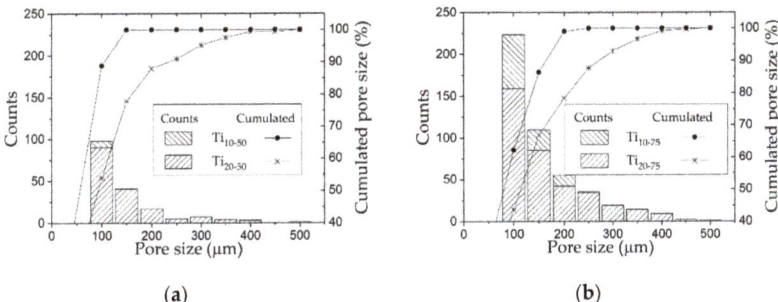

Figure 5. Pore size distribution: (a) Ti$_{10-50}$, Ti$_{20-50}$ (50% milled powder); (b) Ti$_{10-75}$, Ti$_{20-75}$ (75% milled powder).

To complement this analysis of the porosity size, Table 2 presents the image analysis data, showing the porosity percentage, the mean equivalent diameter, and the shape factor values expressed in terms of mean value ± standard error. This value of porosity corresponds to that obtained only with the spacer. To do this, a criterion was applied where the pore size data was filtered, eliminating all the pores that were below 50 µm. Porosity percentage increases with the amount of milled powders, where the highest porosity was achieved for Ti$_{20-75}$ (~36%). Although the porosity obtained is less than the percentage of the added spacer, 50% NH$_4$HCO$_3$, it is important to remember that the hot-pressing technique is used to obtain compacts with high densification. In this case, it is not only intended to introduce porosity in the consolidated titanium matrix, but it also seeks to improve the properties of the matrix and cycle times and reduce the temperature necessary for consolidation. The equivalent pore diameter was found in a range comprised between 84 ± 8 and 141 ± 4 µm. The ANOVA indicates that statistically significant differences exist in the pore size (p-value < 0.05), which confirms the suggested difference observed in the optical micrographs. There are influences of the milling time as well as of the amount of milled powder on the equivalent pore diameter in the specimens. By means of a Tukey's test, it was determined that the greatest significant difference of the equivalent diameter is about 60 µm. The pores shape factor was found in a range comprised between 0.70 and 0.80. In this case, the ANOVA indicates that there are no statistically significant differences (p-value > 0.05). This behavior is attributed to the deformation strengthening mechanism, where the milled powders have lost their ductile behavior; thus, the deformation is carried out mainly by the coarse particles. Another cause is the reduced size of the particles that have a greater surface energy, which results in a faster sintering that hinders the rearrangement of particles along the matrix.

3.3. Mechanical Characterization of Sintered Samples

Vickers microhardness distribution found in the consolidated samples with bimodal microstructure is shown in Figure 6. The mean values of microhardness measurements ranged between 589.4 ± 43.3 HV and 850.6 ± 62.2 HV, where the highest is found for Ti$_{20-75}$. These values exceed the c.p. Ti microhardness mean value (301 HV), being even higher than the reported values for the c.p. titanium components with a nanocrystalline structure or ultrafine-grained components consolidated by SPS [77], high-pressure torsion [78], or multi-pass ECAP [79], which are comprised in a range of 230 to 250 HV. They are closer to the ones reported for titanium components with a bimodal microstructure consolidated by SPS (660 to 853 HV) [45]. As it is observed, the microhardness distribution varies in function of the processing parameters: (i) when the amount of milled powder increases, the frequency of higher microhardness values increases, while (ii) the longer the milling time, the higher the mean value of microhardness. This increase in microhardness with the milling time and quantity of ground powder is mainly attributed to the accumulation of deformation energy [80], but also it could be influenced by the contamination with the milling medium, which in turn depends on the time, intensity, atmosphere of milling, and the difference in the strength/hardness of the powders and milling medium [81]. Table 3 summarizes the statistical analysis of each microhardness distribution. From the ANOVA, it was determined that both parameters have a significant effect (p-value < 0.05) on the microhardness value. However, the amount of ground powder is more significant in increasing the microhardness, as shown in Table 3.

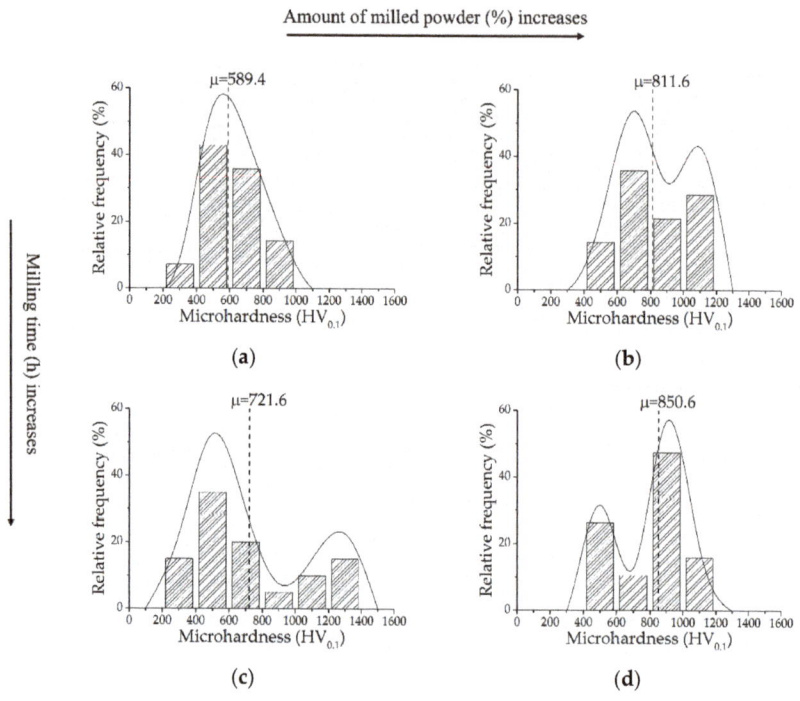

Figure 6. Vickers microhardness distribution for samples: (**a**) Ti_{10-50}; (**b**) Ti_{10-75}; (**c**) Ti_{20-50}; (**d**) Ti_{20-75}.

Table 3. Summary of statistical analysis of microhardness values (HV).

Sample	Milling Time (h)	Milled Powder (%)	Porosity (%)	Max. HV	Min. HV	Mean HV	Std. Error HV
c.p. Ti	0	0	-	430.8	232.1	312.9	11.2
Ti_{10-50}	10	50	1.5 ± 0.4	890.3	374.5	589.4	43.3
Ti_{20-50}	20	50	12.5 ± 3.8	1348.6	227.3	721.6	75.7
Ti_{10-75}	10	75	13.2 ± 0.4	1109.4	516.7	811.6	52.5
Ti_{20-75}	20	75	35.5 ± 1.5	1674.5	456.9	850.6	62.2

In addition, the mechanical properties estimated from the porosity and shape factor obtained from the image analysis are summarized in Table 4. These values are expressed in terms of mean value ± standard error. As it is observed, the elastic modulus decreases as the milling time and/or content of the milled powder increases, following the trend of increasing porosity. Yield strength has similar behavior due to the fact that it depends on the density value, and this is directly related to the porosity level. Nonetheless, the yield strength was estimated from a model which does not consider the strengthening effect due to deformation during the milling stage of the powders, or the presence of a bimodal structure, hence it is necessary to develop models in order to predict this behavior, which can be contrasted with experimental data.

Table 4. Estimated mechanical properties from bulk properties by means of Nielsen's method [56] (E_p) and Jha's correlation [58] ($\sigma_{y,f}$).

Sample	E_p (GPa)	$\sigma_{y,f}$ (MPa)
Ti$_{10-50}$	106.0 ± 0.2	312.8
Ti$_{20-50}$	80.9 ± 8.4	314.5
Ti$_{10-75}$	79.4 ± 1.5	317.8
Ti$_{20-75}$	41.8 ± 4.1	234.1

Previous studies have determined that the optimal porosity for an implant to efficiently stimulate bone ingrowth is in the range of 20–50% [82] with a pore size of 100–400 μm [83]. Furthermore, taking into account that the elastic modulus of the cortical bone is in a range from 20 to 25 GPa and that its ultimate tensile strength is about 195 MPa [3], it is ascertained that the Ti$_{20-75}$ specimen shows the characteristics suitable for bone replacement applications. The fatigue behavior of the specimens presented in this work is planned for future experiments, since metal fatigue [84] is one of the main causes of implants' mechanical failure.

4. Conclusions

In this study, the porous titanium samples with a bimodal microstructure were successfully synthesized by hot pressing with NH_4HCO_3 as a space-holder. The effects of the milling time for obtaining fine powder as well as its amount over the microhardness and porosity were investigated. The conclusions are shown as follows:

- It has been determined that the processing route via the space-holder technique and hot pressing consolidation is effective in producing titanium samples with a porosity of 36% with a bimodal microstructure, whose microhardness has similar values to those obtained in nanocrystalline or ultrafine-grained microstructures synthesized by SPS. The microhardness value depends on the amount of fine powder that constitutes the matrix, and it is the result of deformation strengthening mechanisms and a small grain size;
- The obtained porosity in the titanium samples processed by this route depends on the milling time as well as the amount of fine powder due to the changes in particles size and distribution, deformation/rearrange capability of the powders, and the presence of agglomerates and contamination, all parameters which affect the compaction and sintering processes. However, when the milled powders for 20 h were used, larger pores were reached, whose sizes reached up to 400 μm. The Ti$_{20-75}$ sample presents appropriated mechanical properties for cortical bone replacement applications.

Author Contributions: Conceptualization, S.L., R.V.M. and Y.T.; methodology, C.A. and S.L.; validation, M.R.-V.; formal analysis, S.L., F.J.G., C.A., Y.T., R.C.-V., S.S. and C.S.; investigation, M.R.-V., R.V.M. and S.L.; resources, S.L., R.V.M., F.J.G. and Y.T.; writing—original draft preparation, R.C.-V. and S.S.; writing—review and editing, F.J.G., C.A. and S.L.; supervision, C.S. and R.V.M.; project administration, S.L.; funding acquisition, S.L., R.V.M., F.J.G. and Y.T. All authors have read and agreed to the published version of the manuscript.

Funding: This work was supported by the Agencia Nacional de Investigación y Desarrollo (ANID) of Chile government [grant number Fondecyt 11160865, and FONDEQUIP EQM130103 and EQM150101, and Scholarship Program/DOCTORADO/2021-21211274], and the Ministerio de Ciencia e Innovación of Spain under the grant PID2019-109371GB-I00.

Institutional Review Board Statement: Not applicable.

Informed Consent Statement: Not applicable.

Data Availability Statement: The data presented in this study are available on request from the corresponding author.

Acknowledgments: The authors thank the laboratory technicians Jesus Pinto at Universidad de Sevilla (Spain), Claudio Aravena and Gabriel Cornejo at Universidad Técnica Federico Santa María (Chile) for their support carrying out the microstructure characterization and mechanical testing.

Conflicts of Interest: The authors declare no conflict of interest.

References

1. Gao, C.; Peng, S.; Feng, P.; Shuai, C. Bone biomaterials and interactions with stem cells. *Bone Res.* **2017**, *5*, 17059. [CrossRef]
2. Geetha, M.; Singh, A.; Asokamani, R.; Gogia, A. Ti based biomaterials, the ultimate choice for orthopaedic implants–A review. *Prog. Mater. Sci.* **2009**, *54*, 397–425. [CrossRef]
3. Zhang, L.; Chen, L. A Review on Biomedical Titanium Alloys: Recent Progress and Prospect. *Adv. Eng. Mater.* **2019**, *21*, 1801215. [CrossRef]
4. Pałka, K.; Pokrowiecki, R. Porous Titanium Implants: A Review. *Adv. Eng. Mater.* **2018**, *20*, 1700648. [CrossRef]
5. Lascano, S.; Chávez-Vásconez, R.; Muñoz-Rojas, D.; Aristizabal, J.; Arce, B.; Parra, C.; Acevedo, C.; Orellana, N.; Reyes-Valenzuela, M.; Gotor, F.J.; et al. Graphene-coated Ti-Nb-Ta-Mn foams: A promising approach towards a suitable biomaterial for bone replacement. *Surf. Coat. Technol.* **2020**, *401*, 126250. [CrossRef]
6. Torres-Sanchez, C.; McLaughlin, J.; Bonallo, R. Effect of Pore Size, Morphology and Orientation on the Bulk Stiffness of a Porous Ti35Nb4Sn Alloy. *J. Mater. Eng. Perform.* **2018**, *27*, 2899–2909. [CrossRef]
7. Chino, Y.; Dunand, D.C. Directionally freeze-cast titanium foam with aligned, elongated pores. *Acta Mater.* **2008**, *56*, 105–113. [CrossRef]
8. Yook, S.-W.; Kim, H.-E.; Koh, Y.-H. Fabrication of porous titanium scaffolds with high compressive strength using camphene-based freeze casting. *Mater. Lett.* **2009**, *63*, 1502–1504. [CrossRef]
9. Han, C.; Li, Y.; Wang, Q.; Wen, S.; Wei, Q.; Yan, C.; Hao, L.; Liu, J.; Shi, Y. Continuous functionally graded porous titanium scaffolds manufactured by selective laser melting for bone implants. *J. Mech. Behav. Biomed. Mater.* **2018**, *80*, 119–127. [CrossRef] [PubMed]
10. Zhao, D.; Liang, H.; Han, C.; Li, J.; Liu, J.; Zhou, K.; Yang, C.; Wei, Q. 3D printing of a titanium-tantalum Gyroid scaffold with superb elastic admissible strain, bioactivity and in-situ bone regeneration capability. *Addit. Manuf.* **2021**, *47*, 102223. [CrossRef]
11. Orrù, R.; Licheri, R.; Locci, A.M.; Cincotti, A.; Cao, G. Consolidation/synthesis of materials by electric current activated/assisted sintering. *Mater. Sci. Eng. R Rep.* **2009**, *63*, 127–287. [CrossRef]
12. Ramskogler, C.; Warchomicka, F.; Mostofi, S.; Weinberg, A.; Sommitsch, C. Innovative surface modification of Ti6Al4V alloy by electron beam technique for biomedical application. *Mater. Sci. Eng. C* **2017**, *78*, 105–113. [CrossRef] [PubMed]
13. Sauceda, S.; Lascano, S.; Béjar, L.; Neves, G.O.; Chicardi, E.; Salvo, C.; Aguilar, C. Study of the Effect of the Floating Die Compaction on Mechanical Properties of Titanium Foams. *Metals* **2020**, *10*, 1621. [CrossRef]
14. Yamanoglu, R.; Bahador, A.; Kondoh, K. Fabrication Methods of Porous Titanium Implants by Powder Metallurgy. *Trans. Indian Inst. Met.* **2021**, *74*, 2555–2567. [CrossRef]
15. Niu, W.; Bai, C.; Qiu, G.B.; Wang, Q. Processing and properties of porous titanium using space holder technique. *Mater. Sci. Eng. A* **2009**, *506*, 148–151. [CrossRef]
16. Lascano, S. *Obtención y Caracterización de Ti c.p. Poroso Para Aplicaciones Biomédicas*; Universidad del Norte: Barranquilla, Columbia, 2012.
17. Rodriguez-Contreras, A.; Punset, M.; Calero, J.A.; Gil, F.J.; Ruperez, E.; Manero, J.M. Powder metallurgy with space holder for porous titanium implants: A review. *J. Mater. Sci. Technol.* **2021**, *76*, 129–149. [CrossRef]
18. Torres, Y.; Rodríguez, J.A.; Arias, S.; Echeverry, M.; Robledo, S.; Amigó, V.; Pavón, J.J. Processing, characterization and biological testing of porous titanium obtained by space-holder technique. *J. Mater. Sci.* **2012**, *47*, 6565–6576. [CrossRef]
19. Abhash, A.; Singh, P.; Kumar, R.; Pandey, S.; Sathaiah, S.; Md. Shafeeq, M.; Mondal, D.P. Effect of Al addition and space holder content on microstructure and mechanical properties of Ti2Co alloys foams for bone scaffold application. *Mater. Sci. Eng. C* **2020**, *109*, 110600. [CrossRef]
20. Abhash, A.; Singh, P.; Muchhala, D.; Kumar, R.; Gupta, G.K.; Mondal, D.P. Research into the change of macrostructure, microstructure and compressive deformation response of Ti6Al2Co foam with sintering temperatures and space holder contents. *Mater. Lett.* **2020**, *261*, 126997. [CrossRef]
21. Civantos, A.; Giner, M.; Trueba, P.; Lascano, S.; Montoya-García, M.-J.; Arévalo, C.; Vázquez, M.Á.; Allain, J.P.; Torres, Y. In vitro bone cell behavior on porous titanium samples: Influence of porosity by loose sintering and space holder techniques. *Metals* **2020**, *10*, 696. [CrossRef]
22. Gupta, J.; Ghosh, S.; Aravindan, S. Effect of Mo and space holder content on microstructure, mechanical and corrosion properties in Ti6AlxMo based alloy for bone implant. *Mater. Sci. Eng. C* **2021**, *123*, 111962. [CrossRef] [PubMed]
23. Surace, R.; Filippis, L.A.C.D.; Ludovico, A.D.; Boghetich, G. Influence of processing parameters on aluminium foam produced by space holder technique. *Mater. Des.* **2009**, *30*, 1878–1885. [CrossRef]
24. Torres, Y.; Pavón, J.J.; Rodríguez, J.A. Processing and characterization of porous titanium for implants by using NaCl as space holder. *J. Mater. Proc. Technol.* **2012**, *212*, 1061–1069. [CrossRef]

25. Mansourighasri, A.; Muhamad, N.; Sulong, A.B. Processing titanium foams using taipoca starch as a space holder. *J. Mater. Proc. Technol.* **2012**, *212*, 83–89. [CrossRef]
26. Jakubowicz, J.; Adamek, G.; Dewidar, M. Titanium foam made with saccharose as a space holder. *J. Porous Mater.* **2013**, *20*, 1137–1141. [CrossRef]
27. Esen, Z.; Bor, Ş. Characterization of Ti-6Al-4V alloy foams synthesized by space holder technique. *Mater. Sci. Eng. A* **2011**, *528*, 3200–3209. [CrossRef]
28. Chen, Y.; Frith, J.E.; Dehghan-Manshadi, A.; Attar, H.; Kent, D.; Soro, N.D.M.; Bermingham, M.J.; Dargusch, M.S. Mechanical properties and biocompatibility of porous titanium scaffolds for bone tissue engineering. *J. Mech. Behav. Biomed. Mater.* **2017**, *75*, 169–174. [CrossRef]
29. Lai, T.; Xu, J.-L.; Xiao, Q.-F.; Tong, Y.-X.; Huang, J.; Zhang, J.-P.; Luo, J.-M.; Liu, Y. Preparation and characterization of porous NiTi alloys synthesized by microwave sintering using Mg space holder. *Trans. Nonferrous Met. Soc. China* **2021**, *31*, 485–498. [CrossRef]
30. Manonukul, A.; Muenya, N.; Léaux, F.; Amaranan, S. Effects of replacing metal powder with powder space holder on metal foam produced by metal injection moulding. *J. Mater. Proc. Technol.* **2010**, *210*, 529–535. [CrossRef]
31. Li, B.-q.; Li, Z.-q.; Lu, X. Effect of sintering processing on property of porous Ti using space holder technique. *Trans. Nonferrous Met. Soc. China* **2015**, *25*, 2965–2973. [CrossRef]
32. Chen, Y.; Frith, J.E.; Dehghan-Manshadi, A.; Kent, D.; Bermingham, M.; Dargusch, M. Biocompatible porous titanium scaffolds produced using a novel space holder technique. *J. Biomed. Mater. Res. Part B Appl. Biomater.* **2018**, *106*, 2796–2806. [CrossRef]
33. Guden, M.; Celik, E.; Akar, E.; Cetiner, S. Compression testing of a sintered Ti6Al4V powder compact for biomedical applications. *Mater. Caract.* **2005**, *54*, 399–408. [CrossRef]
34. Bafti, H.; Habibolahzadeh, A. Production of aluminum foam by spherical carbamide space holder technique-processing parameters. *Mater. Des.* **2010**, *31*, 4122–4129. [CrossRef]
35. Arifvianto, B.; Leeflang, M.A.; Zhou, J. The compression behaviors of titanium/carbamide powder mixtures in the preparation of biomedical titanium scaffolds with the space holder method. *Powder Technol.* **2015**, *284*, 112–121. [CrossRef]
36. Lascano, S.; Arévalo, C.; Montealegre-Melendez, I.; Muñoz, S.; Rodriguez-Ortiz, J.A.; Trueba, P.; Torres, Y. Porous Titanium for Biomedical Applications: Evaluation of the Conventional Powder Metallurgy Frontier and Space-Holder Technique. *Appl. Sci.* **2019**, *9*, 982. [CrossRef]
37. Abbasi, N.; Hamlet, S.; Love, R.M.; Nguyen, N.-T. Porous scaffolds for bone regeneration. *J. Sci. Adv. Mater. Devices* **2020**, *5*, 1–9. [CrossRef]
38. Suryanarayana, C.; Al-Aqueeli, N. Mechanically alloyed nanocomposites. *Prog. Mater. Sci.* **2013**, *58*, 383–502. [CrossRef]
39. Misra, D.K.; Rakshit, R.K.; Singh, M.; Shukla, P.K.; Chaturvedi, K.M.; Sivaiah, B.; Gahtori, B.; Dhar, A.; Sohn, S.W.; Kim, W.T.; et al. High yield strength bulk Ti based bimodal ultrafine eutectic composites with enhanced plasticity. *Mater. Des.* **2014**, *58*, 551–556. [CrossRef]
40. Kang, L.; Yang, C. A Review on High-Strength Titanium Alloys: Microstructure, Strengthening, and Properties. *Adv. Eng. Mater.* **2019**, *21*, 1801359. [CrossRef]
41. Long, Y.; Wang, T.; Zhang, H.Y.; Huang, X.L. Enhanced ductility in a bimodal ultrafine-grained Ti-6Al-4V alloy fabricated by high energy ball milling and spark plasma sintering. *Mater. Sci. Eng. A* **2014**, *608*, 82–89. [CrossRef]
42. Okulov, I.V.; Bönisch, M.; Kühn, U.; Skrotzki, W.; Eckert, J. Significant tensile ductility and toughness in an ultrafine-structured Ti68.8Nb13.6Co6Cu5.1Al6.5 bi-modal alloy. *Mater. Sci. Eng. A* **2014**, *615*, 457–463. [CrossRef]
43. Vajpai, S.K.; Sawangrat, C.; Yamaguchi, O.; CIuca, O.P.; Ameyama, K. Effect of bimodal harmonic structure design on the deformation behaviour and mechanical properties of Co-Cr-Mo alloy. *Mater. Sci. Eng. C* **2016**, *58*, 1008–1015. [CrossRef]
44. Zheng, Y.; Yao, X.; Su, Y.; Zhang, D.L. High strength titanium with a bimodal microstructure fabricated by thermomechanical consolidation of a nanocrystalline TiH2 powder. *Mater. Sci. Eng. A* **2017**, *686*, 11–18. [CrossRef]
45. Yang, C.; Ni, S.; Liu, Y.; Song, M. Effects of sintering parameters on the hardness and microstructures of bulk bimodal titanium. *Mater. Sci. Eng. A* **2015**, *625*, 264–270. [CrossRef]
46. Chang, S.; Doremus, R.H.; Schadler, L.S.; Siegel, R.W. Hot-Pressing of Nano-Size Alumina Powder and the Resulting Mechanical Properties. *Int. J. Appl. Ceram. Technol.* **2004**, *1*, 172–179. [CrossRef]
47. Dong, S.; Jiang, D.; Tan, S.; Guo, J. Preparation and characterization of nano-structured monolithic SiC and Si3N4/SiC composite by hot isostatic pressing. *J. Mater. Sci. Lett.* **1997**, *16*, 1080–1083. [CrossRef]
48. McCusker, L.; Dreele, R.V.; Cox, D.; Louër, D.; Scardi, P. Rietveld refinement guidelines. *J. Appl. Crystallogr.* **1999**, *32*, 36–50. [CrossRef]
49. Lopez, E.I.P.; Saint-Laurence, P.I.M.; Ramirez, C.E.A.; Gomez, L.B.; Flores, A.M.; Jimenez, F.D.L.C.; López, I.A. Estudio de perfiles de difracción de rayos X de una aleación Ti-13Ta-3Sn obtenida por aleado mecánico. *Matéria* **2020**, 25. [CrossRef]
50. Hernández-Nava, E.; Mahoney, P.; Smith, C.J.; Donoghue, J.; Todd, I.; Tammas-Williams, S. Additive manufacturing titanium components with isotropic or graded properties by hybrid electron beam melting/hot isostatic pressing powder processing. *Sci. Rep.* **2019**, *9*, 4070. [CrossRef]
51. Petrovskiy, P.; Sova, A.; Doubenskaia, M.; Smurov, I. Influence of hot isostatic pressing on structure and properties of titanium cold-spray deposits. *Int. J. Adv. Manuf. Technol.* **2019**, *102*, 819–827. [CrossRef]
52. *ASTM F1877-16; Standard Practice for Characterization of Particles*; ASTM International: West Conshohocken, PA, USA, 2016.
53. *ASTM E2651-19; Standard Guide for Powder Particle Size Analysis*; ASTM International: West Conshohocken, PA, USA, 2019.

54. ISO 13314:2011; *Mechanical Testing of Metals-Ductility Testing-Compression Test for Porous and Cellular Metals*; ISO: Geneva, Switzerland, 2011.
55. ASTM E9-19; *Standard Test Methods of Compression Testing of Metallic Materials at Room Temperature*; ASTM International: West Conshohocken, PA, USA, 2019.
56. Nielsen, L.F. Elasticity and Damping of Porous Materials and Impregnated Materials. *J. Am. Ceram. Soc.* **1984**, *67*, 93–98. [CrossRef]
57. Li, Y.; Yang, C.; Zhao, H.; Qu, S.; Li, X.; Li, Y. New developments of Ti-based alloys for biomedical applications. *Materials* **2014**, *7*, 1709–1800. [CrossRef] [PubMed]
58. Jha, N.; Mondal, D.P.; Majumdar, J.D.; Badkul, A.; Jha, A.K.; Khare, A.K. Highly porous open cell Ti-foam using NaCl as temporary space holder through powder metallurgy route. *Mater. Des.* **2013**, *47*, 810–819. [CrossRef]
59. Torres, Y.; Lascano, S.; Bris, J.; Pavón, J.; Rodriguez, J.A. Development of porous titanium for biomedical applications: A comparison between loose sintering and space-holder techniques. *Mater. Sci. Eng. C* **2014**, *37*, 148–155. [CrossRef] [PubMed]
60. Soufiani, A.M.; Karimzadeh, F.; Enayati, M.H. Formation mechanism and characterization of nanostructured Ti6Al4V alloy prepared by mechanical alloying. *Mater. Des.* **2012**, *37*, 152–160. [CrossRef]
61. Guzzo, P.L.; Marino, F.B.d.B.; Soares, B.R.; Santos, J.B. Evaluation of particle size reduction and agglomeration in dry grinding of natural quartz in a planetary ball mill. *Powder Technol.* **2020**, *368*, 149–159. [CrossRef]
62. Suryanarayana, C. Mechanical alloying and milling. *Prog. Mater. Sci.* **2001**, *46*, 1–184. [CrossRef]
63. Salvo, C.; Aguilar, C.; Cardoso-Gil, R.; Medina, A.; Bejar, L.; Mangalaraja, R.V. Study on the microstructural evolution of Ti-Nb based alloy obtained by high-energy ball milling. *J. Alloy. Compd.* **2017**, *720*, 254–263. [CrossRef]
64. Chicardi, E.; García-Garrido, C.; Sayagués, M.J.; Torres, Y.; Amigó, V.; Aguilar, C. Development of a novel fcc structure for an amorphous-nanocrystalline Ti-33Nb-4Mn (at.%) ternary alloy. *Mater. Charact.* **2018**, *135*, 46–56. [CrossRef]
65. Aguilar, C.; Pio, E.; Medina, A.; Mangalaraja, R.V.; Salvo, C.; Alfonso, I.; Guzmán, D.; Bejar, L. Structural Study of Novel Nanocrystalline fcc Ti-Ta-Sn Alloy. *Metall. Mater. Trans. A* **2019**, *50*, 2061–2065. [CrossRef]
66. Tejeda-Ochoa, A.; Kametani, N.; Carreño-Gallardo, C.; Ledezma-Sillas, J.E.; Adachi, N.; Todaka, Y.; Herrera-Ramirez, J.M. Formation of a metastable fcc phase and high Mg solubility in the Ti-Mg system by mechanical alloying. *Powder Technol.* **2020**, *374*, 348–352. [CrossRef]
67. Aguilar, C.; Pio, E.; Medina, A.; Martínez, C.; Sancy, M.; Guzman, D. Evolution of synthesis of FCC nanocrystalline solid solution and amorphous phase in the Ti-Ta based alloy by high milling energy. *J. Alloy. Compd.* **2021**, *854*, 155980. [CrossRef]
68. Lange, F.F. Densification of powder compacts: An unfinished story. *J. Eur. Ceram. Soc.* **2008**, *28*, 1509–1516. [CrossRef]
69. Ivasishin, O.M.; Savvakin, D.G.; Froes, F.H.; Mokson, V.C.; Bondareva, K.A. Synthesis of the Ti-6Al-4V alloy having low residual porosity by powder metallurgy method. *Powder Metall. Met. Ceram.* **2002**, *41*, 382–390. [CrossRef]
70. Robertson, I.M.; Schaffer, G.B. Some Effects of Particle Size on the Sintering of Titanium and a Master Sintering Curve Model. *Metall. Mater. Trans. A* **2009**, *40*, 1968–1979. [CrossRef]
71. Qian, M. Cold compaction and sintering of titanium and its alloys for near-net-shape or preform fabrication. *Int. J. Powder Metall.* **2010**, *46*, 29–44.
72. Robertson, I.M.; Schaffer, G.B. Review of densification of titanium based powder systems in press and sinter processing. *Powder Metall.* **2010**, *53*, 146–162. [CrossRef]
73. Lange, F. De-sintering, A phenomenon concurrent with densification within powder compacts: A review. In *Sintering Technology*, 1st ed.; Marcel Dekker, Inc.: New York, NY, USA, 1996; pp. 1–12.
74. Gupta, N.; Basu, B. Hot pressing and spark plasma sintering techniques of intermetallic matrix composites. In *Intermetallic Matrix Composites*, 1st ed.; Mitra, R., Ed.; Elsevier: Duxford, UK, 2018; pp. 243–302.
75. Dirras, G.; Gubicza, J.; Ramtani, S.; Bui, Q.H.; Szilágyi, T. Microstructural and mechanical characteristics of bulk polycrystalline Ni consolidated from blends of powders with different particle size. *Mater. Sci. Eng. A* **2010**, *527*, 1206–1214. [CrossRef]
76. Singer, A.; Barakat, Z.; Mohapatra, S.; Mohapatra, S.S. Chapter 13–Nanoscale Drug-Delivery Systems: In Vitro and In Vivo Characterization. In *Nanocarriers for Drug Delivery*; Mohapatra, S.S., Ranjan, S., Dasgupta, N., Mishra, R.K., Thomas, S., Eds.; Elsevier: Amsterdam, The Netherlands, 2019; pp. 395–419.
77. Ertoter, O.; Topping, T.D.; Li, Y.; Moss, W.; Lavernia, E.J. Nanostructured Ti Consolidated via Spark Plasma Sintering. *Metall. Mater. Trans. A* **2011**, *42*, 964–973. [CrossRef]
78. Edalati, K.; Daio, T.; Arita, M.; Lee, S.; Horita, Z.; Togo, A.; Tanaka, I. High-pressure torsion of titanium at cryogenic and room temperatures: Grain size effect on allotropic phase transformations. *Acta Mater.* **2014**, *68*, 207–213. [CrossRef]
79. Hajizadeh, K.; Eghbali, B.; Topolski, K.; Kurzydlowski, K.J. Ultra-fine grained bulk CP-Ti processed by multi-pass ECAP at warm deformation region. *Mat. Chem. Phys.* **2014**, *143*, 1032–1038. [CrossRef]
80. Suryanarayana, C.; Klassen, T.; Ivanov, E. Synthesis of nanocomposites and amorphous alloys by mechanical alloying. *J. Mater. Sci.* **2011**, *46*, 6301–6315. [CrossRef]
81. Alijani, F.; Amini, R.; Ghaffari, M.; Alizadeh, M.; Okyay, A.K. Effect of milling time on the structure, micro-hardness, and thermal behavior of amorphous/nanocrystalline TiNiCu shape memory alloys developed by mechanical alloying. *Mater. Des.* **2014**, *55*, 373–380. [CrossRef]
82. Vasconcellos, L.; Leite, D.; Oliveira, F.; Carvalho, Y.; Cairo, C. Evaluation of bone ingrowth into porous titanium implants: Histomorphometric analysis in rabbits. *Implantol. Braz. Oral. Res.* **2010**, *24*, 399. [CrossRef]

83. Hollister, S. Scaffold Design and Manufacturing: From Concept to Clinic. *Adv. Mater.* **2009**, *21*, 3330. [CrossRef]
84. Shemtov-Yona, K.; Rittel, D. Fatigue of Dental Implants: Facts and Fallacies. *Dent. J.* **2016**, *4*, 16. [CrossRef] [PubMed]

Article

Gelatin-Based Biofilms with Fe$_x$O$_y$-NPs Incorporated for Antioxidant and Antimicrobial Applications

Johar Amin Ahmed Abdullah [1,*], Mercedes Jiménez-Rosado [1], Antonio Guerrero [1] and Alberto Romero [2]

[1] Departamento de Ingeniería Química, Escuela Politécnica Superior, Universidad de Sevilla, 41011 Sevilla, Spain; mjimenez42@us.es (M.J.-R.); aguerrero@us.es (A.G.)

[2] Departamento de Ingeniería Química, Facultad de Física, Universidad de Sevilla, 41012 Sevilla, Spain; alromero@us.es

* Correspondence: jabdullah@us.es; Tel.: +34-95-455-7179

Citation: Abdullah, J.A.A.; Jiménez-Rosado, M.; Guerrero, A.; Romero, A. Gelatin-Based Biofilms with Fe$_x$O$_y$-NPs Incorporated for Antioxidant and Antimicrobial Applications. *Materials* **2022**, *15*, 1966. https://doi.org/10.3390/ma15051966

Academic Editor: Sandra Maria Fernandes Carvalho

Received: 2 February 2022
Accepted: 3 March 2022
Published: 7 March 2022

Publisher's Note: MDPI stays neutral with regard to jurisdictional claims in published maps and institutional affiliations.

Copyright: © 2022 by the authors. Licensee MDPI, Basel, Switzerland. This article is an open access article distributed under the terms and conditions of the Creative Commons Attribution (CC BY) license (https://creativecommons.org/licenses/by/4.0/).

Abstract: Currently, gelatin-based films are regarded as promising alternatives to non-environmentally friendly plastic films for food packaging. Nevertheless, although they have great biodegradability, their weak mechanical properties and high solubility limit their applications. In this way, the use of nanoparticles, such as Fe$_x$O$_y$-NPs, could improve the properties of gelatin-based biofilms. Thus, the main objective of this work was to include different concentrations of Fe$_x$O$_y$-NPs (0.25 and 1.0%) manufactured by green synthesis (GS) and chemical synthesis (CS) into gelatin-based biofilms in order to improve their properties. The results show that Fe$_x$O$_y$-NPs can be distributed throughout the biofilm, although with a greater concentration on the upper surface. In addition, the incorporation of Fe$_x$O$_y$-NPs into the biofilms improves their physicochemical, mechanical, morphological, and biological properties. Thus, it is possible to achieve suitable gelatin-based biofilms, which can be used in several applications, such as functional packaging in the food industry, antioxidant and antimicrobial additives in biomedical and pharmaceutical biomaterials, and in agricultural pesticides.

Keywords: biofilms; gelatin; nanoparticles; iron oxide; antioxidant activity; antibacterial activity

1. Introduction

A film is a very thin (thickness < 1 mm), transparent, and, in many cases, stretchable plastic with different uses [1]. It is normally made of polyethylene (PE) or polypropylene (PP), which offers great flexibility, making it perfect for wrapping products of different shapes and sizes [2]. The most interesting properties of these films are their transparency (which allows one to see inside the package), their flexibility and adaptation to all kinds of shapes and sizes, and their impermeability, which prevents the passage of air and moisture, acting as a barrier and protecting the interior [3]. These properties are very important in the food industry, as they help to extend the shelf life of food by preventing oxidation–reduction reactions and interaction with microbes [4].

In recent years, the film sector has been affected by an increase in demand due to two fundamental factors: the need for safety in the transport of products, thereby increasing the need for wrapping the merchandise, and the current trend towards better presented products, which has considerably increased the use of films as an element with hygienic and aesthetic characteristics [5]. Nevertheless, the low biodegradability of these packaging materials is generating a great pollution problem, being unsuitable for the food industry. In this way, natural biopolymer-based films are currently being investigated, such as those made of gelatin, chitosan, cellulose, or cellulose derivatives, which confer them good biodegradability without releasing toxic substances [6,7]. Among the raw materials, gelatin has great potential to be used in diverse industrial applications: (1) in the food industry, to protect food from certain factors such as drying, light, and oxidation, since it may be used in biofilms to incorporate a wide range of additives, including antioxidants, antimicrobials, antifungals, nutrients, and flavorings [8]; (2) in the cosmetics sector, e.g., in hair gels, shampoos, and other cosmetic products [9–12]; (3) in the biomedical sector, for intrinsic activities like antidiabetic, antihypertensive, anticancer, antimicrobial, and antioxidant activities, as well as in wound care and healing, tissue engineering, and gene therapy [13,14]; (4) in the pharmaceutical industry and medication delivery, e.g., as a gelling agent for plasma expanders, to manufacture soft and hard gelatin capsule shells, microencapsulation of pharmaceuticals and oils, emulsion stabilization, medicated sponges, scaffoldings, creams and gels, wound care, slow-release, and vaccinations [15,16]; (5) in the photography sector, as a protective coating that extends the life of photographs [12,17]; and (6) other applications such as paints and fertilizers [10]. All this is possible due to the fact that gelatin presents easy processability to form films, high flexibility, suitable gas barrier properties, high availability, and low cost [18,19]. Nevertheless, the greatest problem of gelatin-based biofilms is their high water solubility and vapor permeability, along with weak thermal and mechanical strength [20,21]. Many strategies have been suggested to overcome these problems. However, the most important strategy is the incorporation of nanomaterials as reinforcing fillers [22]. In this way, a variety of metal oxide nanoparticles, nanocellulose, and nanoclays have been incorporated into gelatin-based biofilms, such as zinc oxide nanoparticles [23], gelatin–silver NP antimicrobial composite films [24], titanium dioxide (TiO_2-NPs) [25], gelatin biofilms reinforced with chitosan-NPs [26], montmorillonite [27], chitin NPs [28], and magnetic iron oxide NPs [29]. These nano-sized composites have been used for the manufacture of gelatin-based nanocomposite biofilms. In this sense, researchers are focusing their efforts on discovering alternatives to antibiotic feed additives that do not compromise productivity, since the usage of antibiotic feed additives has been a subject of increasing concern. Nanoparticles have recently been used to replace the high-cost organic source [30].

Furthermore, iron oxide nanoparticles (Fe_xO_y-NPs) have been widely used in biomedical applications due to their particular, unique, and magnetic characteristics, as well as their acceptable biocompatibility and bioavailability [31]. Fe_xO_y-NPs have a high inhibition capacity against the growth of different foodborne pathogens, such as *Staphylococcus aureus*, *Escherichia coli*, and *Pseudomonas aeruginosa* [32]. These NPs have proven to be capable of killing bacteria by producing reactive oxygen species ($^\bullet OH$, $^\bullet O_2^-$), damaging bacterial DNA and proteins, leading to impaired mitochondrial function while keeping non-bacterial cells unharmed [2,18,20]. It is worth mentioning that Fe_xO_y-NPs are nonhazardous and non-cytotoxic at concentrations lower than 0.1 mg/mL [33]. Additionally, Faria et al. reported the oral therapeutic potential of iron oxide NPs to treat iron deficiency anemia [34]. In this context, Fe_xO_y-NPs are considered a suitable additive to incorporate into films and improve their antimicrobial activity.

Fe_xO_y-NPs can be synthesized by chemical or green approaches, which produce different characteristics. Chemically synthesized Fe_xO_y-NPs are more hazardous and have a higher tendency to agglomerate and to show lower stability. As a result of the introduction of green nanotechnology, researchers are focusing more on environmentally beneficial green or biological methods of producing Fe_xO_y-NPs. Thus, the nanoparticles synthesized by green methods are smaller, less agglomerated, more stable, and less toxic than those synthesized by chemical methods [35,36]. In addition, the green synthesis of these NPs can improve their purity and functional properties due to the high presence of active groups coming from polyphenols, which are used for their synthesis.

In this way, the main objective of this work was to develop gelatin-based biofilms with different concentrations of Fe_xO_y-NPs (0.25 and 1.0%). Green (GS) and chemical (CS) Fe_xO_y-NPs were used to compare their influence on the biofilm properties. To this end, the physicochemical, mechanical, microstructural, and functional properties of the different biofilms were evaluated.

2. Materials and Methods

2.1. Materials

The gelatin protein used in this study was food gelatin type B 200/220 g blooms supplied by Manuel Riesgo, S.A. (Madrid, Spain), being a food gelatin that contains sulfur dioxide (<10 ppm). Gallic acid ($C_7H_6O_5$) and DPPH (2,2-diphenyl-1-picrylhydrazyl) were purchased from Sigma Aldrich (Darmstadt, Germany). All the reagents were of analytical grade.

Fe_xO_y-NPs were synthesized according to a previous work with slight modifications [37]. Briefly, it consists of colloidal precipitation in which 20 mL of *Phoenix dactylifera* L. extract, which is rich in polyphenols (green) or NaOH (chemical) (used as reductors) were mixed with 20 mL of $FeCl_3 \cdot 6H_2O$ (used as a precursor). The resulting 40 mL of mixture was heated under continuous stirring for 2 h at 50 °C. Then, the obtained precipitate was filtered, washed, and dried in an oven for 8 h at 100 °C. Finally, they were calcinated in a muffle for 5 h at 500 °C.

CS Fe_xO_y-NPs had a mean size of 49 ± 2 nm, a 2.20 Fe_2O_3:Fe_3O_4 ratio, and 47% crystallinity. GS Fe_xO_y-NPs had a mean size of 32 ± 1 nm, a 0.84 Fe_2O_3:Fe_3O_4 ratio, and 69% crystallinity.

2.2. Biofilm Processing Method

Biofilms were fabricated by the casting procedure [3]. To this end, gelatin was firstly dissolved in distilled water (2% w/v), subjecting it to magnetic stirring for 2 h at 60 °C and 600 rpm. Subsequently, different concentrations of Fe_xO_y-NPs (0.25 and 1.0% w/w) with respect to the initial gelatin were dispersed in the solutions by ultrasound for 0.25 h. Finally, a constant volume (42.7 mL) of the solution was cast into Teflon plates (7.6 cm of diameter) and dried at room conditions (22 ± 1 °C and 35 ± 1% RH) for 3 days. The biofilms were peeled off and kept in a desiccator for further characterization. A reference biofilm was manufactured without the dispersion of NPs. Figure 1 shows a scheme of the different steps of this process.

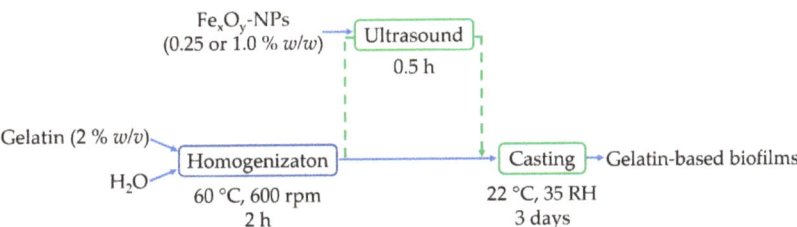

Figure 1. Scheme of the gelatin-based biofilm processing method.

2.3. Physicochemical Properties

2.3.1. Water Solubility

For the determination of this parameter, the samples (2 × 2 cm²) were firstly weighted (W_i) and then placed in an oven at 105 °C for 24 h. Later, the samples were immersed in 50 mL of distilled water for 24 h. Finally, the films were taken out and redried at 105 °C for 24 h to obtain the final dry weight (W_f). The weight loss or water solubility percentage (WS%) was calculated with Equation (1) [26,29]:

$$WS(\%) = \frac{w_i - w_f}{w_i} \cdot 100 \tag{1}$$

2.3.2. Optical Properties

Another essential property of biofilms is their transparency, which restricts light transmission while allowing visibility through the packaging material. To determine this property, UV–vis spectroscopy was used. Thus, 1 × 2 cm^2 samples were measured in a UV–vis spectrophotometer (Model 8451A, Hewlett Packard Co., Santa Clara, CA, USA) at 600 nm. A blank was carried out with air [24,38]. The results were indicated as transmittance (amount of light that can pass through the system).

2.4. Mechanical Properties

Static tensile tests were performed in order to evaluate the mechanical properties of the biofilms. For this, a modification of the ISO 527-3:2019 standard [39] was used. The samples were subjected to an increasing axial force at a speed of 10 mm/min until break in an MTS Insight 10 Universal Testing Machine (Berlin, Germany). During these tests, temperature and relative humidity were constant at 22 ± 1 °C and 35 ± 1%, respectively. The maximum stress (σ_{max}), strain at break (ε_{max}), and Young's modulus of each biofilm were analyzed.

2.5. Morphological Properties

2.5.1. Scanning Electron Microscopy (SEM)

Scanning electron microscopy (SEM) was used to determine the microstructure of the samples. The biofilms were analyzed on both sides (bottom and upper surfaces). In addition, their thickness was measured using ImageJ software. The samples were firstly covered with a thin gold layer to improve their conductivity and, thus, the quality of the micrographs. Later, they were observed in a Zeiss EVO microscope (Pleasanton, CA, USA) with an acceleration voltage of 10 kV and magnification of 3 KX [40].

2.5.2. Energy Dispersive X-ray Spectroscopy (EDX)

The elemental composition distribution of the biofilms was also evaluated. For this, a EDX detector was attached to an SEM microscope [41]. In this way, the Fe distribution was analyzed in both surfaces of each biofilm through mapping (by converting to 8-bit grayscale and then Auto Threshold, where the obtained percentage was representative of the color area of Fe).

2.6. Functional Properties

2.6.1. Antioxidant Activity

Antioxidant activity of the biofilms was determined using the protocol described by Mehmood et al. (2020) with slight modifications [29]. Thus, 1 mL of film solution was mixed with 1 mL of DPPH solution dissolved in methanol (40 ppm). This mixture was kept in the dark for 30 min at 25 °C. Finally, the absorbance of each solution was read at 517 nm in a spectrophotometer. Gallic acid was used as the positive control. DPPH inhibition (*IP*) was calculated using Equation (2).

$$IP\,(\%) = \left(\frac{A-B}{A}\right) \times 100 \qquad (2)$$

where *A* and *B* are the DPPH absorbance without and with antioxidant agent, respectively.

2.6.2. Antimicrobial Activity

The antimicrobial activity of the different biofilms was evaluated using an agar diffusion experiment [29]. In this way, cylindrical biofilms (9 mm of diameter) were firstly sterilized by immersion in 96% (*v/v*) ethanol for 2 min, after which they were rinsed thrice with sterile phosphate buffered saline (PBS). Then, they were placed in agar gels inoculated with *Staphylococcus aureus* (S. au) and *Escherichia coli* (E. col). The antibacterial activity was determined as the inhibition zone (diameters) surrounding the biofilm after 24, 48, and 72 h of incubation at 37 °C using ImageJ software.

2.7. Statistical Analysis

At least 3 replicates of each sample were performed for each measurement. The results were presented as mean values with standard deviations, which were calculated using IBM SPSS statistic software. In addition, the significant differences were evaluated using a one-way ANOVA with 95% confidence level ($p < 0.05$).

3. Results

3.1. Physicochemical Properties

3.1.1. Water Solubility

Water solubility (WS) is a critical parameter for food packaging applications. In this sense, biofilms must be insoluble in water to improve water resistance and product safety [26,29]. Table 1 shows the water solubility (WS) values of the different biofilms. As can be seen, the reference biofilm (neat gelatin) reached the highest WS value (80.9%). Higher WS of neat gelatin biofilms was due to the hydrophilic nature of gelatin [42]. Thus, the incorporation of Fe_xO_y-NPs into the system improved its water resistance. Nevertheless, the increment in the Fe_xO_y-NP concentration did not significantly improve their water resistance. Therefore, it could be concluded that the incorporation of NPs could decrease the solubility in water, regardless of the incorporated concentration. These results could be due to the formation of strong hydrogen bonds between the biopolymer chains and NPs, as has already been reported in previous works [42]. Likewise, Wongphan et al. (2022) reported that the incorporation of enzymes into polymers could cause an interaction via hydrogen bonding that enhances the hydrophobic groups by reducing the contrast, which results in a decrease in the solubility [43]. These results are similar for GS and CS NPs.

Table 1. Physicochemical and mechanical parameters and antioxidant activity values of the biofilms processed with different percentages (0.25 and 1.0%) of green (GS) and chemical (CS) Fe_xO_y nanoparticles (NPs). Neat gelatin-based biofilm without NPs incorporated was used as reference. Different superscript letters (a–d) of each column indicate significant differences ($p < 0.05$).

Sample	C% (w/w)	WS (%)	T_{600} (%)	Thickness (µm)	σ_{max} (MPa)	ε_{max} (mm/mm)	Young's Modulus (MPa)	DPPH Inhibition (%)
GS Fe_xO_y-NPs	0.25%	66.8 ± 2.3 [bc]	45.2 ± 0.3 [b]	98.94 ± 0.8 [c]	14.8 ± 0.8 [a]	0.03 ± 0.01 [a]	494.2 ± 8.9 [a]	78.0 ± 1.9 [ab]
	1%	64.1 ± 2.4 [c]	29.2 ± 0.2 [d]	105.6 ± 0.7 [b]	11.9 ± 1.7 [b]	0.02 ± 0.01 [b]	586.6 ± 148.9 [a]	84.7 ± 3.4 [a]
CS Fe_xO_y-NPs	0.25%	69.6 ± 2.2 [b]	42.0 ± 0.6 [c]	104.9 ± 0.4 [b]	5.2 ± 0.6 [d]	0.03 ± 0.02 [a]	193.9 ± 94.0 [bc]	76.3 ± 2.7 [b]
	1%	67.4 ± 1.5 [bc]	29.3 ± 0.3 [d]	109.1 ± 1.0 [a]	8.4 ± 1.6 [c]	0.02 ± 0.01 [b]	263.5 ± 24.6 [b]	79.6 ± 3.3 [ab]
Neat Gelatin		80.9 ± 3.2 [a]	60.0 ± 0.1 [a]	89.2 ± 1.0 [d]	4.6 ± 0.9 [d]	0.07 ± 0.04 [a]	67.3 ± 33.8 [c]	44.5 ± 0.7 [c]

3.1.2. Optical Properties

The transmittance of each biofilm is also presented in Table 1. The reference system (neat gelatin) had 60% transmittance, which showed the great transparency of these biofilms. The incorporation of Fe_xO_y-NPs into gelatin-based biofilms reduced the transmittance values of the systems. Thus, this decrease was noted linearly with the increase of Fe_xO_y-NP concentration. Moreover, GS NPs led to a lower reduction than CS NPs, although this difference vanished at the highest NP content. The decrease in transparency brought about by NPs could be associated with the increase in solid material, which hindered the mobility of the biopolymer chains. Thus, the dispersion of fillers filled up free space preventing the passage of the light through it. Similar results were obtained in other works with gelatin-based biofilms containing silver or magnetic iron oxide NPs [24,29,44,45].

3.2. Mechanical Properties

Figure 2 shows the tensile profile of the different biofilms. As can be observed, the neat gelatin system presented a short elastic zone followed by a plastic one. This plastic zone decreased as the amount of Fe_xO_y-NPs in the biofilm increased. The comparison of the different systems is better defined through the mechanical parameters shown in

Table 1. In this way, the addition of the Fe_xO_y-NPs generated an increase in Young's modulus and maximum stress (σ_{max}), being more pronounced in GS Fe_xO_y-NPs than in CS Fe_xO_y-NPs. This result is due to the fact that the presence of immiscible particles caused non-homogeneous networks and reduced extensibility of the films [45]. In contrast, no significant differences were found between the incorporated GS and CS-NPs at the same concentration for the strain at break (ε_{max}) of the different films studied. Thus, the incorporation of Fe_xO_y-NPs stiffened the biofilms, probably due to the increasing presence of solid material in the biofilms and the increase of their thickness [29]. The incorporation of GS and CS-NPs may reduce the cohesion between the polymer chains and result in the reduction of strain at break [46]. On the other hand, the GS Fe_xO_y-NPs achieved greater maximum stress and Young's modulus than the CS Fe_xO_y-NPs. This result could be due to the size of the NPs. In this way, smaller NPs generated a better interconnection in the structure, which highlighted this effect [47]. Thus, the incorporation of NPs always improved the mechanical resistance of the biofilms due to the strong network between NPs and biopolymer chains.

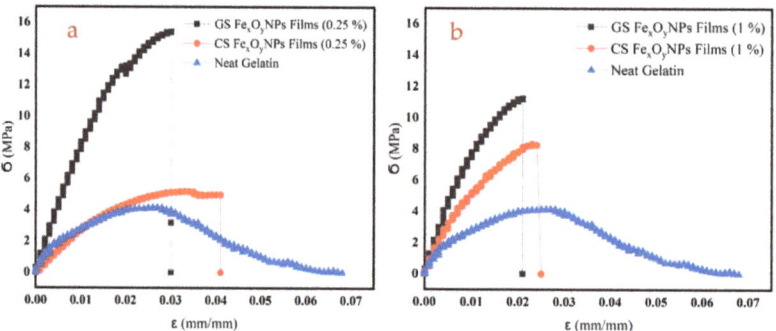

Figure 2. Tensile test profile of the biofilms with different concentrations ((**a**) 0.25 and (**b**) 1.0% *w/w*) of green (GS) and chemical (CS) Fe_xO_y nanoparticles (NPs). Neat gelatin-based biofilm without NPs incorporated was used as the reference system.

Nevertheless, the increase in NP concentration showed much lower effects as compared to their mere incorporation. Thus, no significant increase in Young's modulus could be observed for either of the two types of NPs, while the effect on σ_{max} depended on the NPs used. The maximum stress was reduced when increasing GS NP content or increased when CS was used. These contradictory results could be attributed to the occurrence of two opposite effects. On one hand, an increase in NPs may induce an enhancement of the mechanical properties of the biofilms by promoting NP–protein hydrogen bonding. On the other hand, it may also favor NP agglomeration, which may interfere with the formation of these NP–protein domain arrangements [29]. Likewise, aggregation of nanoparticles in biodegradable polymers was possible with increasing nanofiller concentrations, which may lead to changes in the mechanical and barrier properties of the films [48].

3.3. Morphological Properties

3.3.1. Scanning Electron Microscopy (SEM)

Figure 3 shows the bottom and upper surface of the different biofilms. As can be seen, the biofilm without NP incorporation (neat gelatin) presented a smooth and homogeneous structure on both surfaces. On the other hand, the incorporation of NPs generated rough surfaces with fissures and holes. Thus, the incorporation of NPs seemed to alter the nanostructure of biofilms, being more evident in the case of green nanoparticles (GS Fe_xO_y-NPs). This effect could be due to the functional groups present in GS Fe_xO_y-NPs (as they come from the polyphenols used in their synthesis), which could alter the structure of the biopolymer chains in the biofilms by charge interaction [49]. Furthermore, Fe_xO_y-

NPs seemed to form more aggregates, probably due to the increase in concentration [48]. This effect was more evident in CS Fe$_x$O$_y$-NPs among films, which may be due to the non-availability of stabilizing agents [32].

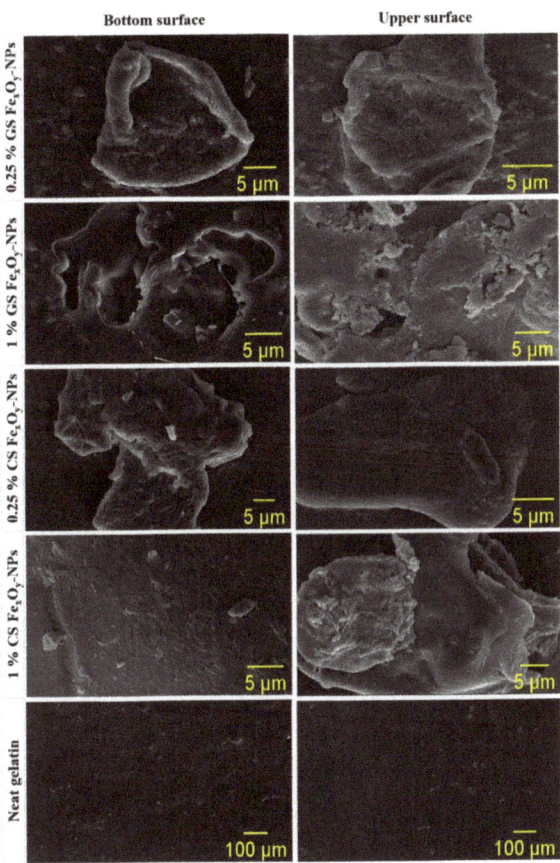

Figure 3. Scanning electron microscopy (SEM) images of the bottom and upper surfaces of the biofilms with different concentrations (0.25 and 1.0% *w/w*) of GS and CS Fe$_x$O$_y$-NPs. Neat gelatin-based biofilm without NPs incorporated was used as the reference system.

3.3.2. Energy Dispersive X-ray Spectroscopy (EDX)

EDX mapping confirmed the purity of biofilms and the synthesized NPs, since only the elemental components of proteins (C, H, N, O) and NPs (Fe, O) were present in the analyses. On the other hand, the distribution of Fe in the biofilms was evaluated through the Fe distribution found in the EDX analyses. Figure 4 shows this distribution in both surfaces (bottom and upper). Nevertheless, the colored area in each image was calculated to improve the comparison between the systems (Table 2). As can be seen, there was a homogeneous distribution of NPs in both surfaces of the biofilms, since Fe could be observed in the whole images. Nevertheless, there was a higher NP concentration on the upper surface than on the bottom surface, being more evident in the higher NP percentage (1%). This behavior can be explained by the hydrophobicity of the NPs, which made them migrate to the surface where they presented less steric repulsion, as well as by their density, which made them float instead of sinking into the film solution [50]. Additionally, this was explained by the strengthened Van Der Waals forces as particle size and concentration increased or that larger particles would exhibit higher surface tension, causing them to settle on the surface [50]. On the other hand, a higher predisposition to the agglomerations or aggregations of NPs could be observed at higher NP concentrations, causing them to fall by gravity. Among the different NPs, CS NPs presented greater precipitate than GS NPs, possibly due to their larger size. In addition, the highest concentration of CS Fe_xO_y-NPs showed agglomeration of NPs in certain areas of the biofilm. This behavior was already reported by Rufus et al. (2017) and Hosseini et al. (2015), who attributed it to an entropic process [26,36].

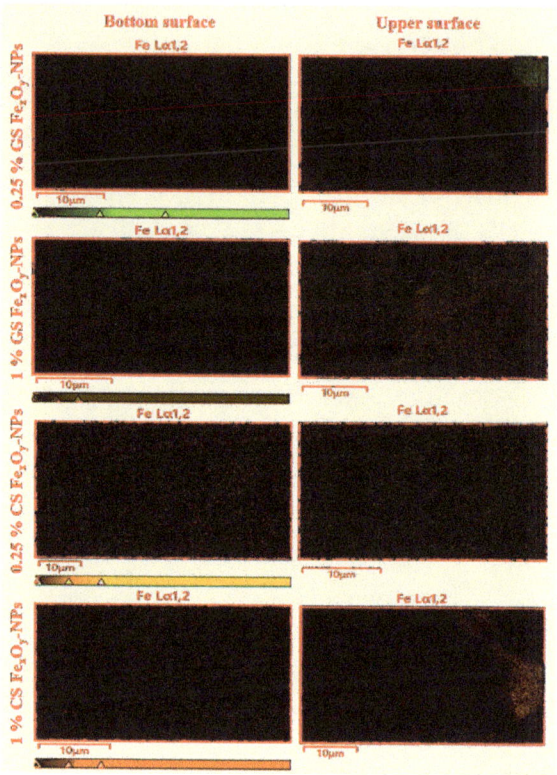

Figure 4. Fe distribution observed by energy dispersive X-ray spectroscopy (EDX) in the bottom and upper surfaces of the biofilms with different concentrations (0.25 and 1.0) of GS and CS Fe_xO_y NPs. Neat gelatin-based biofilm without NPs incorporated was used as the reference system.

Table 2. Fe concentration (%) in the bottom and upper surfaces of the biofilms with different concentrations (0.25 and 1.0) of GS and CS Fe_xO_y NPs.

Biofilms	Bottom	Upper
	Fe (%)	Fe (%)
GS Fe_xO_y-NPs 0.25%	12.0	16.5
GS Fe_xO_y-NPs 1.0%	13.3	30.3
CS Fe_xO_y-NPs 0.25%	12.9	13.6
CS Fe_xO_y-NPs 1.0%	16.7	43.4
Neat Gelatin	-	-

3.4. Functional Properties

3.4.1. Antioxidant Activity

As can be seen in Table 1, the incorporation of Fe_xO_y-NPs produced an increase in antioxidant activity. In addition, the higher the Fe_xO_y-NP concentration, the higher the antioxidant activity. Thus, these nanoparticles act as antioxidants, as was reported in similar articles [29,51]. Regarding the different nanoparticles, the GS Fe_xO_y-NPs presented higher DPPH inhibition than the CS Fe_xO_y-NPs. This behavior could be due to the greater amount of antioxidant groups in them, conferred during their synthesis by polyphenols. In this way, green synthesis did not only display advantages in the synthesis process by reducing toxic waste and lowering costs, but it also allowed the generation of nanoparticles that provided greater functionality in future applications. These results showed that the gelatin-based Fe_xO_y-NPs displayed higher antioxidant activity than others reported in the literature, such as those obtained by Zafar et al. (2020) using iron oxide nanoparticles (IONPs) [29], and Shiv et al. (2019) using melanin nanoparticles [7]. It is worth being mentioned that the *IP%* of the positive control (gallic acid) was from 94.9 to 98.4%.

3.4.2. Antibacterial Activity

The inhibition area produced by the different biofilms is collected in Table 3 through its significant measure (diameter). An image of these inhibition areas can be observed in the supplementary materials. As can be seen, the neat gelatin biofilms showed antibacterial activity against *Staphylococcus aureus*, since a small zone of inhibition could be observed for the first 24 h. This could be due to the presence of sulfur dioxide (<10 ppm), as it is food gelatin. Nevertheless, this effect was lost in time when the bacteria grew. In this way, neat gelatin only had the ability to act on the peptidoglycan layer of bacteria to generate their lysis. However, it did not show the ability to act on the lipid layer, thus it only affected gram-positive bacteria and not gram-negative bacteria [52]. The inclusion of Fe_xO_y-NPs improved the antibacterial activity of biofilms. In this way, higher inhibition areas could be observed, as well as an effect on both types of bacteria. Nevertheless, although the difference in nanoparticle concentration on each of the surfaces of the biofilms was different, the antibacterial activity did not seem to have significant differences on each of the surfaces, making it a fully functional material. Regarding the synthesis of NPs, the GS Fe_xO_y-NPs presented higher antibacterial activity than the CS Fe_xO_y-NPs. This could be due to the smaller size of the green NPs, which conferred the biofilms a better capacity to inhibit the replication of bacterial DNA [53,54], as well as the higher crystallinity shown by GS Fe_xO_y-NPs [30]. Higher crystallinity allowed the released of Fe^{+2}/Fe^{+3} to collide with the negatively charged membranes of bacteria, destroying their protein structure and causing them to die [55]. The incorporated antimicrobial function into biopolymer films could be an efficient method to enhance the shelf-life extension capacity for biodegradable films [56]. On the other hand, all the samples showed a decrease in their inhibitory capacity over time, being more notable in CS Fe_xO_y-NPs.

Table 3. The inhibition areas (represented by their diameter in mm) produced by gelatin-based biofilm with 1.0% GS and CS Fe_xO_y-NPs incorporated against *Staphylococcus aureus* (S. au) and *Escherichia coli* (E. col). Neat gelatin-based biofilm without NPs incorporated was used as the reference system. Different superscript letters (a–e) in a column indicate significant differences ($p < 0.05$).

Test Time (h)	Biofilm	S. Au		E. Col	
		Upper	Bottom	Upper	Bottom
0		9 [b]	9 [c]	9 [c]	9 [b]
24	Neat Gelatin	9.4 ± 0.3 [b]	9.4 ± 0.3 [c]	0.0 [e]	0.0 [c]
	GS-NPs	12.8 ± 1.5 [a]	13.5 ± 1.7 [a]	13.4 ± 2.3 [a]	15.5 ± 2.6 [a]
	CS-NPs	12.9 ± 1.1 [a]	11 ± 1.3 [bc]	12.5 ± 2.0 [ab]	9.3 ± 2.4 [b]
48	Neat Gelatin	0.0 [c]	0.0 [d]	0.0 [e]	0.0 [c]
	GS-NPs	12.3 ± 2.3 [a]	12.6 ± 1.6 [ab]	11.1 ± 0.7 [b]	14.6 ± 1.9 [ab]
	CS-NPs	10.9 ± 2.9 [ab]	10.9 ± 2.5 [bc]	11.5 ± 0.4 [ab]	0.0 [c]
72	Neat Gelatin	0.0 [c]	0.0 [d]	0.0 [e]	0.0 [c]
	GS-NPs	9.1 ± 0 [b]	9.5 ± 0.1 [c]	6.1 ± 0.5 [d]	8.3 ± 0.8 [b]
	CS-NPs	0.0 [c]	0.0 [d]	6.7 ± 1.6 [d]	0.0 [c]

4. Conclusions

The incorporation of Fe_xO_y nanoparticles into gelatin-based biofilms improved their properties even at a low concentration. Thus, the 0.25% (w/w) led to a reduction in the water solubility and to an improvement in the antioxidant activity and mechanical properties. An increase in NPs concentration led to opaquer or less transparent biofilms with greater thickness, but not to a significant improvement in the other properties evaluated, which may be due to the tendency of the nanoparticles to agglomerate at higher concentrations caused by their hydrophobic nature. Therefore, their inclusion reduced the water solubility of the biofilms and improved their mechanical resistance and functional properties. Furthermore, the dispersion of the nanoparticles incorporated by the casting method led to their heterogeneous distribution with differences in the distribution in both upper and bottom surfaces of the biofilms. However, these differences did not alter the antibacterial properties.

Finally, the green nanoparticles (GS Fe_xO_y-NPs) showed better antioxidant and antibacterial activities than the CS Fe_xO_y-NPs, which must improve the added value of these biofilms. This antioxidant activity of (GS Fe_xO_y-NPs) highlighted high antioxidant activity ($PI\% = 84.7\%$ compared to gallic acid as a standard with $PI\% = 96.65$). A control on the distribution of the nanoparticles in the biofilms and the use of the green nanoparticles led to sustainable biofilms with a significant improvement of the functional properties. All this makes these biofilms highly applicable for use as functional packaging to preserve food, antioxidant and antimicrobial additives in multiple applications such as biomedical and pharmaceutical biomaterials, as well as in agricultural pesticides. Nevertheless, further characterization of these biopolymer films, including a biodegradability study, will be required in future works as a previous stage to their use in packaging applications.

Supplementary Materials: The following are available online at https://www.mdpi.com/article/10.3390/ma15051966/s1. Figure S1: Image of inhibition area over time of neat gelatin-based biofilm without Fe_xO_y-NPs incorporated. Figure S2: Image of inhibition area over time of gelatin-based biofilm with 1.0% GS Fe_xO_y-NPs incorporated. Figure S3: Image of inhibition area over time of gelatin-based biofilm with 1.0% CS Fe_xO_y-NPs incorporated.

Author Contributions: Conceptualization, J.A.A.A., A.R. and A.G.; methodology, J.A.A.A. and M.J.-R.; software, J.A.A.A.; validation, J.A.A.A. and A.R.; formal analysis, M.J.-R.; investigation, J.A.A.A. and M.J.-R.; resources, A.G.; data curation, J.A.A.A.; writing—original draft preparation, J.A.A.A. and M.J.-R.; writing—review and editing, A.R. and A.G.; visualization, A.R. and A.G.; supervision, A.R. and A.G.; project administration, A.G.; funding acquisition, A.G. All authors have read and agreed to the published version of the manuscript.

Funding: Grant RTI2018-097100-B-C21 funded by MCIN/AEI/10.13039/501100011033 and by "ERDF A way of making Europe".

Institutional Review Board Statement: Not applicable.

Informed Consent Statement: Not applicable.

Data Availability Statement: The data presented in this study are available upon request from the corresponding author.

Acknowledgments: The authors acknowledge the MCI/AEI/FEDER, EU project (Ref. RTI2018-097100-B-C21) that supports this work. In addition, the authors thank the predoctoral grant from Johar Amin Ahmed Abdullah (Universidad de Sevilla, CODE 810) and Mercedes Jiménez-Rosado (Ministerio de Educación y Formación Profesional, FPU2017/01718). The authors also thank CITIUS for granting access to and their assistance with the DRX area characterization and microscopy services.

Conflicts of Interest: The authors declare no conflict of interest.

References

1. AL-Assadi, Z.I.; AL-Assadi, F.I. Enhancing the aesthetic aspect of the solar systems used as facades for building by designing multi-layer optical coatings. *Tech. Rom. J. Appl. Sci. Technol.* **2021**, *3*, 1–10. [CrossRef]
2. Glaser, T.K.; Plohl, O.; Vesel, A.; Ajdnik, U.; Ulrih, N.P.; Hrnčič, M.K.; Bren, U.; Zemljič, L.F. Functionalization of polyethylene (PE) and polypropylene (PP) material using chitosan nanoparticles with incorporated resveratrol as potential active packaging. *Materials* **2019**, *12*, 2118. [CrossRef]
3. Roy, S.; Rhim, J.-W.W. Gelatin-Based Film Integrated with Copper Sulfide Nanoparticles for Active Packaging Applications. *Appl. Sci.* **2021**, *11*, 6307. [CrossRef]
4. Nur Hanani, Z.A.; Roos, Y.H.; Kerry, J.P. Use and application of gelatin as potential biodegradable packaging materials for food products. *Int. J. Biol. Macromol.* **2014**, *71*, 94–102. [CrossRef]
5. Zhang, X.-L.; Zhao, Y.-Y.; Zhang, X.-T.; Shi, X.-P.; Shi, X.-Y.; Li, F.-M. Re-used mulching of plastic film is more profitable and environmentally friendly than new mulching. *Soil Tillage Res.* **2022**, *216*, 105256. [CrossRef]
6. Samsi, M.S.; Kamari, A.; Din, S.M.; Lazar, G. Synthesis, characterization and application of gelatin–carboxymethyl cellulose blend films for preservation of cherry tomatoes and grapes. *J. Food Sci. Technol.* **2019**, *56*, 3099–3108. [CrossRef]
7. Shankar, S.; Wang, L.-F.; Rhim, J. Effect of melanin nanoparticles on the mechanical, water vapor barrier, and antioxidant properties of gelatin-based films for food packaging application. *Food Packag. Shelf Life* **2019**, *21*, 100363. [CrossRef]
8. Etxabide, A.; Uranga, J.; Guerrero, P.; de la Caba, K. Development of active gelatin films by means of valorisation of food processing waste: A review. *Food Hydrocoll.* **2017**, *68*, 192–198. [CrossRef]
9. Abd Elgadir, M.; Mirghani, M.E.S.; Adam, A. Fish gelatin and its applications in selected pharmaceutical aspects as alternative source to pork gelatin. *J. Food Agric. Environ.* **2013**, *11*, 73–79.
10. Rawdkuen, S.; Thitipramote, N.; Benjakul, S. Preparation and functional characterisation of fish skin gelatin and comparison with commercial gelatin. *Int. J. Food Sci. Technol.* **2013**, *48*, 1093–1102. [CrossRef]
11. Gómez-Guillén, M.C.; Pérez-Mateos, M.; Gómez-Estaca, J.; López-Caballero, E.; Giménez, B.; Montero, P. Fish gelatin: A renewable material for developing active biodegradable films. *Trends Food Sci. Technol.* **2009**, *20*, 3–16. [CrossRef]
12. Da Trindade Alfaro, A.; Balbinot, E.; Weber, C.I.; Tonial, I.B.; Machado-Lunkes, A. Fish Gelatin: Characteristics, Functional Properties, Applications and Future Potentials. *Food Eng. Rev.* **2015**, *7*, 33–44. [CrossRef]
13. Kim, S.-K.; Ngo, D.-H.; Vo, T.-S. Marine Fish-Derived Bioactive Peptides as Potential Antihypertensive Agents. *Adv. Food Nutr. Res.* **2012**, *65*, 249–260. [PubMed]
14. Gudipati, V. Fish Gelatin: A Versatile Ingredient for the Food and Pharmaceutical Industries. In *Marine Proteins and Peptides*; John Wiley & Sons, Ltd.: Chichester, UK, 2013; pp. 271–295.
15. Jeevithan, E.; Qingbo, Z.; Bao, B.; Wu, W. Biomedical and Pharmaceutical Application of Fish Collagen and Gelatin: A Review. *J. Nutr. Ther.* **2013**, *2*, 218–227. [CrossRef]
16. Liu, L.S.; Liu, C.K.; Fishman, M.L.; Hicks, K.B. Composite films from pectin and fish skin gelatin or soybean flour protein. *J. Agric. Food Chem.* **2007**, *55*, 2349–2355. [CrossRef]
17. Jamilah, B.; Harvinder, K.G. Properties of gelatins from skins of fish—Black tilapia (Oreochromis mossambicus) and red tilapia (Oreochromis nilotica). *Food Chem.* **2002**, *77*, 81–84. [CrossRef]
18. Shiao, W.C.; Wu, T.C.; Kuo, C.H.; Tsai, Y.H.; Tsai, M.L.; Hong, Y.H.; Huang, C.Y. Physicochemical and antioxidant properties of gelatin and gelatin hydrolysates obtained from extrusion-pretreated fish (Oreochromis sp.) scales. *Mar. Drugs* **2021**, *19*, 275. [CrossRef]
19. Babayevska, N.; Przysiecka, Ł.; Nowaczyk, G.; Jarek, M.; Järvekülg, M.; Kangur, T.; Janiszewska, E.; Jurga, S.; Iatsunskyi, I. Fabrication of gelatin-zno nanofibers for antibacterial applications. *Materials* **2021**, *14*, 103. [CrossRef]
20. Taokaew, S.; Seetabhawang, S.; Siripong, P.; Phisalaphong, M. Biosynthesis and characterization of nanocellulose-gelatin films. *Materials* **2013**, *6*, 782–794. [CrossRef]

21. Mousazadeh, S.; Ehsani, A.; Moghaddas Kia, E.; Ghasempour, Z.; Moghaddas, E.; Ghasempour, Z. Zinc oxide nanoparticles and periodate oxidation in developing pH-sensitive packaging film based on modified gelatin. *Food Packag. Shelf Life* **2021**, *28*, 100654. [CrossRef]
22. Liff, S.; Mckinley, G.H.; Mehmood, Z.; Sadiq, M.B.; Khan, M.R.; Hanani, Z.A.N.; Sahraee, S.; Milani, J.M.; Ghanbarzadeh, B.; Hamishehkar, H.; et al. High-performance elastomeric nanocomposites via solvent-exchange processing. *Food Packag. Shelf Life* **2007**, *21*, 100363. [CrossRef] [PubMed]
23. Mohammadi, A.; Moradpour, M.; Saeidi, M.; Karim, A. LWT—Food Science and Technology Effects of nanorod-rich ZnO on rheological, sorption isotherm, and physicochemical properties of bovine gelatin films. *LWT—Food Sci. Technol.* **2014**, *58*, 142–149. [CrossRef]
24. Kanmani, P.; Rhim, J.W. Physicochemical properties of gelatin/silver nanoparticle antimicrobial composite films. *Food Chem.* **2014**, *148*, 162–169. [CrossRef] [PubMed]
25. He, Q.; Zhang, Y.; Cai, X.; Wang, S. International Journal of Biological Macromolecules Fabrication of gelatin—TiO_2 nanocomposite film and its structural, antibacterial and physical properties. *Int. J. Biol. Macromol.* **2016**, *84*, 153–160. [CrossRef]
26. Hosseini, S.F.; Rezaei, M.; Zandi, M.; Farahmandghavi, F. Fabrication of bio-nanocomposite films based on fish gelatin reinforced with chitosan nanoparticles. *Food Hydrocoll.* **2015**, *44*, 172–182. [CrossRef]
27. Flaker, C.H.C.; Lourenço, R.V.; Bittante, A.M.Q.B.; Sobral, P.J.A. Gelatin-based nanocomposite films: A study on montmorillonite dispersion methods and concentration. *J. Food Eng.* **2015**, *167*, 65–70. [CrossRef]
28. Sahraee, S.; Milani, J.M.; Ghanbarzadeh, B.; Hamishehkar, H. International Journal of Biological Macromolecules Physicochemical and antifungal properties of bio-nanocomposite film based on gelatin-chitin nanoparticles. *Int. J. Biol. Macromol.* **2017**, *97*, 373–381. [CrossRef]
29. Mehmood, Z.; Sadiq, M.B.; Khan, M.R. Gelatin nanocomposite films incorporated with magnetic iron oxide nanoparticles for shelf life extension of grapes. *J. Food Saf.* **2020**, *40*, . [CrossRef]
30. Mohan, P.; Mala, R. Comparative antibacterial activity of magnetic iron oxide nanoparticles synthesized by biological and chemical methods against poultry feed pathogens. *Mater. Res. Express* **2019**, *6*, 115077. [CrossRef]
31. Tran, N.; Mir, A.; Mallik, D.; Sinha, A.; Nayar, S.; Webster, T.J. Bactericidal effect of iron oxide nanoparticles on Staphylococcus aureus. *Int. J. Nanomed.* **2010**, *5*, 277–283. [CrossRef]
32. Junaid, M.; Dowlath, H.; Anjum, S.; Khalith, S.B.M.; Varjani, S.; Kumar, S.; Munuswamy, G.; Woong, S.; Jin, W.; Ravindran, B. Comparison of characteristics and biocompatibility of green synthesized iron oxide nanoparticles with chemical synthesized nanoparticles. *Environ. Res.* **2021**, *201*, 111585. [CrossRef]
33. Singh, N.; Jenkins, G.J.S.; Asadi, R.; Doak, S.H. Potential toxicity of superparamagnetic iron oxide nanoparticles (SPION). *Nano Rev.* **2010**, *1*, 5358. [CrossRef] [PubMed]
34. Faria, N.; Pereira, D.; Mergler, B.; Powell, J.; Synthesis, A. Ligand doping of iron oxide nanoparticles as an approach to novel oral iron therapeutics. In Proceedings of the 2011 11th IEEE International Conference on Nanotechnology, Portland, OR, USA, 15–18 August 201; pp. 837–840.
35. Mukherjee, P.; Ahmad, A.; Mandal, D.; Senapati, S.; Sainkar, S.R.; Khan, M.I.; Parishcha, R.; Ajaykumar, P.V.; Alam, M.; Kumar, R.; et al. Fungus-Mediated Synthesis of Silver Nanoparticles and Their Immobilization in the Mycelial Matrix: A Novel Biological Approach to Nanoparticle Synthesis. *Nano Lett.* **2001**, *1*, 515–519. [CrossRef]
36. Rufus, A.; Sreeju, N.; Vilas, V.; Philip, D. Biosynthesis of hematite (α-Fe2O3) nanostructures: Size effects on applications in thermal conductivity, catalysis, and antibacterial activity. *J. Mol. Liq.* **2017**, *242*, 537–549. [CrossRef]
37. Abdullah, J.A.A.; Salah Eddine, L.; Abderrhmane, B.; Alonso-González, M.; Guerrero, A.; Romero, A.; Ahmed, J.A.; Salah, L.; Abderrhmane, B. Green synthesis and characterization of iron oxide nanoparticles by pheonix dactylifera leaf extract and evaluation of their antioxidant activity. *Sustain. Chem. Pharm.* **2020**, *17*, 100280. [CrossRef]
38. Felix, M.; Perez-Puyana, V.; Romero, A.; Guerrero, A. Production and Characterization of Bioplastics Obtained by Injection Moulding of Various Protein Systems. *J. Polym. Environ.* **2017**, *25*, 91–100. [CrossRef]
39. AERNOR UNE-EN ISO 527-3; Plásticos. Determinación de las Propiedades en Tracción. Parte 3: Condiciones de Ensayo Para Películas y Hojas. EUROPEAN STANDARD: Pilsen, Czech Republic, 2019.
40. El, S.; Koraichi, S.; Latrache, H.; Hamadi, F. Scanning Electron Microscopy (SEM) and Environmental SEM: Suitable Tools for Study of Adhesion Stage and Biofilm Formation. In *Scanning Electron Microscopy*; InTech: London, UK, 2012.
41. Barkarmo, S.; Longhorn, D.; Leer, K.; Johansson, C.B.; Stenport, V.; Franco-Tabares, S.; Kuehne, S.A.; Sammons, R. Biofilm formation on polyetheretherketone and titanium surfaces. *Clin. Exp. Dent. Res.* **2019**, *5*, 427–437. [CrossRef]
42. Voon, H.C.; Bhat, R.; Easa, A.M.; Liong, M.T.; Karim, A.A. Effect of Addition of Halloysite Nanoclay and SiO 2 Nanoparticles on Barrier and Mechanical Properties of Bovine Gelatin Films. *Food Bioprocess Technol.* **2012**, *5*, 1766–1774. [CrossRef]
43. Wongphan, P.; Khowthong, M.; Supatrawiporn, T.; Harnkarnsujarit, N. Novel edible starch films incorporating papain for meat tenderization. *Food Packag. Shelf Life* **2022**, *31*, 100787. [CrossRef]
44. Šupová, M.; Martynková, G.S.; Barabaszová, K. Effect of nanofillers dispersion in polymer matrices: A review. *Sci. Adv. Mater.* **2011**, *3*, 1–25. [CrossRef]
45. Leelaphiwat, P.; Pechprankan, C.; Siripho, P.; Bumbudsanpharoke, N.; Harnkarnsujarit, N. Effects of nisin and EDTA on morphology and properties of thermoplastic starch and PBAT biodegradable films for meat packaging. *Food Chem.* **2022**, *369*, 130956. [CrossRef] [PubMed]

46. Klinmalai, P.; Srisa, A.; Laorenza, Y.; Katekhong, W. Antifungal and plasticization effects of carvacrol in biodegradable poly (lactic acid) and poly (butylene adipate terephthalate) blend films for bakery packaging. *LWT* **2021**, *152*, 112356. [CrossRef]
47. An, L.; Zhang, D.; Zhang, L.; Feng, G. Effect of nanoparticle size on the mechanical properties of nanoparticle assemblies. *Nanoscale* **2019**, *11*, 9563–9573. [CrossRef] [PubMed]
48. Phothisarattana, D.; Wongphan, P.; Promhuad, K.; Promsorn, J. Biodegradable Poly (Butylene Adipate-Co-Terephthalate) and Thermoplastic Starch-Blended TiO_2 Nanocomposite Blown Films as Functional Active Packaging of Fresh Fruit. *Polymers* **2021**, *13*, 4192. [CrossRef]
49. Villasante, J.; Martin-Lujano, A.; Almajano, M.P. Characterization and application of gelatin films with pecan walnut and shell extract (Carya illinoiensis). *Polymers* **2020**, *12*, 1424. [CrossRef]
50. Tanvir, S.; Qiao, L. Surface tension of Nanofluid-type fuels containing suspended nanomaterials. *Nanoscale Res. Lett.* **2012**, *7*, 226. [CrossRef]
51. Paul, D.R.; Robeson, L.M. Polymer nanotechnology: Nanocomposites. *Polymer* **2008**, *49*, 3187–3204. [CrossRef]
52. Salton, M.R.J. Studies of the bacterial cell wall. *Biochim. Biophys. Acta* **1953**, *10*, 512–523. [CrossRef]
53. Armijo, L.M.; Wawrzyniec, S.J.; Kopciuch, M.; Brandt, Y.I.; Rivera, A.C.; Withers, N.J.; Cook, N.C.; Huber, D.L.; Monson, T.C.; Smyth, H.D.C.; et al. Antibacterial activity of iron oxide, iron nitride, and tobramycin conjugated nanoparticles against Pseudomonas aeruginosa biofilms. *J. Nanobiotechnology* **2020**, *18*, 1–27. [CrossRef]
54. Lee, C.; Jee, Y.K.; Won, I.L.; Nelson, K.L.; Yoon, J.; Sedlak, D.L. Bactericidal effect of zero-valent iron nanoparticles on Escherichia coli. *Environ. Sci. Technol.* **2008**, *42*, 4927–4933. [CrossRef]
55. Shuai, C.; Wang, C.; Qi, F.; Peng, S.; Yang, W.; He, C.; Wang, G.; Qian, G. Enhanced Crystallinity and Antibacterial of PHBV Scaffolds Incorporated with Zinc Oxide. *J. Nanomater.* **2020**, *2020*, 1–12. [CrossRef]
56. Laorenza, Y.; Harnkarnsujarit, N. Carvacrol, citral and α-terpineol essential oil incorporated biodegradable films for functional active packaging of Pacific white shrimp. *Food Chem.* **2021**, *363*, 130252. [CrossRef] [PubMed]

Article

Effect of Solution Properties in the Development of Cellulose Derivative Nanostructures Processed via Electrospinning

Pablo Sánchez-Cid [1,*], José Fernando Rubio-Valle [2], Mercedes Jiménez-Rosado [3], Víctor Pérez-Puyana [1] and Alberto Romero [1]

[1] Departamento de Ingeniería Química, Facultad de Química, Universidad de Sevilla, 41012 Sevilla, Spain; vperez11@us.es (V.P.-P.); alromero@us.es (A.R.)
[2] Pro2TecS—Chemical Process and Product Technology Research Centre, Department Ingeniería Química, ETSI, Campus de "El Carmen", Universidad de Huelva, 21071 Huelva, Spain; josefernando.rubio@diq.uhu.es
[3] Departamento de Ingeniería Química, Escuela Politécnica Superior, Universidad de Sevilla, 41011 Sevilla, Spain; mjimenez42@us.es
* Correspondence: psanchezcid@us.es; Tel.: +34-954557179

Abstract: In the last few years, electrospinning has proved to be one of the best methods for obtaining membranes of a micro and nanometric fiber size. This method mainly consists in the spinning of a polymeric or biopolymeric solution in solvents, promoted by the difference in the electric field between the needle and collector, which is finally deposited as a conjunction of randomly oriented fibers. The present work focuses on using cellulose derivatives (namely cellulose acetate and ethylcellulose), based on the revaluation of these byproducts and waste products of biorefinery, to produce nanostructured nanofiber through electrospinning with the objective of establishing a relation between the initial solutions and the nanostructures obtained. In this sense, a complete characterization of the biopolymeric solutions (physicochemical and rheological properties) and the resulting nanostructures (microstructural and thermal properties) was carried out. Therefore, solutions with different concentrations (5, 10, 15, and 20 wt%) of the two cellulose derivatives and different solvents with several proportions between them were used to establish their influence on the properties of the resulting nanostructures. The results show that the solutions with 10 wt% in acetic acid/H$_2$O and 15 wt% in acetone/N,N-dimethylformamide of cellulose acetate and 5 wt% of ethylcellulose in acetone/N,N-dimethylformamide, exhibited the best properties, both in the solution and nanostructure state.

Keywords: electrospinning; cellulose acetate; ethylcellulose; nanostructures; rheological properties; thermal properties; microstructure

1. Introduction

Electrospinning is defined as the spinning of polymeric solutions or molten polymers in the presence of strong electric fields. Thus, this technique allows the fabrication of fibers with diameters between micro and nanometers. The application of a high electric force, enough to overcome the surface tension of the polymeric solution, leads to the generation of fibers from the solution that are finally deposited randomly over the collector, while the solvent evaporates. These fibers move along the direction of the electric field, thus they are unstable and may undergo elongation, depending on the set parameters [1,2]. There are numerous variables that should be considered when designing an electrospinning process, such as compositional parameters (solution concentration, viscosity, conductivity, surface tension, molecular weight of the polymer, etc.), processing parameters (flow rate, voltage, distance between the needle and the collector, etc.), and environmental parameters, such as temperature and relative humidity [1–4]. In addition, electrospinning is a versatile process that allows obtaining structured nanofibers, which are formed by micro- and nano-sized fibers. Such nanofibrous membranes offer countless advantages over traditional

fibers, such as an enormous surface area/volume ratio, surface flexibility, high porosity, porosity control, pore interconnectivity, and superior mechanical performance compared to other known forms of the material. In addition, a wide range of raw materials, such as synthetic polymers, natural polymers, proteins, polysaccharides, etc. can be used to obtain electrospun nanostructures [5]. Due to these characteristics, the use of nanofibers and therefore electrospinning has increased in recent years, making them optimal candidates for a wide variety of applications, including tunable hydrophobicity and water adhesion, scaffolds for tissue engineering, air filtration media, controlled drug release, biosensors, textiles, wound dressings, special membranes, and antimicrobial activities [6–8].

Cellulose $(C_6H_{10}O_5)_n$ is the most important and abundant structural polysaccharide and biomolecule in the world [9]. Cellulose was first synthesized in 1992 by Kobayashi and Shoda, without using any enzyme of a biological origin [10,11]. Regarding its structure, cellulose is a glucose polymer formed by β-type glucose molecules attached by β (1→4) glycosidic bonds. Cellulose chains present a linear structure, united by hydrogen bonds. These units are not located exactly in the plane of the structure, but adopt a saddle conformation with the successive glucose residues rotated at an angle of 180° with respect to the molecular axis and the hydroxyl groups in the equatorial position, which provide high mechanical resistance [12,13]. Due to its structure, cellulose is susceptible to considerable modifications to give rise to new derivative compounds that may be more appropriate than cellulose for certain applications [14]. There is a large variety of cellulose derivatives, such as methylcellulose, cellulose acetate, and ethylcellulose, among others, with the latter two being the most widely used [8,14–17]. These organic compounds are obtained from cellulose, modifying its original structure by substituting external hydroxyl groups with methyl, acetyl, or ethoxy groups, respectively [12].

The electrospinning of cellulose derivatives has been extensively studied in the last decade. However, there have been enormous advances, particularly in the preparation of composite materials based on the electrospinning of these derivatives, which have enormous potential to turn several industrial sectors around [18]. Many researchers have studied electrospun cellulose derivatives nanofibers using different solvent systems [19]. The different solvents used for electrospinning were: Acetone, dimethylacetamide, dimethylformamide, acetic acid, chloroform, methanol, water, or their mixtures in different proportions, with acetone being the most commonly used solvent [20–22]. The main problem is that the boiling point (56 °C) of acetone is low, and it evaporates quickly, which hinders the long-term electrospinning required for the large-scale production of nanostructures for various applications.

In recent years, cellulose derivatives electrospun nanostructures have been used as scaffolds for tissue regeneration [23,24], as filter membranes, for catalytic processes, due to their high surface area [25]. Cellulose derivatives fibers have also been utilized in the textile industry, combining different types of fibers, with polymeric coatings [26]. However, the main applications of cellulose derivatives nanostructures are medical applications [27,28], such as the previously mentioned TE, bandage fabrication, drug-controlled release systems, medical implants, and artificial organs [1,12]. Although cellulose derivatives nanostructures potential applications have been investigated, no correlation between the properties of the previous solutions and the properties of electrospun nanostructures has been established yet. Thus, in the present work it is intended to establish a correlation between the physicochemical properties of the solution with the microstructural properties of the electrospun nanostructures, for this purpose it will be evaluated how it affects the concentration of natural and biodegradable polymers, as well as how it affects the solvent used.

Therefore, the main objective of this study was to relate the characteristics of the initial solutions with those of the obtained nanostructures, using cellulose derivatives as raw materials. For this purpose, solutions of ethylcellulose and cellulose acetate were prepared using different weight concentrations as well as different solvents. They were rheological and physically characterized. On the other hand, a complete microstructural, thermal, and chemical characterization of the electrospun fiber mats was carried out. Properties of the initial solution and nanostructures have been correlated to establish relationships between them.

2. Materials and Methods

2.1. Materials

The polysaccharides chosen in this study were cellulose acetate (AC) (M_w = 30,000 g·mol^{-1}; acetyl groups percentage = 39.8%; DS = 2.45) and ethylcellulose (EC) (M_w = 45,000 g·mol^{-1}; ethoxy groups percentage = 48%; DS = 2.45), which are low molecular weight cellulose derivatives. Both materials were provided by SIGMA ALDRICH S.A. (Taufkirchen, Germany). The selected solvents were acetone, provided by Honeywell (Offenbach am Main, Germany), N,N-dimethyl-formamide (DMF), supplied by EMSURE (Darmstadt, Germany), acetic acid provided by Panreac Química S.A. (Barcelona, Spain), and distilled water. These solvents were chosen based on the great solubility of both cellulose derivatives in them.

2.2. Nanofabrication of Cellulose Derivatives

2.2.1. Solution Preparation

AC solutions were prepared by dissolving different concentrations of polymer (5, 10, 15 and 20 wt.%) into a 2:1 mixture of acetone/DMF or acetic acid/water, due to its high solubility in these solvents [14], to evaluate the influence of the solvents in the procedure and the properties of the systems. Each system underwent an agitation step at 500 rpm, using a magnetic stirrer for 4 h.

In the same way, EC was dissolved at different concentrations (2.5, 5, 10, and 15 wt.%) in a 2:1 acetone/DMF solvent, using the same protocol, in order to evaluate the differences between both polymers. It is worth mentioning that acetic acid/water solutions were not prepared for EC due to its low solubility in this solvent [15].

2.2.2. Electrospinning

The fabrication of the nanostructures of AC and EC was developed with a Fluidnatek LE-50, electrospinning equipment (Bioinicia, Valencia, Spain). A total of 5 mL of each solution were introduced into a syringe, equipped with a 21 G needle. The syringe was fixed to the support, using the vertical setting, with a distance of 15 cm from the needle to the collector. The high-voltage power source supplied 17 kV. Each experiment was carried out at room temperature (22 ± 1 °C) and a controlled relative humidity of 45 ± 1%.

2.3. Characterization

2.3.1. Characterization of the Solutions

Once the solutions of each system were prepared, rheological properties, as well as physical properties, such as density, surface tension, and electric conductivity were characterized and analyzed.

Rheological Properties

Rheological measurements were carried out to determine the viscosity of the solutions as a function of the shear rate. The measurements were carried out with a stress-controlled rheometer AR 2000 (TA Instruments, New Castle, DE, USA). In the measurements, the samples were placed into a cone/plate measuring geometry with 1.016° and a diameter of 60 mm, in order to minimize possible inertia effects and to avoid gliding. Each measurement was taken at room temperature and the shear rate range studied was 0.02 to 100 s^{-1}. In order to establish a relationship between rheological and microstructural properties, specific viscosity (η_{sp}) was determined using Equation (1):

$$\eta_{sp} = \frac{\eta - \eta_0}{\eta_0}, \quad (1)$$

where η (Pa·s) is the viscosity of the solution and η_0 (Pa·s) is the viscosity of the pure solvent.

In addition, extensional viscosity measurements were carried out using a HAAKE CaBER 1 (Thermo Haake GmbH, Germany) at room temperature. One bead of solution was taken between two plates (4 mm of diameter) concentrically set. The plates were separated vertically with a gap of 1 mm and a traction stress was applied to form a fluid unstable filament, observing the evolution of the filament diameter (D_{min}) with time, using a micrometer and its interaction with a laser, which provides a way to estimate the extensional viscosity (η_{ext}).

Physical Properties

Density measurements were carried out using a Densito 30P digital densimeter (Mettler Toledo, Sevilla, Spain). Moreover, surface tension was analyzed using a force tensiometer Sigma 703D (Biolin Scientific, Shanghai, China). The measurements of each system were taken by temporal stability, utilizing a Wilhelmy platinum plate (39.24 mm wide and 0.1 mm thick). Each measurement was recorded at room temperature. In addition, conductivity measurements of the solutions were performed with an EC-Meter BASIC 30+ digital conductometer (Crison Instruments, Barcelona, Spain) by electrical stability.

2.3.2. Characterization of the Nanostructures

Chemical Properties

Chemical bonds were analyzed by Fourier-transform infrared spectroscopy (FTIR), using a Hyperion 1000 spectrophotometer (Bruker, Santa Clara, CA, USA). The samples were placed in an ATR diamond sensor to obtain their infrared profile. The measurements were obtained between 4000 and 400 cm^{-1} with an opening of 4 cm^{-1} and an acquisition of 200 scans. Baseline correction was performed by measuring without a sample.

Thermal Properties

Thermogravimetric analyses were carried out using TGA Discovery equipment (TA Instruments, USA), evaluating the loss of weight of the systems with temperature, in order to observe the thermal events that may take place. Sample quantities between 4 and 7 mg were placed on platinum pans and heated from 25 °C to 600 °C, at 10 °C·min^{-1}, under a N_2 atmosphere with a flow rate of 60 mL/h. Moreover, differential scanning calorimetry analyses were also carried out, in order to determine the heat flow associated with the different thermal transitions, using a SDT Q600 (TA Instruments, USA). The tests were performed between −50 and 220 °C, with a heating rate of 10 °C·min^{-1} and N_2 atmosphere with a flow rate of 60 mL/h. AC and EC raw materials were also thermally characterized in order to evaluate if electrospinning process produces changes in the thermal transition of the final nanostructures in respect to the raw material.

Microstructural Properties

Prior to microscopy examination, samples (2–3 mm) were cut and treated with osmium vapor (1%) for 8 h to fix the samples and facilitate their observation under the microscope. Then, the fixed samples were covered by a thin film of Au to improve the quality of the micrograph (improving sample conductivity). The microscopy examination was performed using a Zeiss EVO scanning electron microscope (Stuttgart, Germany) with a secondary electron detector at an acceleration voltage of 10 kV. A digital processing free software, FIJI Image-J (National Institutes of Health, Bethesda, MD, USA), was used to determine the pore size distribution and the mean pore size of the nanostructures.

2.4. Statistical Analysis

At least three replicates were carried out for each measurement. Statistical analyses were performed with t-tests and one-way analysis of variance ($p < 0.05$), using PASW Statistics for Windows (Version18: SPSS Inc., Endicott, NY, USA). Standard deviations were calculated for selected parameters. The significant differences were established with a confidence level of 95% ($p < 0.05$), which are indicated with different letters in the different tables.

3. Results

3.1. Characterization of the Solutions

As was previously mentioned, the electrospinning process depends on many variables and physicochemical properties of solutions, such as viscosity, surface tension, and electric conductivity. In particular, the viscosity of the solution can be adapted by modifying the polymer concentration [29–31] and in fact, some correlations were applied in order to find the minimal concentration that leads to adequate viscosity values to perform the electrospinning process. Table 1 displays the values of shear viscosity of different solutions of AC and EC, as well as other physicochemical parameters, namely, extensional viscosity, density, surface tension, and conductivity. It is important to mention that every system exhibited a Newtonian behavior in the applied shear rate range.

Table 1. Rotational viscosity (η), extensional viscosity (η_{ext}), density (ρ), surface tension (γ), and conductivity (σ) values of the different AC and EC solutions. Different letters in each column present significant differences between the parameters ($p < 0.05$).

Systems		η (Pa·s)	η_{ext} (Pa·s)	ρ (g·cm^{-3})	γ (mN·m^{-1})	σ (µS·cm^{-1})
AC + Acetic acid/H$_2$O (2:1)	5 wt.%	0.004 [a]	0.011 [A]	1.067 [α]	29.75 [I]	184.0 [a]
	10 wt.%	0.011 [a]	0.035 [B]	1.074 [α]	30.15 [II]	197.7 [b]
	15 wt.%	0.185 [b]	0.575 [C]	1.087 [β]	31.25 [III]	174.3 [c]
	20 wt.%	0.311 [c]	0.983 [D]	1.095 [β]	30.75 [II]	157.7 [d]
AC + Acetone/DMF (2:1)	5 wt.%	0.006 [a]	0.015 [A]	0.862 [γ]	30.35 [II]	128.1 [e]
	10 wt.%	0.105 [d]	0.335 [E]	0.881 [δ]	31.54 [III]	113.1 [f]
	15 wt.%	0.385 [e]	1.105 [F]	0.892 [δ]	32.15 [IV]	96.71 [g]
	20 wt.%	0.523 [f]	1.529 [G]	0.902 [ε]	31.74 [III]	73.13 [h]
EC + Acetone/DMF (2:1)	2.5 wt.%	0.009 [a]	0.029 [B]	0.859 [γ]	30.58 [II]	144.5 [i]
	5 wt.%	0.132 [g]	0.356 [H]	0.869 [γ]	32.51 [IV]	97.57 [g]
	10 wt.%	0.678 [h]	2.104 [I]	0.877 [γ]	33.58 [V]	69.27 [h]
	15 wt.%	1.115 [i]	3.395 [J]	0.886 [δ]	33.47 [V]	25.20 [j]

From these results, it can be observed that an increase of the polymer concentration used to make the solution leads to an increase of viscosity, regardless of the solvents used in the process. On the other hand, comparing the results obtained for the AC systems with both combination of solvents, the solvents appear to influence viscosity, obtaining higher viscosities when using acetone/DMF, due to the interaction between the solvents and polymer. Finally, when comparing the effect of the polymer itself, i.e., using different polymers (AC and EC) with the same mixture of solvents, it can be observed that the viscosities of the EC systems are considerably higher, due to the structural properties of each polymer [32]. EC + Acetone/DMF 20 wt.% was not included in Table 1, as it did not provide relevant information, conversely to the 2.5 wt.% solution. The differences in the physicochemical properties of the systems prepared with EC with respect to AC, using in both cases as solvents a DMF/Ac mixture is mainly due to two factors. On the one hand, the main factor is due to the molecular weight, since EC has a molecular weight of 45,000 g.mol^{-1}, while CA has a molecular weight of less than 30,000 g.mol^{-1}. As is well known, systems made with a polymer with a higher molecular weight have higher viscosities, in addition to causing the displacement towards lower concentrations of the critical concentration (Ce) from which these systems are in the semi-dilute entangled regime and can generate micro and nanofibers during the electrospinning process [31,33].

On the other hand, the viscosity of the systems is affected by the length of the polymer chain. In the case of systems made by EC, the chain of the ethyl functional group is slightly longer than that of systems made with CA (ethoxy group). It is also important to mention that the content of ethyl groups in the raw material from which the systems were prepared is higher than that of acetyl groups (48% of ethyl groups compared to 39.8% of acetyl groups in AC), as can be seen in Section 2.1 [21,22].

As discussed above, the solution viscosity has often been adapted by modifying the polymer concentration [29,30] and, in fact, some correlations were applied to find the minimum polymer concentration (C_e) to reach the appropriate viscosity values to perform the electrospinning process. For example, Aslanzadeh et al. [34] determined that a certain viscosity threshold corresponding to the critical polymer overlap concentration is required to obtain relatively uniform nanofibers. To obtain this C_e, specific viscosity η_{sp} was plotted vs. polymer concentration (Figure 1).

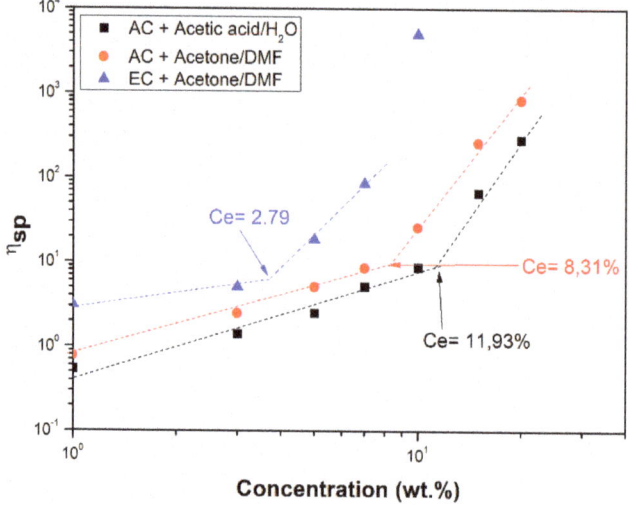

Figure 1. Concentration dependence of the specific viscosity for different solutions of AC and EC with different solvents.

Figure 1 shows the relationship between the specific viscosity (η_{sp}) and derivates celluloses concentrations. The critical entanglement concentration (C_e) delimiting the semi-diluted unentangled and the semi-diluted entangled regimes can be obtained as the change in the slope of this plot. It must be highlighted that C_e strongly depends on the polymer and the solvents used. On the semi-dilute unentangled, strong polymer-solvent interactions predominate, which means that in this regime, no electrospun fibers would be obtained; whereas at higher concentrations (C > C_e), the predominant process would be electrospinning, thus obtaining nanofibers [29,30].

In addition to rotational viscosity, Table 1 displays the values of extensional viscosity, density, surface tension, and conductivity obtained for each system. Extensional viscosity (η_{ext}) reflects the same behavior as the one shown by rotational viscosity, increasing with polymer concentration regardless of the solvent mixture used. Moreover, when comparing each system, similar results are observed, obtaining higher extensional viscosity values for AC when using acetone/DMF as solvents; with the EC systems exerting higher values than the AC systems. Density values are determined by the combination of solvents used, obtaining higher values for AC + acetic acid/H_2O and similar values for systems with different polymers, however the same solvents. Regardless of the solvents used, increasing the polymer concentration slightly increased density values [35,36]. On the other hand, surface tension results did not show a clear deviation pattern for the studied systems, with the ones obtained for the EC being slightly higher. Finally, conductivity values generally decreased when the polymer concentration was increased, except for the case of the low polymer concentration AC + acetic acid/H$_2$O solutions. This could be due to the fact that, when increasing the polymer concentration in the solutions, the number of electric charges also increases up to a maximum concentration, from which, despite the increase in the number of charges, conductivity decreases due to the fact that the movement of the charges becomes progressively more limited [29,30].

3.2. Characterization of the Nanostructures

3.2.1. Chemical Properties

Figure 2 shows the resulting FTIR spectra of both raw materials (AC and EC) without electrospinning and a representative nanostructure processed with each solvent, namely AC (10 wt.%) + acetic acid/H$_2$O, AC (15 wt.%) + acetone/DMF and EC (5 wt.%) + acetone/DMF. These systems are selected as representative because they are the concentrations of each biopolymer immediate in each case to the overcoming of the threshold of the critical concentration (C_e), from which each system is expected to present microstructural properties formed by micro and nanofibers, being able to speak already of nanostructures that are composed by fibers.

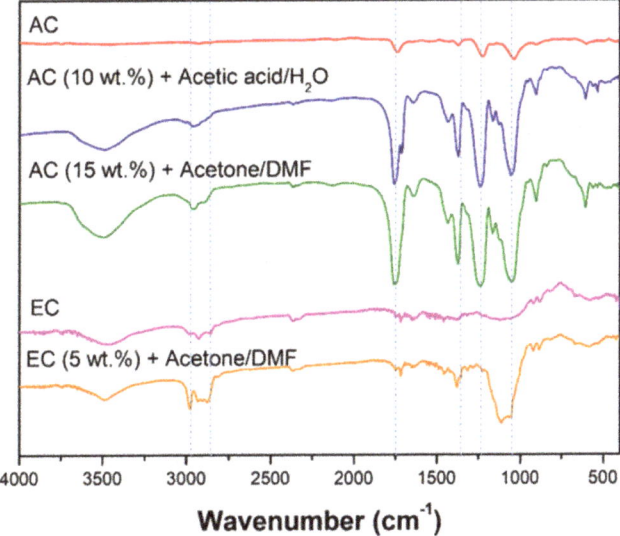

Figure 2. FTIR spectra of both raw materials (AC and EC) and selected nanostructures of the different systems. The dashed lines marked the most representative peaks that have been analyzed.

The AC spectrum shows the characteristic bands of esters at 1735 cm^{-1}, due to the existence of double bonds conjugated with the carbonyl group, at 1300 cm^{-1} and 1000 cm^{-1}. These bands can also be observed in the AC nanostructures, both using acetic acid/H$_2$O and acetone/DMF solvents combinations. The differences with the raw AC spectrum are the intensity of the common bands, which are higher for the nanostructures, with respect to the raw material, and the presence of the characteristic bands of the solvents used, namely the ones of carboxylic acids in the case of using acetic acid and water as solvents or the characteristic bands of ketones and amides when using the acetone/DMF mixture. On the other hand, in the case of EC systems, characteristic bands of ethers, ketones, and amides (2950–2800 cm^{-1}) can be observed, due to the molecular structure of the polymer itself and the solvents used [37–42]. The same increase in the common bands can be observed in this case.

3.2.2. Thermal Properties

Table 2 displays the values of onset temperatures (T_{onset}), maximum temperatures (T_{max}), weight loss, and residues of the thermal events observed on each processed nanostructure and in both raw materials, as well as the glass transition temperatures (T_g). The thermal profile of TGA and DSC of the different systems are incorporated as supplementary material (Figures S2 and S3). It is worth mentioning that AC (5 wt.%) + acetone/DMF and EC (2.5 wt.%) + acetone/DMF were not included in Table 2, since it was not possible to carry out TGA/DSC assays, due to the fact that neither of the nanostructures could be removed appropriately. This was due to the fact that the nanostructures could not be correctly separated from the aluminum foil enveloping the aluminum collecting plate. This is an experimental limitation since the structure of these two systems consisted entirely of micrometer-sized particles, which formed randomly arranged agglomerates, and when attempting to remove these nanostructures, part of the aluminum foil was always extracted, which could distort the thermal measurements.

Table 2. Onset temperature (T_{onset}), maximum temperature (T_{max}), weight loss, residue percentage, and glass transition temperature (T_g) values obtained for the different raw materials and nanostructured systems of AC and EC.

Systems		T_{onset} (°C)	T_{max} (°C)	Weight Loss (%)	Residue (%)	T_g (°C)
AC		272.0/345.4	285.6/367.9	81.9	18.1	171.1
AC + Acetic acid/H$_2$O (2:1)	5 wt.%	258.7/330.8	305.8/356.7	88.7	11.3	–
	10 wt.%	259.9/335.3	307.2/358.8	88.1	11.9	190.1
	15 wt.%	260.8/337.4	309.9/358.8	87.5	12.5	–
	20 wt.%	261.6/337.7	311.9/359.8	87.3	12.7	–
AC + Acetone/DMF (2:1)	10 wt.%	257.2/322.8	314.4/356.8	87.9	12.1	–
	15 wt.%	259.0/329.0	314.8/357.2	87.8	12.2	188.2
	20 wt.%	260.0/334.7	315.4/359.3	87.5	12.5	–
EC		261.2/322.9	283.1/354.2	95.2	4.8	181.5
EC + Acetone/DMF (2:1)	5 wt.%	253.1/337.2	308.3/359.8	95.5	4.5	183.0

On the one hand, it must be highlighted that both T_{onset} and T_{max} columns include two values, which correspond to two different thermal events (Figure S1). The first one appears at temperatures of 250–300 °C and corresponds to the degradation of acetyl and ethoxyl groups of AC and EC, respectively. The other one occurs at about 350 °C and corresponds to the degradation of cellulose [35,36]. Thus, the first values of these two columns, the T_{onset} and T_{max}, of the first thermal event and the second ones correspond to values obtained for the second thermal event.

On the other hand, it can be noted that an increase of polymer concentration causes a slight increment of both T_{onset} and T_{max}. Moreover, this increment of polymer concentration also provokes a higher generation of residues.

Furthermore, comparing the effect of the solvents on the different AC systems, it can be observed that, using acetic acid/water as a solvent, higher T_{onset}, T_{max}, and residue values were obtained with respect to the values obtained for the acetone/DMF combination (Figures S2 and S3). A similar behavior was observed when analyzing the effect of the polymer, obtaining similar values for the AC and EC systems and the same tendency for T_{onset}. Additionally, it can be noted that the residue percentages are lower for the processed nanostructures than for the raw materials. This effect could be due to the fact that the electrospinning process enables the formation of fibers with the polymer without dragging the inorganic impurities (i.e., salts) of the raw materials The tendencies observed for the EC systems were similar to those of the AC systems.

Moreover, the T_g values of the processed nanostructures were generally higher than the ones obtained for the raw materials, possibly due to the higher crosslinking degree, as the new nanostructure is formed by an entangled nanofiber, whose movement is more limited. Only selected nanostructures had their T_g measured, which were AC (10 wt.%) + acetic acid/H_2O, AC (15 wt.%) + acetone/DMF, and EC (5 wt.%) + acetone/DMF, since although there were no significant differences with the other nanostructures of each system, the ones selected had comparatively better properties than the rest.

3.2.3. Microstructural Properties

Figures 3–5 shows the analysis of the microstructural properties of every different nanostructured system, displaying both SEM image and the distribution of the microstructural properties for AC + Acetic acid/H_2O, AC + Acetone/DMF, and EC + Acetone/DMF systems, respectively.

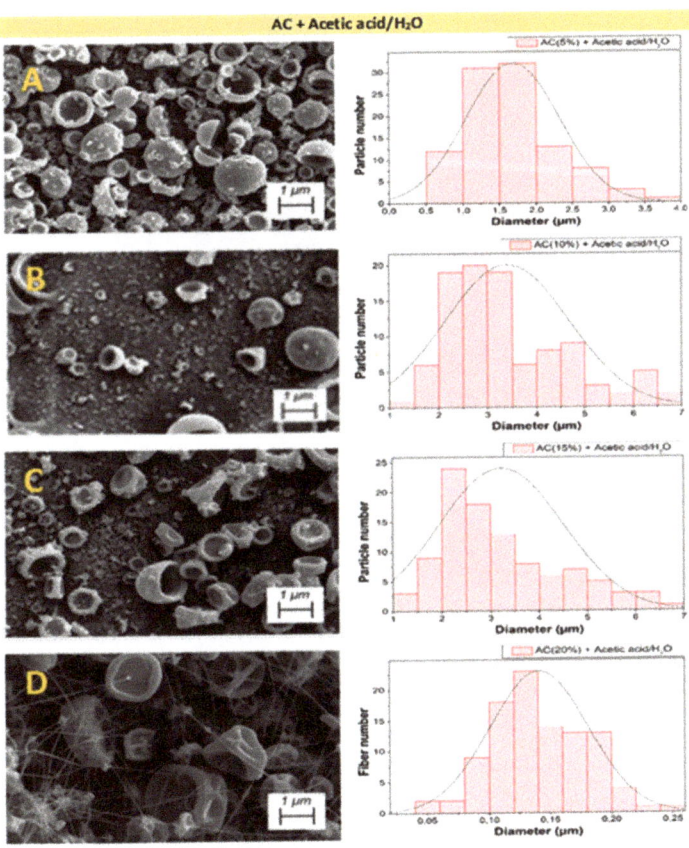

Figure 3. SEM images and distribution of microstructural properties of the nanostructures made with different AC concentrations: (**A**) 5 wt.%, (**B**) 10 wt.%, (**C**) 15 wt.%, and (**D**) 20 wt.%, using acetic acid and water as solvents.

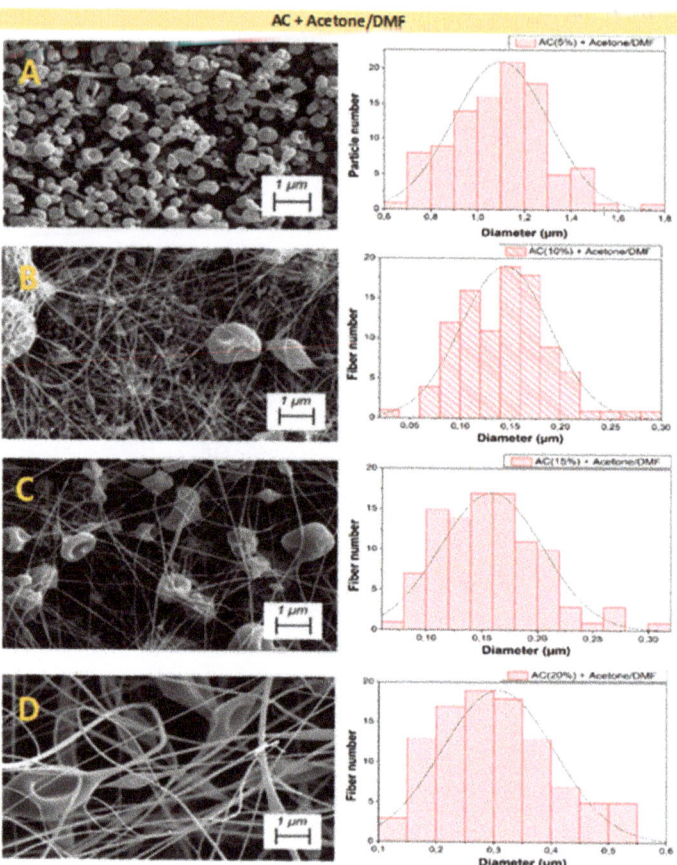

Figure 4. SEM images and distribution of microstructural properties of the nanostructures made with different AC concentrations: (**A**) 5 wt.%, (**B**) 10 wt.%, (**C**) 15 wt.%, and (**D**) 20 wt.%, using acetone and DMF as solvents.

Figure 3A–C shows an increase in the size of nanoparticles when increasing polymer concentration, which means that these results are in line with the obtained rheological properties of the solutions of each system, as the systems in Figure 3A,B were below C_e and consequently, as expected, no fibers were obtained in these systems. Conversely, Figure 3C is slightly above C_e, which is why, in these systems, both particles and fibers with very small diameters were obtained in these systems. Finally, Figure 3D shows a distribution of interconnected particles and fibers, with the latter being larger than the fibers obtained in Figure 3C.

Figure 4 shows the results of microstructural analyses carried out for the AC systems using acetone/DMF as solvents. Once again, the obtained results are consistent with the previous rheological characterization of the solutions, as only particles can be observed in Figure 4A, since this concentration of AC with this combination of solvents is lower than C_e, whereas Figure 4B–D, whose concentrations are in the regime predominated by the electrospinning, shows predominates and fibers with an increasing size as the polymer concentration is increased.

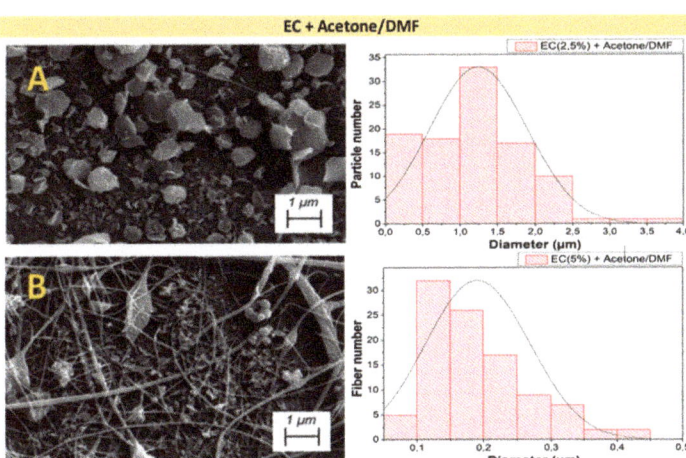

Figure 5. SEM images and distribution of microstructural properties of the nanostructures made with different EC concentrations (**A**) 2.5 wt.% and (**B**) 5 wt.%, using acetone and DMF as solvents.

When comparing the influence of the selected solvents for AC systems, it can be observed that, in solutions with acetic acid/H$_2$O, particle formation is favored rather than fiber spinning, due to the difficulty that the solution presented for the electrospinning process. Conversely, AC solutions with acetone/DMF achieved a more abundant fiber formation with irregular morphology. This is due to the difference in the physicochemical properties of the solvents used, which leads to different interactions with CA, in this case, therefore obtaining different physicochemical properties for the resulting solutions, which explains the difference in the C_e when using different solvents [43,44]. This fact bears out the importance of making a previous rheological characterization in order to obtain C_e and to determine both regimes.

Figure 5 shows the results of the microstructural characterization results of the EC systems. In Figure 5A, only irregular particles can be observed, without fiber formation. Nevertheless, in Figure 5B, when C_e was exceeded, a heterogeneous distribution of fibers was obtained. The rest of the EC systems were not included in this analysis due to the fact that it was not possible of electrospun solutions with higher concentrations, since maintaining the electrospinning parameters used in the other systems, resulted in the high viscosity of the solutions made the formation of the Taylor cone impossible [45].

By comparing both raw materials (AC and EC) using the same solvents, it can be concluded that, for the same polymer concentration, fibers and particles with larger diameters were obtained for the EC systems (Table 3), highlighting the influence that the selected polymer exerts on both the rheological properties of the solutions and the properties of the final nanostructures. Moreover, the values of Table 3 corroborate those results observed in diameter distribution for every system in Figures 3–5, that is the increment of particle and fiber diameter when increasing the polymer concentration.

From the particle and fiber diameters of the different systems, it is possible to correlate the specific viscosity of electrospun solutions with the mean diameter of both particles and fibers (Figure 6). This correlation can be established from specific viscosities in the semi-diluted unentangled regimes for particles (in other words, specific viscosities from concentrations below C_e) (Figure 6A) and specific viscosities of solutions with concentrations above C_e for fibers (Figure 6B). These correlations allow one to determine the values of specific viscosity required to obtain certain diameters of particles or fibers.

Table 3. Mean values of particles diameter, fiber diameter, and porosity obtained for the different AC and EC systems. Different letters in each column represent significant differences between the parameters ($p < 0.05$).

Systems		Particle Diameter (µm)	Fiber Diameter (µm)	Porosity (%)
AC + Acetic acid/H$_2$O (2:1)	5 wt.%	1.70 ± 0.12 [a]	–	24.29 ± 0.18 [α]
	10 wt.%	3.40 ± 0.09 [b]	–	22.72 ± 0.16 [β]
	15 wt.%	3.05 ± 0.14 [b]	0.06 ± 0.01	28.19 ± 0.27 [γ]
	20 wt.%	5.00 [c] ± 0.27	0.14 ± 0.01 [A]	25.30 ± 0.25 [α,γ]
AC + Acetone/DMF (2:1)	5 wt.%	1.10 ± 0.04 [d]	–	30.42 ± 0.28 [δ]
	10 wt.%	3.10 ± 0.11 [e]	0.15 ± 0.01 [A]	26.65 ± 0.15 [γ]
	15 wt.%	3.90 ± 0.19 [f]	0.16 ± 0.01 [A]	33.32 ± 0.38 [ε]
	20 wt.%	–	0.31 ± 0.03 [B]	25.67 ± 0.41 [γ]
EC + Acetone/DMF (2:1)	2.5 wt.%	1.25 ± 0.03 [d]	–	35.45 ± 0.21 [ζ]
	5 wt.%	2.75 ± 0.08 [g]	0.19 ± 0.03 [C]	26.21 ± 0.26 [γ]

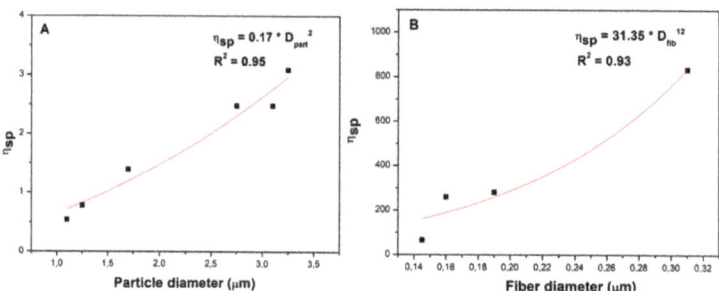

Figure 6. Empirical correlation between specific viscosity of solutions with (**A**) particle diameter and (**B**) fiber diameter obtained for the AC and EC systems.

4. Conclusions

As a general conclusion of the study, a relationship between the initial solutions and final nanostructures was established. It was corroborated that the chemical, thermal, and microstructural properties (fiber size, particle size, and porosity) of the different cellulose derivatives strongly depended on the physicochemical and rheological properties of the solutions, which are provided for the different molecular weight of the raw materials and the solvent used.

The results obtained show that solutions could be prepared with different cellulose derivatives at different concentrations and different solvent ratios. In general, the solutions of AC at 10 wt.% in acetic acid/H$_2$O and 15 wt.% in acetone/DMF, and those of EC at 5 wt.% in acetone/DMF are the ones that present the best compositional and final properties of the nanostructures obtained. In addition, the processing parameters established in the electrospinning allow for obtaining AC nanostructures without difficulties in the process. However, the EC solutions generated problems for electrospinning at higher concentrations (10, 15, and 20 wt.%) whose viscosity values were too high to be processed using the selected electrospinning parameters. Furthermore, an empirical relationship could be established between particle and fiber diameters and the specific viscosity of solutions, which allowed for determining the required value of specific viscosity (and, therefore, polymer concentration) to obtain a certain particle or fiber size. Thus, these results open the possibility of estimating the properties obtained in the nanofibers processed by electrospinning through the characterization of the biopolymeric solutions. However, future studies will characterize the nanostructures in depth to evaluate their functionality.

Supplementary Materials: The following supporting information can be downloaded at: https://www.mdpi.com/article/10.3390/polym14040665/s1. Figure S1: (A) TGA profile and (B) DSC analysis of EC (5 %) + acetone/DMF membrane as representative to show the selected parameters. (1) and (2) refers to the thermal events that occurs during the heating process, namely the degradation of acetyl and ethoxy groups of AC and EC, respectively (1) and the degradation of cellulose (2); Figure S2: TGA analysis of electrospun nanostructures for systems made with: A) AC + Acetic acid/H2O, B) AC + Acetone/DMF and C) EC + acetone/DMF; Figure S3: DSC analysis of AC, AC (10 wt.%) + Acetic acid/H2O, AC (15 wt.%) + Acetone /DMF, EC and EC (5 %) + acetone/DMF.

Author Contributions: Conceptualization, M.J.-R., V.P.-P., and A.R.; methodology, P.S.-C.; validation, V.P.-P., M.J.-R., J.F.R.-V., and A.R.; formal analysis, P.S.-C.; investigation, P.S.-C.; resources, A.R.; data curation, P.S.-C., J.F.R.-V., and V.P.-P.; writing—original draft preparation, P.S.-C. and V.P.-P.; writing—review and editing, M.J.-R., J.F.R.-V., V.P.-P., and A.R.; visualization, P.S.-C.; supervision, A.R. and V.P.-P.; project administration, A.R.; funding acquisition, A.R. All authors have read and agreed to the published version of the manuscript.

Funding: This work is part of a research project sponsored by the "Ministerio de Ciencia e Innovación-Agencia Estatal de Investigación" (MINECO/AEI/FEDER, EU) from the Spanish government (Ref. RTI2018-097100-B-C21 and Ref. RTI2018-096080-B-C21). This work is also possible thanks to the predoctorals grants of Mercedes Jiménez Rosado (FPU2017/01718, Ministerio de Educación y Formación Profesional), José Fernando Rubio Valle (PRE2019-090632, Ministerio de Educación y Formación Profesional), and the postdoctoral grant of Víctor Manuel Pérez Puyana (Convocatoria 2019-20, Junta de Andalucía and European Social Fund, DOC_00586).

Institutional Review Board Statement: Not applicable.

Informed Consent Statement: Not applicable.

Data Availability Statement: The data presented in this study are available on request from the corresponding author.

Acknowledgments: The authors acknowledge CITIUS for granting access to and their assistance with the functional characterization and microscopy services.

Conflicts of Interest: The authors declare no conflict of interest.

References

1. Duque Sánchez, L.M.; Rodriguez, L.; López, M. Electrospinning: The Nanofibers Age. *Rev. Iberoam. Polímeros Vol. Iber. Polímeros* **2014**, *14*, 10–27.
2. Pérez-Puyana, V.; Guerrero, A.; Romero, A. *Biomateriales*; Gandulfo Impresores S.L.: Sevilla, Spain, 2020; ISBN 978-84-617-5006-1.
3. Gritsch, L.; Liverani, L.; Lovell, C.; Boccaccini, A.R. Polycaprolactone Electrospun Fiber Mats Prepared Using Benign Solvents: Blending with Copper(II)-Chitosan Increases the Secretion of Vascular Endothelial Growth Factor in a Bone Marrow Stromal Cell Line. *Macromol. Biosci.* **2020**, *20*, 1900355. [CrossRef]
4. Jiménez-Rosado, M.; Rubio-Valle, J.F.; Perez-Puyana, V.; Guerrero, A.; Romero, A. Comparison between pea and soy protein-based bioplastics obtained by injection molding. *J. Appl. Polym. Sci.* **2021**, *138*, 50412. [CrossRef]
5. Jiménez-Rosado, M.; Bouroudian, E.; Perez-Puyana, V.; Guerrero, A.; Romero, A. Evaluation of different strengthening methods in the mechanical and functional properties of soy protein-based bioplastics. *J. Clean. Prod.* **2020**, *262*, 121517. [CrossRef]
6. El-Aswar, E.I.; Ramadan, H.; Elkik, H.; Taha, A.G. A comprehensive review on preparation, functionalization and recent applications of nanofiber membranes in wastewater treatment. *J. Environ. Manage.* **2022**, *301*, 113908. [CrossRef]
7. Elnaggar, M.; Shalaby, E.; Abd-Al-Aleem, A.-A.-A.; Youssef, A. Nanomaterials and nanofibers as wound dressing mats: An overview of the fundamentals, properties and applications. *Egypt. J. Chem.* **2021**, *64*, 7447–7473. [CrossRef]
8. Zhang, Y.; Zhang, C.; Wang, Y. Recent progress in cellulose-based electrospun nanofibers as multifunctional materials. *Nanoscale Adv.* **2021**, *3*, 6040–6047. [CrossRef]
9. Brown, R.M. Cellulose Structure and Biosynthesis: What is in Store for the 21st Century? *J. Polym. Sci. Part A Polym. Chem.* **2004**, *42*, 487–495. [CrossRef]
10. Rafael, V. Bacterial degradation of lignin. *Enzyme Microb. Technol.* **1988**, *10*, 646–655.
11. Kobayasi, S.; Kashiwa, K.; Shimada, J.; Kawasaki, T.; Shoda, S.-I. Enzymatic polymerization: The first in vitro of cellulose via non biosynthetic path catalyzed by cellulase. *Makromol. Chemie. Macromol. Symp.* **1992**, *54–55*, 509–518. [CrossRef]
12. Habibi, Y. Key advances in the chemical modification of nanocelluloses. *Chem. Soc. Rev.* **2014**, *43*, 1519–1542. [CrossRef] [PubMed]
13. Mascal, M.; Nikitin, E.B. Comment on Processes for the Direct Conversion of Cellulose or Cellulosic Biomass into Levulinate Esters. *ChemSusChem* **2010**, *3*, 1349–1351. [CrossRef]
14. Muqeet, M.; Mahar, R.B.; Gadhi, T.A.; Ben Halima, N. Insight into cellulose-based-nanomaterials—A pursuit of environmental remedies. *Int. J. Biol. Macromol.* **2020**, *163*, 1480–1486. [CrossRef] [PubMed]

15. Teixeira, M.A.; Paiva, M.C.; Amorim, M.T.P.; Felgueiras, H.P. Electrospun nanocomposites containing cellulose and its derivatives modified with specialized biomolecules for an enhanced wound healing. *Nanomaterials* **2020**, *10*, 557. [CrossRef]
16. Huang, C.K.; Zhang, K.; Gong, Q.; Yu, D.G.; Wang, J.; Tan, X.; Quan, H. Ethylcellulose-based drug nano depots fabricated using a modified triaxial electrospinning. *Int. J. Biol. Macromol.* **2020**, *152*, 68–76. [CrossRef]
17. Ahmadi, P.; Jahanban-Esfahlan, A.; Ahmadi, A.; Tabibiazar, M.; Mohammadifar, M. Development of Ethyl Cellulose-based Formulations: A Perspective on the Novel Technical Methods. *Food Rev. Int.* **2020**, *10*, 1–48. [CrossRef]
18. Konwarh, R.; Karak, N.; Misra, M. Electrospun cellulose nanofibers: The present status and gamut of biotechnological applications. *Biotechnol. Adv.* **2013**, *31*, 421–437. [CrossRef] [PubMed]
19. Wsoo, M.A.; Shahir, S.; Mohd Bohari, S.P.; Nayan, N.H.M.; Razak, S.I.A. A review on the properties of electrospun cellulose acetate and its application in drug delivery systems: A new perspective. *Carbohydr. Res.* **2020**, *491*, 107978. [CrossRef] [PubMed]
20. Zhang, X.; Wang, B.; Qin, X.; Ye, S.; Shi, Y.; Feng, Y.; Han, W.; Liu, C.; Shen, C. Cellulose acetate monolith with hierarchical micro/nano-porous structure showing superior hydrophobicity for oil/water separation. *Carbohydr. Polym.* **2020**, *241*, 116361. [CrossRef] [PubMed]
21. Crabbe-Mann, M.; Tsaoulidis, D.; Parhizkar, M.; Edirisinghe, M. Ethyl cellulose, cellulose acetate and carboxymethyl cellulose microstructures prepared using electrohydrodynamics and green solvents. *Cellulose* **2018**, *25*, 1687–1703. [CrossRef]
22. Um-i-Zahra, S.; Shen, X.X.; Li, H.; Zhu, L. Study of sustained release drug-loaded nanofibers of cellulose acetate and ethyl cellulose polymer blends prepared by electrospinning and their in-vitro drug release profiles. *J. Polym. Res.* **2014**, *21*, 602. [CrossRef]
23. Souza, P.R.; de Oliveira, A.C.; Vilsinski, B.H.; Kipper, M.J.; Martins, A.F. Polysaccharide-based materials created by physical processes: From preparation to biomedical applications. *Pharmaceutics* **2021**, *13*, 621. [CrossRef]
24. Oprea, M.; Voicu, S.I. Recent advances in applications of cellulose derivatives-based composite membranes with hydroxyapatite. *Materials* **2020**, *13*, 2481. [CrossRef]
25. Gopiraman, M.; Bang, H.; Yuan, G.; Yin, C.; Song, K.-H.; Lee, J.S.; Chung, I.M.; Karvembu, R.; Kim, I.S. Noble metal/functionalized cellulose nanofiber composites for catalytic applications. *Carbohydr. Polym.* **2015**, *132*, 554–564. [CrossRef]
26. Morsi, R.E.; Elsawy, M.; Manet, I.; Ventura, B. Cellulose acetate fabrics loaded with rhodamine B hydrazide for optical detection of Cu(II). *Molecules* **2020**, *25*, 3751. [CrossRef]
27. Fatima, A.; Yasir, S.; Khan, M.S.; Manan, S.; Ullah, M.W.; Ul-Islam, M. Plant extract-loaded bacterial cellulose composite membrane for potential biomedical applications. *J. Bioresour. Bioprod.* **2021**, *6*, 26–32. [CrossRef]
28. Deeksha, B.; Sadanand, V.; Hariram, N.; Rajulu, A.V. Preparation and properties of cellulose nanocomposite fabrics with in situ generated silver nanoparticles by bioreduction method. *J. Bioresour. Bioprod.* **2021**, *6*, 75–81. [CrossRef]
29. Bekturov, E.A.; Bakauova, Z.K. *Synthetic Water-Soluble Polymers in Solution*, 6th ed.; Hüthig and Wepf Verlag: Mainz, Germany, 1986.
30. Painter, P.C.; Coleman, M.M. *Fundamentals of Polymer Science: An Introductory Text*, 3rd ed.; Technocomic Publishing Company Inc.: Lancaster, PA, USA, 1994.
31. Rubio-Valle, J.F.; Sánchez, M.C.; Valencia, C.; Martín-Alfonso, J.E.; Franco, J.M. Electrohydrodynamic Processing of PVP-Doped Kraft Lignin Micro- and Nano-Structures and Application of Electrospun Nanofiber Templates to Produce Oleogels. *Polymers* **2021**, *13*, 2206. [CrossRef]
32. Agarwal, S.; Greiner, A.; Wendorff, J.H. *Kirk-Othmer Encyclopedia of Chemical Technology*; Wiley-VCH: Weinheim, Germany, 2014.
33. Gupta, P.; Elkins, C.; Long, T.E.; Wilkes, G.L. Electrospinning of linear homopolymers of poly(methyl methacrylate): Exploring relationships between fiber formation, viscosity, molecular weight and concentration in a good solvent. *Polymer* **2005**, *46*, 4799–4810. [CrossRef]
34. Aslanzadeh, S.; Ahvazi, B.; Boluk, Y.; Ayranci, C. Carbon Fiber Production from Electrospun Sulfur Free Softwood Lignin Precursors. *J. Eng. Fiber. Fabr.* **2017**, *12*, 155892501701200. [CrossRef]
35. Khatri, M.; Ahmed, F.; Jatoi, A.W.; Mahar, R.B.; Khatri, Z.; Kim, I.S. Ultrasonic dyeing of cellulose nanofibers. *Ultrason. Sonochem.* **2016**, *31*, 350–354. [CrossRef]
36. Wang, X.; He, G.; Liu, H.; Zheng, G.; Sun, D. Fabrication and morphological control of electrospun ethyl cellulose nanofibers. In Proceedings of the 8th Annual IEEE International Conference on Nano/Micro Engineered and Molecular Systems, Suzhou, China, 7–10 April 2013; pp. 324–327.
37. Arribas-Jimeno, S.; Burriel-Martí, F.; Hernández-Méndez, J.; Lucena-Conde, F. *Química Analítica Cualitativa*, 1st ed.; Paraninfo: Madrid, Spain, 2006.
38. Jiménez-Álvarez, M.D.; Gómez-del Río, M.I. *Guía Didáctica Química Analítica II*, 14th ed.; UNED: Madrid, Spain, 1999.
39. Bermejo-Barrera, M.P. *Química Analítica General, Cuantitativa e Instrumental*, 7th ed.; Paraninfo: Madrid, Spain, 1990.
40. Blanco, M.; Cerdá, V.; Sanz-Medel, A. *Espectroscopía Atómica Analítica*, 1st ed.; Universidad Autónoma de Barcelona: Barcelona, Spain, 1990.
41. Burriel-Martí, F. *Química Analítica Cualitativa*, 6th ed.; Paraninfo: Madrid, Spain, 2003.
42. Burriel-Martí, F.; Lucena-Conde, C.F.; Arribas-Jimeno, S. *Química Analítica Cuantitativa*, 18th ed.; Revolucionaria: Madrid, Spain, 1978.
43. Pal, S.; Srivastava, R.K.; Nandan, B. Effect of spinning solvent on crystallization behavior of confined polymers in electrospun nanofibers. *Polym. Cryst.* **2021**, *4*, e10209. [CrossRef]

44. Nuamcharoen, P.; Kobayashi, T.; Potiyaraj, P. Influence of volatile solvents and mixing ratios of binary solvent systems on morphology and performance of electrospun poly(vinylidene fluoride) nanofibers. *Polym. Int.* **2021**, *70*, 1465–1477. [CrossRef]
45. Nayak, R.; Padhye, R.; Kyratzis, I.L.; Truong, Y.B.; Arnold, L. Effect of viscosity and electrical conductivity on the morphology and fiber diameter in melt electrospinning of polypropylene. *Text. Res. J.* **2013**, *83*, 606–617. [CrossRef]

Article

Rice Bran-Based Bioplastics: Effects of Biopolymer Fractions on Their Mechanical, Functional and Microstructural Properties

María Alonso-González [1,2,*], Manuel Felix [2] and Alberto Romero [2]

1. Departamento de Ingeniería Química, Facultad de Química, Universidad de Sevilla, 41012 Sevilla, Spain
2. Departamento de Ingeniería Química, Escuela Politécnica Superior, Universidad de Sevilla, 41011 Sevilla, Spain; mfelix@us.es (M.F.); alromero@us.es (A.R.)
* Correspondence: maralonso@us.es; Tel.: +34-954557179

Abstract: Rice bran is an underutilized by-product of rice production, containing proteins, lipids and carbohydrates (mainly starches). Proteins and starches have been previously used to produce rice bran-based bioplastics, providing a high-added-value by-product, while contributing to the development of biobased, biodegradable bioplastics. However, rice bran contains oil (18–22%), which can have a detrimental effect on bioplastic properties. Its extraction could be convenient, since rice bran oil is becoming increasingly attractive due to its variety of applications in the food, pharmacy and cosmetic industries. In this way, the aim of this work was to analyze the effect of the different components of rice bran on the final properties of the bioplastics. Rice bran refining was carried out by extracting the oil and fiber fractions, and the effects of these two procedures on the final properties were addressed with mechanical, functional and microstructural measures. Results revealed that defatted rice bran produced bioplastics with higher viscoelastic moduli and better tensile behavior while decreasing the water uptake capacity and the soluble matter loss of the samples. However, no significant improvements were observed for systems produced from fiber-free rice bran. The microstructures observed in the SEM micrographs matched the obtained results, supporting the conclusions drawn.

Keywords: bioplastics; rice bran; rice bran oil; valorization; starch; injection molding

1. Introduction

Environmental pollution derived from conventional plastics produced from fossil resources has become a global concern. Consequently, the production of environmentally sustainable materials as an alternative is drawing the attention of the scientific community [1–3]. Great effort has been made to develop biodegradable plastics from renewable natural resources with the aim of producing biodegradable materials that resemble the behavior of fossil-based polymers. It is estimated that, between 1950 and 2015, 8.3 billion tons of plastic were produced worldwide and only 21% were recycled or incinerated, the remaining 79% accumulated in landfills and surrounding areas [4]. In addition, conventional plastics are not biodegradable, thus they remain in the environment for many years, physically breaking into smaller particles occurs, releasing microplastics in landfills and marine environments, entering food chains and, consequently, animal bodies, eventually causing different diseases [5,6].

In this context, bioplastics, which can be either biobased or biodegradable, appear as a promising alternative to replace or at least reduce the extensive use of conventional plastics and their harmful waste [7,8]. Biodegradable plastics are usually made from biopolymers. Furthermore, biopolymers derived from renewable sources, such as animals or plants, can play an essential role in overcoming the challenges derived from the depletion of oil resources and the environmental problems related to the increasing use of petroleum-based plastics. These biopolymers can be natural fibers, cellulose, polysaccharides, proteins, lipopolysaccharides, etc. [9,10].

Natural polymers such as polysaccharides (starch, cellulose, pectin, hemicellulose) and proteins (casein, zein, gluten, gelatin) are generally able to form intramolecular and intermolecular interactions and cross-linking between polymeric constituents, forming a semi-rigid three-dimensional network [11]. Furthermore, these biopolymers are present in wastes and by-products from the agro-food industry. For example, rice (*Oryza sativa* L.) is the major cereal crop grown in the world [12] and is generally processed by shelling and polishing to remove the bran from the grain prior to commercialization. Although rice bran constitutes about one-tenth of the rice weight, it is still underutilized. It is a low-added-value by-product mainly used for animal feeding or as an organic fertilizer [12]. However, the proteins and starches present in rice bran have previously been used to develop rice bran-based bioplastics; their mechanical and functional properties depend on the processing parameters and composition [13,14]. This valorization turns rice bran into a high-added-value by-product, a new raw material, such as gluten or soybean flours [15,16], that can be used to produce biobased, biodegradable plastic, benefiting both the environment and the producing companies. However, there are no studies that evaluate the effect of the different fractions of rice bran on their mechanical and functional properties.

Furthermore, rice bran is emerging as a potential by-product of rice processing also as a result of the increasing demand for rice bran oil (RBO). Rice bran contains 18–22% oil, similar to other edible vegetable oil sources such as soybean (15–20%) or tung (16–18%) [17]. Rice bran oil is unique among edible vegetable oils due to its composition (i.e., fatty acids, phenolic compounds and vitamin-E) [18]. Among the health benefits of RBO, it reduces oxidative stress and hypertension, has anti-cancer and anti-diabetic activities, can act as anti-inflammatory or anti-allergic agent, etc. In this way, it has a variety of food and non-food applications in pharmacy and cosmetics [18–20].

In this context, this research work aimed to analyze the effect of oil extraction, which can have its own valorization route, on the properties of defatted rice bran-based bioplastics. In addition, the fibers were also extracted to evaluate the influence of the different biopolymer fractions on the properties of the bioplastics obtained. In this way, different systems were obtained from virgin rice bran, defatted rice bran, and fiber-free rice bran, each of them plasticized with a mixture of water with either glycerol or sorbitol. By these means, the plasticizer effect was also analyzed for the different active matters employed. The other processing parameters, raw materials proportion, mixing temperature and injection pressure and temperature, were kept constant, selected based on previous studies. Through these means, the effects of both the biopolymer fractions and the selected plasticizer were successfully analyzed, achieving a better understanding of the mechanisms involved in the development of protein- and starch-based bioplastics.

2. Materials and Methods

2.1. Materials

Vaporized indica rice bran (RB) was obtained from Herba Ingredients (San José de la Rinconada, Spain). Deionized-grade water, sorbitol and glycerol were employed as plasticizers. Both sorbitol and glycerol were provided by PANREAC S.A. (Barcelona, Spain). All other reagents were supplied by Sigma-Aldrich (St. Louis, MI, USA).

2.2. Preparation of Defatted Rice Bran (DRB) and Fiber-Free Rice Bran (FRB)

DRB was prepared by suspending sieved RB (<500 µm) in hexane (1:10 *w/v*). The mixture was vigorously stirred at room temperature for 24 h and, after this time, it was centrifuged at 5000 rpm for 10 min. The supernatant containing both hexane and lipids was carefully separated from the solid fraction, which was dried in a fume hood to remove any residual hexane. This process was carried out twice, ensuring that >90% of the lipids were removed.

Fiber removal was carried out following the methodology used by Singh et al. [12] by first soaking DRB in deionized grade water (1:40 *w/v*) for 2 h. The pH was then adjusted to 9.5 using a NaOH solution and the mixture was stirred at room temperature for 1 h.

After this time, the slurry was sieved through a 125 µm mesh to separate the fibers and the remaining mixture was centrifuged. The supernatant (containing some soluble protein) was discarded and the solid fraction was washed with distilled water and lyophilized.

2.3. Chemical Composition

The chemical composition of RB was already characterized by Alonso-González et al. [13]. The approximate composition of DRB and FRB flours was determined following the approved methods of A.O.A.C. [21]. The water content was determined by mass difference after placing 3 g of sample in a conventional oven (Memmert B216.1126, Schwabach, Germany) at 105 °C for 24 h. The lipid content was quantified using the Soxhlet extraction method [22], where hexane was used as a solvent in contact with the sample. The lipids were dragged in subsequent cycles until the whole lipid content was removed and quantified by mass difference. The ash content was determined by heating a small amount of sample at 550 °C in a muffle furnace (Hobersal HD-230, Barcelona, Spain) for 5 h in air atmosphere. The sample was then cooled to room temperature in a desiccator before being weighed again to calculate the mass difference. Protein content was determined as% N × 6.25 [23] using a LECO TRUSPEC CHNS-932 nitrogen microanalyzer (Leco Corporation, St. Joseph, MI, USA). Finally, the starch and fiber content in the RB sample was determined by analytical methods in an external laboratory and it was assumed that their ratios were constant after defatting (i.e., DRB system) while their contents in the FRB were determined by mass difference.

2.4. Sample Preparation

Samples were prepared according to the methodology followed by Alonso-González et al. [14]. In this way, the active matter (RB, DRB and FRB) was introduced along with plasticizers into a HAAKE POLYLAB QC mixer-rheometer (ThermoScientific, Waltham, MA, USA), equipped with counter-rotating rotors, obtaining homogeneous blends. Two different plasticizers, glycerol (G) and sorbitol (S), were evaluated in combination with water (W) for each system. All blends contained 55% active matter and 45% total plasticizer, maintaining the proportion of 2:1 water-G/S. The proportions were selected according to previous studies [13,14]. The blends were mixed at 200 rpm and 80 °C for 1 h. The systems are identified with the corresponding active matter (RB, DRB and FRB) and the plasticizer used (G or S). In this way, a system developed from defatted rice bran using a mixture of water and sorbitol would be DRBS.

Once mixed, the doughs were kept inside a desiccator until the moisture content was adequate for further processing (between 10 and 40 wt.%). The final moisture content, calculated during the drying process following the A.O.A.C. methods, varied for each system. The doughs were finally processed by injection molding using a Haake pneumatic piston injection molding equipment (MiniJet ThermoScientific, Waltham, MA, USA) to obtain the bioplastic samples. The temperatures for the injection cylinder and the mold were 50 °C and 150 °C, respectively. The injection pressure was 500 bar, which was applied for 15 s, while the post-injection time and pressure were 200 s and 500 bar, respectively. These conditions were selected according to previous studies [13,14]. By these means, rectangular probes (60 mm × 10 mm × 1 mm) were obtained and employed for mechanical, functional and microstructural characterization.

2.5. Bioplastics Characterization

2.5.1. Dynamic Mechanical Thermal Analysis (DMTA)

DMTA tests were carried out with a DMA850 rheometer (TA Instruments, New Castle, DE, USA) on the rectangular probes using the film clamp in tension mode. First, the linear viscoelastic range was determined by strain sweep tests. Subsequently, a strain within the linear viscoelastic range was selected for the frequency sweep tests (between 0.01 and 20 Hz) performed at room temperature and the temperature ramp tests (between −10 and 160 °C). Temperature tests were carried out at constant frequency (1 Hz) at the heating rate of 5 °C/min. By these means, the elastic modulus (E'), viscous modulus (E") and loss tangent (tan δ = E"/E') were obtained for the whole studied range.

2.5.2. Tensile Tests

Tensile tests were performed with RSA3 equipment (TA Instruments, New Castle, DE, USA) according to a modification of the ISO 527-2 method for the tensile properties of plastics [24] using rectangular probes and measuring three replicates. Stress-strain curves were obtained for all evaluated systems and Young's modulus, maximum stress and deformation at break were successfully determined with a deformation rate of 1 mm/min at room temperature.

2.5.3. Water Uptake Capacity and Soluble Matter Loss

Water uptake capacity (WUC) was measured according to the ASTM D570 method [25] using one-third of the rectangular samples mentioned above, that is, probes measuring 20 mm × 10 mm × 1 mm. The samples were subjected to a dehydrothermal treatment in a conventional oven at 50 °C (Memmert UN 55, Schwabach, Germany) for 24 h to determine the dry weight (*Initial dry weight*). Subsequently, they were immersed in distilled water and weighted after 24 h of immersion (*Wet weight*). Finally, they were frozen at −40 °C before lyophilization at −80 °C and vacuum atmosphere (0.125 bar) using a LyoQuest freeze-dryer with a Flask M8 head (Telstar, Barcelona, Spain) and subsequently weighted (*Final dry weight*). According to the methodology used, WUC and soluble matter loss (SML) were determined by the following equations:

$$WUC\ (\%) = \frac{Wet\ weight - Initial\ dry\ weight}{Initial\ dry\ weight} \cdot 100 \quad (1)$$

$$SML\ (\%) = \frac{Initial\ dry\ weight - Final\ dry\ weight}{Initial\ dry\ weight} \cdot 100 \quad (2)$$

2.5.4. Scanning Electron Microscopy (SEM)

Selected bioplastics samples after water uptake and subsequent freeze-drying (1 mm thick) were observed by SEM examination, showing the micrographs their microstructural appearance. The equipment used was a ZEISS EVO microscope (Carl Zeiss Microscopy, White Plains, NY, USA) at a voltage of 10 kV and a magnification of 500×. These samples were previously coated by Pd/Au sputtering (13 nm) using a Leica AC600 metalizer. Finally, the porosity of the samples was calculated using ImageJ software.

2.6. Statistical Analyses

At least three replicates of each measurement were carried out. Statistical analyses were performed using t-test and one-way analysis of variance (ANOVA) (significance value ρ < 0.05) using the STATGRAPHICS 18 software (Statgraphics Technologies, Old Tavern Rd, The Plains, VA, USA). Standard deviations from some selected parameters were calculated. Significant differences are indicated by different letters, that is, all mean values labeled with the same letter did not show significant differences.

3. Results

3.1. Chemical Composition

The chemical composition of this variety of rice bran is similar to those obtained by different authors in their research, the fiber and starch content being those with the highest differences depending on the variety [26,27]. The effects of lipid and fiber extractions on the chemical composition of RB are gathered in Table 1. First, the lipid extraction produced a DRB system with only 1.8% lipids, which caused a proportional increase in the rest of the components as happened in a similar study directed by [12]. Thus, the chemical characterization revealed a composition of 15.9% moisture, 13.4% ashes, 16.8% proteins and 27.9 and 24.2% fiber and starch, respectively. Previous studies revealed that the protein content consists mainly of variable fractions of glutelin, globulin, albumin, and prolamin, while the starch content is the mixture of two biopolymers: amylose and amylopectin [12,28]. However, the fiber removal process caused the lipid fraction to increase proportionally again to 5.8%. Consequently, the rest of the fractions also increased containing 16.8% moisture, 24.6% ashes, and 34.1% starch, leaving the remaining of 3.0% fibers. The only exception would be the protein content, which appeared to decrease to 15.7% due to the removal of some soluble protein during the fiber extraction procedure.

Table 1. Chemical composition of the different active matters employed: rice bran (RB), defatted rice bran (DRB) and fiber-free rice bran (FRB).

Composition	RB	DRB	FRB
Moisture (%)	12.5 ± 5.0	15.9 ± 5.0	16.8 ± 3.4
Ashes (%)	10.5 ± 0.3	13.4 ± 0.5	24.6 ± 0.5
Lipids (%)	22.8 ± 1.3	1.8 ± 0.6	5.8 ± 1.6
Proteins (%)	13.2 ± 0.5	16.8 ± 0.5	15.7 ± 1.0
Fiber (%)	22.0 ± 1.0	27.9 ± 1.0	3.0 ± 1.0
Starches (%)	19.0 ± 1.0	24.2 ± 1.0	34.1 ± 1.0

3.2. Dynamic Mechanical Thermal Analyses (DMTA)

The frequency sweep tests of rice bran-based bioplastics are shown in Figure 1. As can be seen, E' is higher than E'' for all processed systems, which confirms that the samples processed by the injection molding process exhibit mainly elastic behavior. At the same time, the values obtained for the viscoelastic moduli present a certain frequency dependence, increasing as the frequency increases for all studied systems. Regarding the active matter, the lowest recorded values were observed for RB systems, processed with either glycerol or sorbitol as plasticizers (that is, RBG and RBS, respectively). However, this analysis shows that the lipid and fiber extractions resulted in bioplastics with better rheological behavior than the virgin RB, with the viscoelastic moduli increasing for all DRB- and FRB-based systems, with no significant differences between them. In this way, the values of the viscoelastic moduli overlay for the DRBS and FRBS systems, as well as for the DRBG and FRBG. Finally, it can be observed that sorbitol produced bioplastics with higher viscoelastic moduli than those processed with glycerol as a plasticizer. In fact, the two glycerol systems with better rheological behavior, DRBG and FRBG, present practically similar E' and E' with respect to RBS.

Figure 2a,b show the temperature dependence of viscoelastic moduli (E' and E'') and tan δ, respectively. As can be seen in Figure 2a, all systems exhibit the same behavior, where, again, the elastic component is always above the viscous one within the whole frequency range studied. Furthermore, in all cases, both E' and E'' decrease with increasing temperature; the difference between them is the abruptness of this drop towards the end of the experiment. First, the recorded values followed the same trend as in the previous case: the two virgin systems (RBS and RBG) presented the lowest E' and E'' values, while the two extraction methods appear to enhance their rheological behavior. In this way, the DRBS and FRBS systems are above the RBS one, and the DRBG and FRBG systems are above the RBG samples. However, although in Figure 1 both extraction methods seem to give similar bioplastics, the DRB and FRB samples differ from each other in this case, since the FRBS and FRBG underwent a less accused drop of the viscoelastic moduli with temperature being above DRBS and DRBG during the last part of the studied range (T > 90 °C). In addition, the same tendency as in Figure 1 is maintained with the sorbitol systems which exhibited higher E' and E'' than the glycerol ones; however, in this case, there is a great difference in their behavior depending on the plasticizer used. Thus, the viscoelastic moduli of the glycerol systems seem to stabilize at the end of the experiment, maintaining their values, whereas for the effect of the plasticizer used (sorbitol or glycerol), since the sorbitol-based bioplastics continue to decrease for the whole temperature range studied.

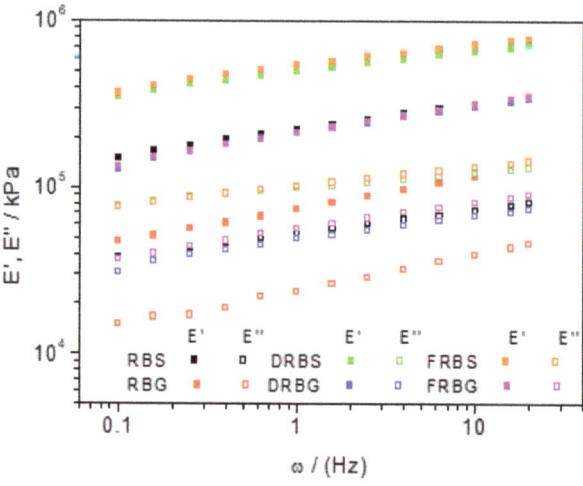

Figure 1. Frequency sweep tests performed between 0.1 and 20 Hz of the different studied systems (RBS, RBG, DRBS, DRBG, FRBS, and FRBG) carried out at room temperature.

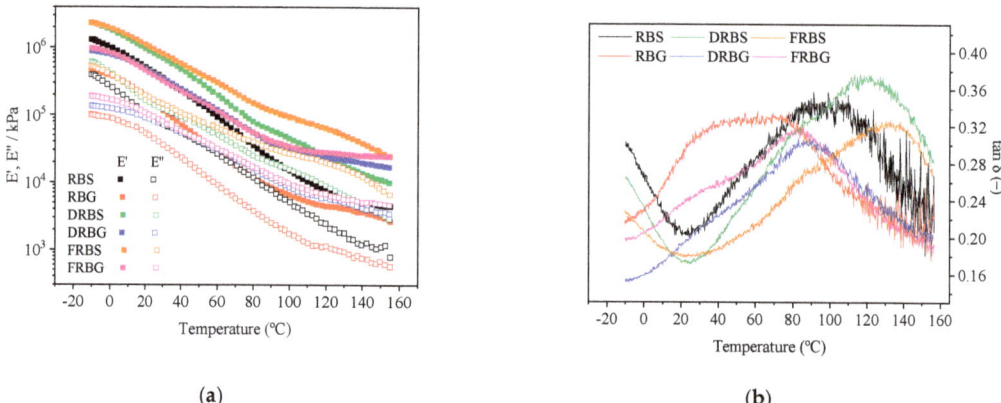

Figure 2. Temperature sweep tests between −10 and 160 °C performed on the different evaluated systems at 1 Hz; (**a**) Elastic (E′) and viscous (E″) modulus (**b**) Loss tangent (tan δ).

Figure 2b shows the evolution of the loss tangent with temperature. The observed maximum values correspond to the glass transition temperature (T_g) of the plasticized RB, DRB and FRB [29]. In this way, the influence of the biopolymer fraction as well as the plasticizer used on the T_g of the systems can be analyzed. First, it can be observed that glycerol led to a Tg earlier than sorbitol. In this way, the T_g of the RBG appeared around 50 °C, while for the RBS is around 100 °C. A similar behavior was shown by DRBG and FRBG, whose transition temperatures were nearly 80–90 °C, although their equivalent sorbitol systems underwent the glass transition at 120–130 °C, approximately. In addition, the effect of the biopolymer fractions can be observed in both the widths of the peaks and the higher T_g with respect to the virgin RB. In this way, the two extraction procedures, oil and fiber removal, led to narrower peaks with respect to virgin RB. The RBG and RBS systems showed wider transitions. Additionally, it is possible to distinguish the beginning of another peak (at the lowest temperatures studied) for the samples plasticized with the water-sorbitol mixture that is related to the glass transition of the plasticizer mix [30]. For the samples plasticized with water-glycerol, this peak would be expected at lower temperatures.

3.3. Tensile Tests

Figure 3 shows the stress-strain curves obtained for the different samples. As can be seen, both extraction procedures enhanced the tensile response of the obtained bioplastics. In this way, the response corresponding to the RBS samples is below those of DRBS and FRBS, and the DRBG and FRBG systems are above the RBG curve. Moreover, when comparing the three different active matters employed in this study, the samples with better mechanical properties are those obtained from the defatted rice bran, since the fiber-free samples did not induce further improvements in the mechanical properties. On the other hand, it can be observed that the sorbitol systems exhibited higher stiffness and withstood higher stresses, while the glycerol-based probes presented improved elasticity, achieving, by these means, greater elongations.

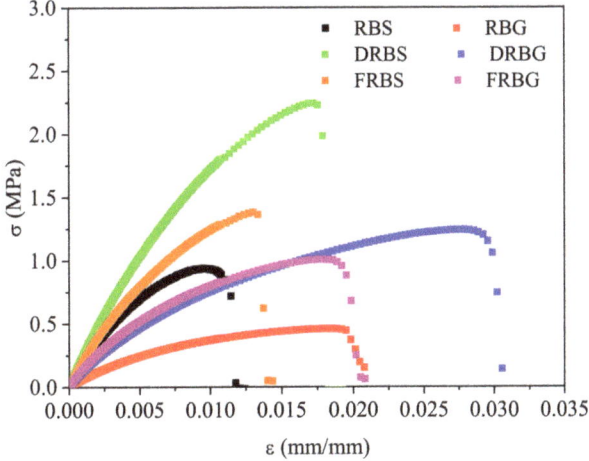

Figure 3. Stress-strain curves obtained for the different evaluated systems at 1 mm/min and room temperature.

The three parameters obtained from the strain-stress curves (that is, Young's modulus (YM), maximum stress (MS), and deformation at break (DB)) are gathered in Table 2, which allows for a more accurate evaluation of the tensile properties. First, the values obtained for the YM improved for the treated RB samples and for the systems containing sorbitol. Therefore, the lowest value is associated with the RBG sample 53 ± 1 MPa, followed by its two treated systems, namely DRBG and FRBG, which did not show significant differences exhibiting 104 ± 2 and 106 ± 9 MPa, respectively. The systems plasticized with sorbitol exhibited improved stiffness compared to the previous ones, beginning with the similar values recorded for the RBS and FRBS systems (156 ± 6 and 171 ± 17 MPa), which were exceeded by the DRBS, obtaining the highest value (227 ± 11 MPa). MS values followed a similar trend with the lowest and highest values presented by RBG and DRBS, respectively. The second-lowest was the FRBG system, followed closely by the RBS and DRBG systems and, finally, the FRBS system. The last parameter, that is, DB, differed from the other two. Although defatted samples induced better elongations (DRBS and DRBG), no significant differences were observed for fiber-free systems. The RBS and FRBS values did not show significant differences, either RBG nor FRBG. In addition, the elastic deformation of the bioplastics was enhanced by glycerol instead of the improvements observed in the other parameters for the sorbitol-containing samples.

Table 2. Young's modulus (YM), maximum stress (MS) and deformation at break (DB) of the different evaluated systems obtained from the stress-strain curves.

System	YM (MPa)	MS (MPa)	DB (mm/mm)
RBS	156 ± 6 [C*]	1.00 ± 0.10 [FG]	1.26 ± 0.18 [J]
RBG	53 ± 1 [A]	0.42 ± 0.04 [E]	2.03 ± 0.13 [K]
DRBS	227 ± 11 [D]	2.30 ± 0.30 [I]	1.82 ± 0.20 [K]
DRBG	104 ± 2 [B]	1.17 ± 0.03 [GH]	3.04 ± 0.26 [L]
FRBS	171 ± 17 [C]	1.33 ± 0.06 [H]	1.36 ± 0.23 [J]
FRBG	106 ± 9 [B]	0.90 ± 0.09 [F]	2.11 ± 0.13 [K]

* Different superscript letters within the column indicate significant differences ($p < 0.05$). Significant differences are indicated by different letters, all mean values labeled with the same letter did not show significant differences.

3.4. Water Uptake Capacity and Soluble Matter Loss

The water uptake capacity (WUC) and soluble matter loss (SML) of the obtained systems are shown in Figure 4a,b, respectively. This figure indicates that the removal of lipids and fibers has a detrimental effect on WUC, since treated samples exhibited lower absorption capacities (especially those without fiber). The only exception is the DRBS sample, with a value of 148 ± 5%, while its native sample showed a lower value (131 ± 10%), although the decreasing trend was again followed by the FRBS sample, with only 116 ± 3%. This drop in the absorption capacities was more progressive for glycerol-containing samples (RBG, DRBG, and FRBG systems). Thus, the highest value is associated with the RBG sample, with 162 ± 24%, followed by the DRBG sample, with 122 ± 13%, and finally the FRBG sample, with 108 ± 4%, which are the lowest results of all evaluated systems. The WUC values obtained are similar to those observed by López-Castejón et al. [30] for albumen/tragacanth-based bioplastics.

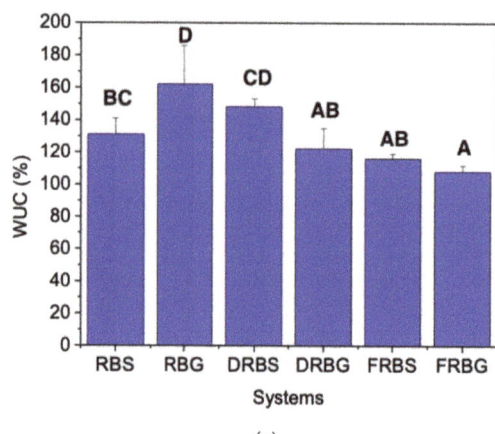
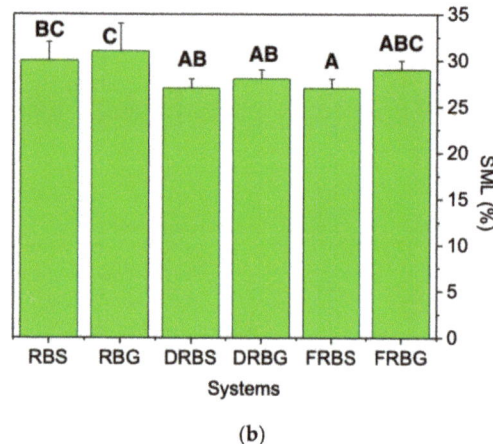

Figure 4. (a) Water uptake capacity (WUC) and (b) soluble matter loss (SML) of the different evaluated systems. Different letters above columns indicate significant differences ($p < 0.05$).

However, the effects were less remarkable for the SML values. It seems that higher WUC is associated with higher SML, although there were no significant differences. In this way, the lowest values were those of the FRBS and DRBS systems with 27 ± 1%, followed by the DRBG and FRBG systems, with 28 ± 1% and 29 ± 1%, respectively. Finally, the two systems with the highest losses are the RBS and RBG systems, with 30 ± 2% and 31 ± 3%, respectively.

3.5. Scanning Electron Microscopy (SEM)

Figure 5 shows the SEM micrographs obtained for the different bioplastics. Figure 5A,B correspond to the samples obtained with the raw RB containing sorbitol and glycerol, respectively. These two samples present microstructures with great flaws, large pores and cracks with no continuous surface, showing 15.8 and 25.8% porosity, respectively. On the contrary, the effect of employing treated rice bran is perfectly noticeable. In this way, the defatted rice bran produces more continuous surfaces with smaller pores, which present a more homogeneous distribution, as can be observed in Figure 5C,D. Although the pores are smaller, the total porosity increases for these two samples to 40.4% for DRBS and to 39.4% for DRBG. On the other hand, the effect is more pronounced in the last two micrographs, i.e., E and F, corresponding to the fiber-free samples, which again exhibit homogeneous surfaces with smaller pores, maintaining the total porosity in 38.4 and 38.1%, respectively. Finally, when the two plasticizers used are compared, both sorbitol and glycerol seem to produce similar porosities.

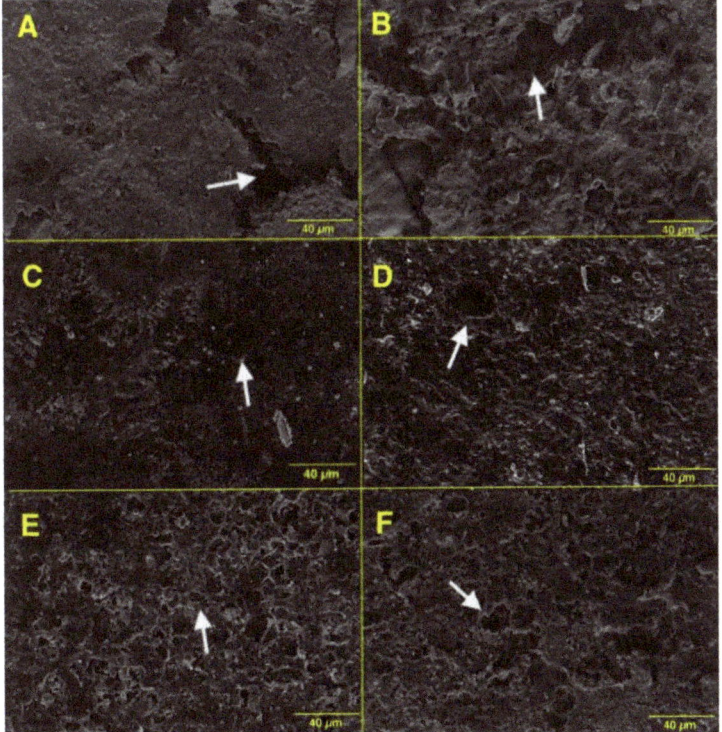

Figure 5. SEM micrographs of the different lyophilized bioplastics observed (**A**) RBS; (**B**) RBG; (**C**) DRBS; (**D**) DRBG; (**E**) FRBS; (**F**) FRBG.

4. Discussion

4.1. Chemical Composition

The chemical composition of RB, DRB, and FRB gathered in Table 1 allows analyzing the efficiency of the extraction methods. The lipid extraction was successfully carried out, decreasing the fat content from 22.8% for the RB sample to 1.8% for the DRB sample. Therefore, within 48 h after suspension in hexane, 92.11 ± 0.6% lipids were extracted at room temperature. Although the lipid extraction could be optimized using the Soxhlet device, removing most of the fat content in hours, the sample would reach higher temperatures, which could imply certain thermal modifications (i.e., earlier gelatinization) that are avoided with the methodology used. Note that lipids have been reported to act as plasticizer, increasing the mechanical properties of the final material generated [31]. Moreover, the hydrophobic character of lipids may also affect the water uptake capacity of the bioplastics obtained, this will be discussed in Section 4.4. With respect to the removal of the fibers, the methodology employed allowed the removal of the fiber fraction along with some soluble protein. This removal of fiber must affect the final properties of the bioplastic generated. It should be noticed that fiber used as filler can reinforce the structure of polymers. However, an excessive amount also may affect the mechanical properties of the final bioplastics [32]. The different systems studied will allow evaluating the effect of the fiber fraction on the bioplastic formed.

4.2. Dynamic Mechanical Thermal Analyses (DMTA)

Figure 1 shows the frequency sweep tests of predominantly elastic probes, indicating a frequency dependence. This dependence was already observed in previous studies conducted on RB-based bioplastics [13]. This frequency dependence has been previously related for uncross-linked polymers of high molecular weight [33], which is the case of these polymers, where the biopolymer chains (i.e., proteins or carbohydrates), are not specifically linked to each other. Thus, the systems plasticized with water/sorbitol (RBS, DRBS and FRBS) led to higher E' and E'' than water/glycerol ones (RBG, DRBG and FRBG), which according to the "lubricating" theory, the glycerol has a higher lubricating effect [34]. With respect to the active matter, the removal of the lipid fraction in the DRBS and DRBG-based bioplastics produced probes with improved rheological properties, that is, higher viscoelastic moduli. Thus, these results confirm that rice bran oil could act as a plasticizer, which increases the biopolymer chain mobility, leading to reduced viscoelastic moduli. Lipid removal appears to be beneficial for rheological behavior, since the viscoelastic moduli increased. The improvement in rheological behavior could also be attributed to the relative increase in the number of proteins and starches, that is, the active matter that produces most biopolymers. In this way, although fiber removal could have a similar effect, increasing total interactions, fibers are known to act as reinforcement in the biopolymeric matrix [26], therefore, the two effects counteract each other, leaving bioplastics without significant improvements in the rheological properties, as can be observed in the E' and E'' values of the FRBS and FRBG samples. In this experiment, the DRB and FRB systems maintained a similar behavior.

Some of the effects described above can also be observed in Figure 2a. First, the removal of the lipids again led to enhanced rheological behavior, with the DRBS and DRBG systems exhibiting viscoelastic moduli above the RBS and RBG systems, respectively, which are associated with a higher proportion of active matter, producing more interactions among the biopolymer chains. Moreover, the lack of RB oil reduced the mobility of the biopolymer chain as a result of its plasticizer role. However, although no significant differences were observed between the DRB and FRB systems when evaluating the frequency dependence, fiber removal enhanced the rheological behavior with temperature, especially for the FRBS samples, which were above the DRBS samples above room temperature, while the FRBG system is only superior to the DRBG towards the end of the temperature interval studied. Since fibers do not contribute to creating significant interactions in protein- and starch-based biopolymers [35,36], the lack of fibers allows creating a more compact structure with better behavior with temperature for higher protein and starch proportions. Finally, in terms of the plasticizer effect, although sorbitol contributes to the production of bioplastics with higher viscoelastic moduli, E' and E'', these systems experienced a more longer temperature dependence than the systems containing glycerol (which even stabilized above 100 °C until the end of the experiment). This behavior was previously observed in preliminary studies (results not published yet).

In addition, analysis of the loss tangent (tan δ) behavior was performed in order to assess the effect of the two extraction methods and the plasticizer used on the glass transition temperature of the bioplastic obtained. Figure 2b reveals that glycerol induced an earlier transition compared to the sorbitol-containing system, indicating a higher plasticizer efficiency for glycerol, which was also deduced from the biopolymer chain mobility above-mentioned. This effect is also reflected in the delayed peaks deduced at the lowest temperatures for the sorbitol-based systems. This higher plasticizer effect of glycerol was also observed when these compounds plasticized whey-based films [37]. The narrow peaks associated with the T_g of the DRB and FRB bioplastic using either glycerol or sorbitol as the plasticizer might be related to the purer composition obtained with the two extraction methods followed. As is well-known, purer substances undergo the glass transition in small temperature ranges, while complex compounds exhibit wider transition ranges [38]. In addition, the two extraction procedures exhibited higher T_g compared to the virgin RB, which might be due to the lack of oil in their composition, which also reflects its role as plasticizer. Moreover, this plasticizer effect is also observed in the

4.3. Tensile Tests

The results collected in Figure 3 and Table 2 confirm the beneficial effects of lipid removal on the mechanical properties of the final bioplastics. In this way, the plasticizer behavior of rice bran oil was lost, while, the proportion of proteins and starches increased, leading to stiffer, more resistant and more elastic samples (i.e., higher E, σ and ε for the DRBS and DRBG samples). This effect has been previously reported when for crayfish-PCL composites when increasing the active matter, which also led to an increase in the tensile parameters [39]. As for the fiber content, its removal also resulted in improved tensile behavior compared to virgin rice bran (RBS and RBG samples), the results for the FRBS and FRBG systems are below those of the defatted ones (DRBS and DRBG). In this case, as previously described, although there was a higher protein and starch content, the reinforcing effect of the fibers was lost and the samples obtained showed poorer tensile-strength properties. With respect to the plasticizer effect of sorbitol and glycerol, it seems that glycerol forms a closer mixture with the active matter, having a more efficient plasticizer effect and producing more elastic but less stiff samples. On the other hand, as a solid material, sorbitol can also act as a filler, not achieving such high elongations but leading to very stiff and resistant samples [40]. These results are in agreement with those obtained from DMA tests, confirming the different roles played by each plasticizer (glycerol or sorbitol). Although the tensile tests assume that the deformation of the material is related to its elasticity, whereas the DMA tests consider that the response is a combination of the elastic and viscous components, both assumptions reach the same conclusion for the plasticizing effect. Thus, this agreement between oscillatory and continuous deformation test has been also found for synthetic polymers such as epoxy foams [41].

4.4. Water Uptake Capacity and Soluble Matter Loss

The detrimental effect on WUC and the lower SML associated with defatted and fiber-free samples may be related to the higher proportions of proteins and starches present in these systems. In these samples, the interactions within the active matter were favored in greater proportions, leading to samples with high physical integrity (low soluble matter loss) that hold their structure during water immersion, although with a lower tendency to absorb and retain water. As can be seen in Section 3.4, the effect is observed for the DRBS and DRBG samples, but it was more pronounced in the FRBS and FRBG samples. Regarding the effect of the plasticizer, glycerol seems to produce bioplastics with higher WUC and, consequently, higher SML, as previously indicated, both parameters are related [42]. The effect of the plasticizer can also be related to the higher hydrophilic character of the glycerol, this higher affinity to water favors its penetration, and it also can justify the higher values obtained for the SML parameter, since once solubilized, it can be released in the medium, where the concentration difference is the driving force [43,44]. Moreover, it should be also noted that rice-bran oil can be extracted with both polar (i.e., acetone) and non-polar solvents (e.g., hexane, chloroform) [45]. Although we are not assuming that water extracts rice bran oil, it seems that polar solvent may interact with rice bran oil, facilitating water diffusion through the biopolymer structure, which is required for water uptake [46].

4.5. Scanning Electron Microscopy (SEM)

The microstructures observed in Figure 5 can be related to the results obtained in the previous sections regarding mechanical properties and the WUC. In this sense, the improved mechanical properties associated with the purified samples (i.e., the defatted rice bran, Figure 5C,D, and fiber-free rice bran, Figure 5E,F), correspond to more homogeneous microstructures with smaller pores and cracks that also present a more continuous distribution compared to those of Figure 5A,B. Thus, the higher homogeneity found in the micrographs evidenced a lower water uptake, since probe swelling created fewer pores and cracks when more active matter was present. The higher interactions led to enhanced mechanical properties observed. Thus, systems that usually have higher mechanical properties generally have a detrimental effect on water uptake [13], which was also obtained in this study, where the higher porosities observed in the DRB and FRB samples were not related to higher water absorption. Moreover, the glycerol-based samples (Figure 5B,D,F) also seem to exhibit more pores than those obtained with sorbitol (Figure 5A,C,E), which can be related to the above-mentioned of glycerol compared to sorbitol. Higher probe porosity after water uptake as glycerol content increases was previously observed for bioplastics [47].

5. Conclusions

This work evidences the beneficial effect of the removal of the lipid fraction on the mechanical properties of bioplastic samples. The lipids imply a certain plasticizing effect, which is lost in the DRB, also leading to a higher proportion of active matter and consequently to better rheological and tensile properties. The higher amount of active matter was later reflected in the WUC of the probes generated. Where also was observed the hydrophilic character of glycerol since it was released in a larger extend than sorbitol (SML values). However, although the removal of the fibers produces a higher proportion of active matter, the reinforcing effect of fibers is eliminated, which causes either a detrimental or no significant effect on the final properties. The only exception is the rheological behavior at high temperatures that was enhanced for these last samples. Finally, among the two plasticizers used, glycerol exhibited a higher plasticizing efficiency, leading to a lower glass transition temperature and producing bioplastics with higher deformability. On the contrary, the sorbitol-containing systems exhibited higher stiffness and maximum stress values.

Furthermore, the chemical characterization of the DRB and FRB revealed great quantities of ash content, which could also be removed via dialysis to obtain an optimized and pure biopolymer. In this way, there are still many modifications and parameters to be evaluated to develop suitable rice bran-based bioplastics. In addition, according to their mechanical and functional properties, RB-based bioplastics could be selected as promising candidates for the substitution of conventional fossil plastics in certain applications in the future.

Author Contributions: Conceptualization, M.A.-G. and M.F.; methodology, M.A.-G. and M.F.; software, M.A.-G.; validation, M.A.-G. and M.F.; formal analysis, M.A.-G. and A.R.; investigation, M.A.-G.; resources, M.F. and A.R.; data curation, M.A.-G.; writing—original draft preparation, M.A.-G.; writing—review and editing, M.F. and A.R.; visualization, M.A.-G. and A.R.; supervision, M.F. and A.R.; project administration, A.R.; funding acquisition, A.R. All authors have read and agreed to the published version of the manuscript.

Funding: This research was funded by the "FEDER/Ministerio de Ciencia, Innovación-Agencia Estatal de Investigación" through the project RTI2018-097100-B-C21.

Acknowledgments: The authors acknowledge the University of Seville for the VPPI-US grant (Ref.-II.5) to Manuel Felix and the VI-PPITUS grant (Ref.-II.2A) to María Alonso-González and the financial support of the Spanish Government "MCI/AEI/FEDER, UE". The authors also thank CITIUS for granting access to and their assistance with the Microscopy service. Finally, the authors also thank Herba Ingredients for providing the raw material used in this study.

Conflicts of Interest: The authors declare no conflict of interest. The funders had no role in the design of the study; in the collection, analyses, or interpretation of data; in the writing of the manuscript, or in the decision to publish the results.

References

1. Barnes, S.J. Understanding plastics pollution: The role of economic development and technological research. *Environ. Pollut.* **2019**, *249*, 812–821. [CrossRef] [PubMed]
2. Khalil, H.P.S.A.; Davoudpour, Y.; Saurabh, C.K.; Hossain, M.S.; Adnan, A.S.; Dungani, R.; Paridah, M.T.; Mohamed, Z.I.S.; Fazita, M.R.N.; Syakir, M.I.; et al. A review on nanocellulosic fibres as new material for sustainable packaging: Process and applications. *Renew. Sustain. Energy Rev.* **2016**, *64*, 823–836. [CrossRef]
3. Sabbagh, F.; Muhamad, I.I. Production of poly-hydroxyalkanoate as secondary metabolite with main focus on sustainable energy. *Renew. Sustain. Energy Rev.* **2017**, *72*, 95–104. [CrossRef]
4. Law, K.L. Plastics in the Marine Environment. *Ann. Rev. Mar. Sci.* **2017**, *9*, 205–229. [CrossRef] [PubMed]
5. Posnack, N.G. Plastics and cardiovascular disease. *Nat. Rev. Cardiol.* **2020**, *18*, 69–70. [CrossRef]
6. Wright, S.L.; Kelly, F.J. Plastic and Human Health: A Micro Issue? *Environ. Sci. Technol.* **2017**, *51*, 6634–6647. [CrossRef]
7. George, A.; Sanjay, M.R.; Srisuk, R.; Parameswaranpillai, J.; Siengchin, S. A comprehensive review on chemical properties and applications of biopolymers and their composites. *Int. J. Biol. Macromol.* **2020**, *154*, 329–338. [CrossRef]
8. Muhamad, I.I.; Sabbagh, F.; Karim, N.A. Polyhydroxyalkanoates: A Valuable Secondary Metabolite Produced in Microorganisms and Plants. In *Plant Secondary Metabolites*; Apple Academic Press: Oakville, ON, Canada, 2016; pp. 185–214. [CrossRef]
9. Kalia, S.; Avérous, L. *Biopolymers: Biomedical and Environmental Applications*; Wiley: Hoboken, NJ, USA, 2011. [CrossRef]
10. Omran, A.A.B.; Mohammed, A.A.B.A.; Sapuan, S.M.; Ilyas, R.A.; Asyraf, M.R.M.; Koloor, S.S.R.; Petrů, M. Micro- and Nanocellulose in Polymer Composite Materials: A Review. *Polymers* **2021**, *13*, 231. [CrossRef]
11. Guilbert, S.; Gontard, N.; Cuq, B. Technology and applications of edible protective films. *Packag. Technol. Sci.* **1995**, *8*, 339–346. [CrossRef]
12. Singh, T.P.; Sogi, D.S. Comparison of Physico-Chemical Properties of Starch Isolated From Bran and Endosperm of Rice (*Oryza sativa* L.). *Starch/Staerke* **2018**, *70*, 1700242. [CrossRef]
13. Alonso-González, M.; Felix, M.; Guerrero, A.; Romero, A. Effects of Mould Temperature on Rice Bran-Based Bioplastics Obtained by Injection Moulding. *Polymers* **2021**, *13*, 398. [CrossRef]
14. Alonso-González, M.; Felix, M.; Guerrero, A.; Romero, A. Rice bran-based bioplastics: Effects of the mixing temperature on starch plastification and final properties. *Int. J. Biol. Macromol.* **2021**, *188*, 932–940. [CrossRef]
15. Jiménez-Rosado, M.; Alonso-González, M.; Rubio-Valle, J.F.; Perez-Puyana, V.; Romero, A. Biodegradable soy protein-based matrices for the controlled release of zinc in horticulture. *J. Appl. Polym. Sci.* **2020**, *137*, 49187. [CrossRef]
16. Jiménez-Rosado, M.; Zarate-Ramírez, L.S.; Romero, A.; Bengoechea, C.; Partal, P.; Guerrero, A. Bioplastics based on wheat gluten processed by extrusion. *J. Clean. Prod.* **2019**, *239*, 117994. [CrossRef]
17. Karmakar, A.; Karmakar, S.; Mukherjee, S. Properties of various plants and animals feedstocks for biodiesel production. *Bioresour. Technol.* **2010**, *101*, 7201–7210. [CrossRef]
18. Punia, S.; Kumar, M.; Siroha, A.K.; Purewal, S.S. Rice Bran Oil: Emerging Trends in Extraction, Health Benefit, and Its Industrial Application. *Rice Sci.* **2021**, *28*, 217–232. [CrossRef]
19. Orthoefer, F.T. Rice Bran Oil. *Bailey's Ind. Oil Fat Prod.* **2020**, *68*, 1–25. [CrossRef]
20. Wang, Y. Applications of Rice Bran Oil. In *Rice Bran and Rice Bran Oil Chemistry, Processing and Utilization*; Elsevier: Amsterdam, The Netherlands; pp. 159–168. [CrossRef]
21. AOAC International. *Official Methods of Analysis of AOAC International*; AOAC International: Rockville, MD, USA, 2005.
22. López-Bascón-Bascon, M.A.; de Castro, M.D.L. Soxhlet extraction in Liquid-Phase Extraction. In *Liquid Extraction*; Elsevier: Amsterdam, The Netherlands, 2019; pp. 327–354. [CrossRef]
23. Mariotti, F.; Tomé, D.; Mirand, P.P. Converting nitrogen into protein—Beyond 6.25 and Jones' factors. *Crit. Rev. Food Sci. Nutr.* **2008**, *48*, 177–184. [CrossRef]
24. ISO. *ISO 527:2012 Plastics—Determination of Tensile Properties*; International Organization for Standardization—ISO: Geneva, Switzerland, 2006. [CrossRef]
25. D570, ASTM D 570—98—Standard Test Method for Water Absorption of Plastics. *ASTM Stand.* 1985. [CrossRef]
26. Klanwan, Y.; Kunanopparat, T.; Menut, P.; Siriwattanayotin, S. Valorization of industrial by-products through bioplastic production: Defatted rice bran and kraft lignin utilization. *J. Polym. Eng.* **2016**, *36*, 529–536. [CrossRef]
27. Silventoinen, P.; Rommi, K.; Holopainen-Mantila, U.; Poutanen, K.; Nordlund, E. Biochemical and Techno-Functional Properties of Protein- and Fibre-Rich Hybrid Ingredients Produced by Dry Fractionation from Rice Bran. *Food Bioprocess Technol.* **2019**, *12*, 1487–1499. [CrossRef]

28. Kunanopparat, T.; Menut, P.; Srichumpoung, W.; Siriwattanayotin, S. Characterization of Defatted Rice Bran Properties for Biocomposite Production. *J. Polym. Environ.* **2014**, *22*, 559–568. [CrossRef]
29. Gibbs, J.H.; DiMarzio, E.A. Nature of the glass transition and the glassy state. *J. Chem. Phys.* **1958**, *28*, 373. [CrossRef]
30. López-Castejón, M.L.; Bengoechea, C.; García-Morales, M.; Martínez, I. Effect of plasticizer and storage conditions on thermomechanical properties of albumen/tragacanth based bioplastics. *Food Bioprod. Process.* **2015**, *95*, 264–271. [CrossRef]
31. Chiralt, A.; González-Martínez, C.; Vargas, M.; Atarés, L. Edible films and coatings from proteins. In *Proteins Food Process*, 2nd ed; Elsevier: Amsterdam, The Netherlands, 2018; pp. 477–500. [CrossRef]
32. Yang, J.; Ching, Y.C.; Chuah, C.H. Applications of Lignocellulosic Fibers and Lignin in Bioplastics: A Review. *Polymers* **2019**, *11*, 751. [CrossRef]
33. Ferry, J.D. *Viscoelastic Properties of Polymers*; Wiley: Hoboken, NJ, USA, 1980.
34. Daniels, P.H.; Cabrera, A. Plasticizer compatibility testing: Dynamic mechanical analysis and glass transition temperatures. *J. Vinyl Addit. Technol.* **2015**, *21*, 7–11. [CrossRef]
35. Gurunathan, T.; Mohanty, S.; Nayak, S.K. A review of the recent developments in biocomposites based on natural fibres and their application perspectives. *Compos. Part A Appl. Sci. Manuf.* **2015**, *77*, 1–25. [CrossRef]
36. Satyanarayana, K.G.; Arizaga, G.G.C.; Wypych, F. Biodegradable composites based on lignocellulosic fibers—An overview. *Prog. Polym. Sci.* **2009**, *34*, 982–1021. [CrossRef]
37. Thomazine, M.; Carvalho, R.A.; Sobral, P.J.A. Physical Properties of Gelatin Films Plasticized by Blends of Glycerol and Sorbitol. *J. Food Sci.* **2005**, *70*, E172–E176. [CrossRef]
38. Pacáková, V.; Virt, J. Plastics. In *Encyclopedia of Analytical Science*, 2nd ed.; Worsfold, P., Townshend, A., Poole, C., Eds.; Elsevier: Oxford, UK, 2005; pp. 180–187. [CrossRef]
39. Félix, M.; Romero, A.; Martín-Alfonso, J.E.E.; Guerrero, A. Development of crayfish protein-PCL biocomposite material processed by injection moulding. *Compos. Part B Eng.* **2015**, *78*, 291–297. [CrossRef]
40. Felix, M.; Carpintero, V.; Romero, A.; Guerrero, A. Influence of sorbitol on mechanical and physico-chemical properties of soy protein-based bioplastics processed by injection molding. *Polímeros* **2016**, *26*, 277–281. [CrossRef]
41. Hu, G.; Yu, D. Tensile, thermal and dynamic mechanical properties of hollow polymer particle-filled epoxy syntactic foam. *Mater. Sci. Eng. A* **2011**, *528*, 5177–5183. [CrossRef]
42. Felix, M.; Martínez, I.; Aguilar, J.M.; Guerrero, A. Development of Biocomposite Superabsorbent Nanomaterials: Effect of Processing Technique. *J. Polym. Environ.* **2018**, *26*, 4013–4018. [CrossRef]
43. Orliac, O.; Rouilly, A.; Silvestre, F.; Rigal, L. Effects of additives on the mechanical properties, hydrophobicity and water uptake of thermo-moulded films produced from sunflower protein isolate. *Polymer (Guildf)* **2002**, *43*, 5417–5425. [CrossRef]
44. Fernández-Espada, L.; Bengoechea, C.; Cordobés, F.; Guerrero, A. Thermomechanical properties and water uptake capacity of soy protein-based bioplastics processed by injection molding. *J. Appl. Polym. Sci.* **2016**, *133*, 43524. [CrossRef]
45. Ruen-Ngam, D.; Thawai, C.; Nokkoul, R.; Sukonthamut, S. Gamma-oryzanol extraction from upland rice bran. *Int. J. Biosci. Biochem. Bioinforma.* **2014**, *4*, 252. [CrossRef]
46. Chaiyasat, A.; Jearanai, S.; Christopher, L.P.; Alam, M.N. Novel superabsorbent materials from bacterial cellulose. *Polym. Int.* **2019**, *68*, 102–109. [CrossRef]
47. Chen, P.; Tian, H.; Zhang, L.; Chang, P.R. Structure and Properties of Soy Protein Plastics with ε-Caprolactone/Glycerol as Binary Plasticizers. *Ind. Eng. Chem. Res.* **2008**, *47*, 9389–9395. [CrossRef]

Article

Processing and Characterization of Bioplastics from the Invasive Seaweed *Rugulopteryx okamurae*

Ismael Santana, Manuel Félix, Antonio Guerrero and Carlos Bengoechea *

Higher Polytechnic School, University of Seville, Calle Virgen de África, 7, 41011 Sevilla, Spain; isantana@us.es (I.S.); mfelix@us.es (M.F.); aguerrero@us.es (A.G.)
* Correspondence: cbengoechea@us.es; Tel.: +34-954-557-179

Abstract: The seaweed *Rugulopteryx okamurae,* from the Pacific Ocean, is considered an invasive species in the Mediterranean Sea. In this work, the use of this seaweed is proposed for the development of bio-based plastic materials (bioplastics) as a possible solution to the pollution produced by the plastic industry. The raw seaweed *Rugulopteryx okamurae* was firstly blended with glycerol (ratios: 50/50, 60/40 and 70/30), and subsequently, they were processed by injection molding at a mold temperature of 90, 120 and 150 °C. The rheological properties (frequency sweep tests and temperature ramp tests) were obtained for blends before and after processing by injection molding. The functional properties of the bioplastics were determined by the water uptake capacity (WUC) values and further scanning electron microscopy (SEM). The results obtained indicated that E′ was always greater than E″, which implies a predominantly elastic behavior. The 70/30 ratio presents higher values for both the viscoelastic moduli and tensile properties than the rest of the systems (186.53 ± 22.80 MPa and 2.61 ± 0.51 MPa, respectively). The WUC decreased with the increase in seaweed in the mixture, ranging from 262% for the 50/50 ratio to 181% for the 70/30 ratio. When carrying out the study on molded bioplastic 70/30 at different temperatures, the seaweed content did not exert a remarkable influence on the final properties of the bioplastics obtained. Thus, this invasive species could be used as raw material for the manufacture of environmentally friendly materials processed by injection molding, with several applications such as food packaging, control–release, etc.

Keywords: *Rugulopteryx okamurae*; bioplastics; DMA; injection molding; seaweed

Citation: Santana, I.; Félix, M.; Guerrero, A.; Bengoechea, C. Processing and Characterization of Bioplastics from the Invasive Seaweed *Rugulopteryx okamurae*. *Polymers* **2022**, *14*, 355. https://doi.org/10.3390/polym14020355

Academic Editor: Jean-Marie Raquez

Received: 23 December 2021
Accepted: 10 January 2022
Published: 17 January 2022

Publisher's Note: MDPI stays neutral with regard to jurisdictional claims in published maps and institutional affiliations.

Copyright: © 2022 by the authors. Licensee MDPI, Basel, Switzerland. This article is an open access article distributed under the terms and conditions of the Creative Commons Attribution (CC BY) license (https:// creativecommons.org/licenses/by/ 4.0/).

1. Introduction

Rugulopteryx okamurae (RO), also known as *Dictyota marginata, Dilophus marginatus,* or *Dictyota okamurae* [1,2], is a species of brown algae belonging to the *Dictyotaceae* family, originally from the coasts of the warm and temperate northwestern Pacific Ocean (Korea, Japan, China, Taiwan and the Philippines) [3]. This alga has been introduced from the Pacific Ocean to the Mediterranean through the Strait of Gibraltar. The Spanish Mediterranean coasts and those of the Strait of Gibraltar present a highly favorable environment for the species, favoring its expansion and an increase in the derived impacts. Consequently, on 1 December 2020 it was included in the Spanish Catalog of Invasive Exotic Species since it represents one of the main threats to biodiversity in the Mediterranean, due to its ability to spread and adapt [4].

In 2020, García-Gómez et al. [5] reported that more than 3,000,000 m² of the seabed of the Strait Natural Park was occupied by this invasive seaweed at different depths. The highest proportion of coverage occurs between 5 and 30 m, reaching 85–96%, while the smallest was found at greater depths or practically on the shores, 30–40 or 0–5 m (around 42–45%). This invasive species not only has effects on the seagrass, but it should also be noted that 5000 tons of this Asian seaweed were dislodged from the beaches of Ceuta in 2015, and 400 tons from the beaches of Tarifa only in July 2020 [6].

The great technological development from the twentieth century to the present has caused an increase in environmental pollution. Among these advances, the industrial production of non-biodegradable plastic materials stands out [7]. At the beginning of the 21st century, the average consumption of plastic was 15 kg/year per person [8]. This also accounts for a large part of the pollution that is produced, both by the emission of greenhouse gases during their manufacture and disposal, as well as by their accumulation in landfills and a high presence in the oceans due to their low biodegradability [9]. Since the 1950s, more than 7800 million metric tons have been produced, of which more than 10% ends up in the oceans once disposed. This produces worrying impacts on the seabed, its biodiversity and people due to the consumption of fish contaminated with microplastics [10]. Thus, an annual consumption of 39,000 to 52,000 microplastic particles is estimated that can negatively affect health [11].

Nowadays, an important part of this development is inconceivable without taking care of the environment, and that is why there are increasingly more restrictive regulations and a greater and growing interest in finding biodegradable materials with properties similar to those of conventional plastics. The solution of conventional plastics is the use of bioplastics, which can be defined as those plastics made from a renewable source or those which are biodegradable [12,13]. This demand for bio-based raw materials will demand a huge amount of resources, which could be supplied by both agri-food wastes or invasive species with no applications in the local market. Underutilized invasive species would provide adequate biomass for processing, which would reduce their environmental impact. The present work proposes the use of the invasive seaweed RO from the bay of Algeciras (located in the Strait of Gibraltar) as a raw material to produce bioplastics. The use of the invasive seaweed RO to generate biodegradable materials would serve both to reduce the impact of algae in the Mediterranean biodiversity and find an alternative to fossil-based plastics that pollute the planet. Depending on their properties, algae are used for various applications, such as in cosmetic or bioplastic packaging industries [14].

One of the problems when replacing conventional plastics with bioplastics are the poorer mechanical properties of the latter during their processing and end use, which can limit their potential. As a solution, a plasticizer is added to improve its mechanical properties. According to IUPAC (International Union of Pure and Applied Chemistry), a plasticizer is defined as "a substance or material incorporated into another material to improve its flexibility, compliance or viability" [15]. Plasticizers generally have a high boiling point and carbon chains of between 14 and 40 carbons, with a molecular weight of between 300 and 600 [16]. Thanks to these characteristics, plasticizers can be inserted into the structure of the polymers that make up plastics, reorganizing their three-dimensional structure, reducing the intramolecular forces of the polymer and allowing its mobility [17]. Due to this function, the choice of plasticizer for the plastic to be developed is important, as it will affect its final properties. Other parameters to consider when choosing a plasticizer are the boiling or melting temperature, polarity or solvation. Moreover, little or no toxicity, such as that of fatty acid esters or vegetable glycerol, is desired when selecting a plasticizer [18].

The selection of the technique and conditions used when processing bioplastics is also highly influential in their final properties. It should be noticed that Mass Flow Rate (MFR) is also of extreme importance when selecting a processing technique for bioplastics. Some authors have pointed out that some biodegradable plastics, such as polylactide (PLA), experience a greater increase in MFR when increasing the processing speed than conventional plastics, such as low-density polyethylene (LDPE) [19]. This clearly affects the energy efficiency of the processing techniques. There are different techniques for processing bioplastics, such as compression molding, extrusion or injection molding, among others. Compression molding consists of two metal molds that apply a certain force on the sample to produce the bioplastic. As the sample flows into the mold, it acquires its shape [20]. Extrusion is the most widely used technique for processing polymers, and consists of a continuous process, where the polymer is transported through a barrel where it is heated to enhance its flowability through a final, conveniently shaped orifice. The polymer cools as it exits and results in a constant section solid [21]. Injection molding consists of the softening of a plastic material under the appropriate conditions and its subsequent introduction under pressure into the cavities of a mold, in which it is brought to a temperature at which the pieces can be extracted without deforming. The injection molding process can be divided into two main stages: injection and compaction. First, the substance to be molded is introduced into a tank that is heated to a certain temperature that allows the material to flow for subsequent injection into the mold, by means of a piston which exerts a certain pressure. This is typically called the injection stage. The compaction stage is the second part of the process and occurs with the sample already inside the cavities of the mold, which is heated to the molding temperature. In the case of biopolymers, the molding temperature can be higher than that of the cylinder, so as to promote crosslinking and fix the structure. During this stage, the injected homogeneous mixture is subjected to a certain stable pressure by the piston until its extraction [20]. Among all the different techniques available for dry bioplastic processing, this work selected injection molding due to its versatility and scalability.

This processing technique has been already used to manufacture bioplastics from gluten [22], soy protein [23] or blood plasma from the meat industry [24], among others. Even if different examples of the development of bioplastics have already been reported [14]. There is no such information from the invasive macroalgae RO. The qualities that macroalgae offer (low cost, low toxicity, proper mechanical properties) make them a good candidate for producing bioplastics [25].

The aim of this work has been the development and characterization of bioplastic materials based on RO seaweed. To this end, RO was firstly blended with glycerol (ratios: 50/50, 60/40 and 70/30), and subsequently, bioplastics were processed at 120 °C. Subsequently, the effect of mold temperature was also analyzed by sample processing at 90 and 150 °C (RO/GLY ratio: 70/30). The rheological properties (frequency sweep tests and temperature ramp tests) were obtained for blends before and after processing by injection molding, whereas the functional properties of the bioplastics obtained were assessed by water uptake capacity (WUC) and electron scan microscopy (SEM).

2. Materials and Methods

2.1. Materials and Sample Preparation

The seaweed RO used in the present study was gently supplied by the Andalusian Institute for Agricultural, Fisheries, Food and Organic Production Research and Training (IFAPA, Puerto Real, Spain). The RO was hand-picked from the bay of Algeciras, and subsequently freeze dried. Freeze-dried algae samples were ground in a kitchen blender (Mambo10070, CECOTEC, Valencia, Spain) at maximum speed, obtaining a flour where the diameters of most particles (~90%) were within the 10 to 100 µm range. GLY was used as plasticizer (Panreac Química S.A (Castellar del Vallès, Spain). The proximate composition of the seaweed is 18.47 ± 0.35% ashes, 13.48 ± 0.26% water, 9.76 ± 0.16% proteins (conversion factor N-protein: 4.92 [26]) and 11.63 ± 0.22% lipids.

Bioplastics were obtained in a two-step method: (i) RO and glycerol (GLY) was mixed in a two-blade counter-rotating batch mixer Haake Polylab QC (ThermoHaake, Karlsruhe, Germany) at room temperature and 50 rpm until system homogeneity (60, 10 and 5 min for RO/GLY ratios 50/50, 60/40 and 70/30, respectively); (ii) RO/GLY blends were injected in the lab-scale Minijet Piston Injection Molding System (ThermoHaake, Karlsruhe, Germany), obtaining $1 \times 10 \times 60$ mm^3 probes. The samples were processed at 60 °C (cylinder temperature), whereas the mold temperature was set at 90, 120 or 150 °C. The injection pressure was 500 bar (20 s), whereas the post-injection pressure was 200 bar (150 s). These processing conditions were similar to those from plasma porcine protein [24].

Figure 1 shows the schematic process for the processing and characterization of bioplastics.

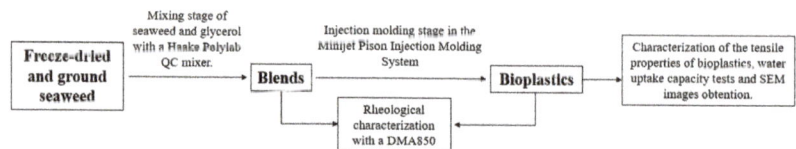

Figure 1. Scheme of the production process and characterization of bioplastics.

2.2. Methods

2.2.1. Rheological Characterization of RO/GLY Blends

Rheological Measurements

The RO/GLY blends obtained after the mixing stage were analyzed by dynamic mechanical analysis (DMA) in compression mode. Frequency sweep tests (from 0.1 to 10 Hz) and temperature ramp tests (from −50 to 150 °C) were performed using a DMA850 (TA Instruments, Wakefield, MA, USA). All DMA tests were performed within the linear viscoelastic region, which was determined prior to any measurement by strain sweep tests at 1 Hz. Frequency sweep tests were performed at constant temperature (25 °C), whereas temperature ramp tests were performed at a constant frequency (1 Hz). DMA tests were performed using the 15 mm diameter parallel plates geometry.

2.2.2. Characterization of the RO-Based Bioplastics

Rheological Measurements

Final bioplastics were characterized by DMA tests with a tension clamp as geometry. For these tests, rectangular probes (10 mm × 1 mm × 15 mm) were analyzed in the DMA 850 (TA Instruments, MA, USA). Frequency sweep tests (from 0.1 to 10 Hz) and temperature ramp tests (from −30 to 180 °C) were performed using a DMA850 (TA Instruments, MA, USA). All DMA tests were performed within the linear viscoelastic region, which was determined prior to any measurement by strain sweep tests at 1 Hz. Frequency sweep tests were performed at constant temperature (25 °C), whereas temperature ramp tests were performed at a constant frequency (1 Hz).

Tensile Properties

Stress–strain curves were obtained in uniaxial tensile tests until fracture using a RSA3 (TA Instruments, MA, USA). From these tests, three parameters were obtained: Young's modulus (E), maximum stress (σ_{max}) and maximum strain (ε_{max}). These tests were performed using rectangular probes (60 mm × 10 mm × 1 mm) according to the standard ISO 527-2 [27]. All tensile tests were carried out at a constant elongation rate of 1 mm·min^{-1} and room temperature.

Water Uptake Capacity (WUC)

Water uptake capacity (WUC) was determined as follows [28]: bioplastic samples were first dried at 50 °C for 24 h and weight (w_1); then, dried samples were immersed for 24 h in 100 mL of deionized water and weighed (w_2). Finally, samples were lyophilized for 24 h and weighed (w_3). WUC and soluble matter loss (SML) were calculated using Equations (1) and (2):

$$\text{WUC (\%)} = \frac{(w_2 - w_3)}{w_3} \cdot 100 \qquad (1)$$

$$\text{SML (\%)} = \frac{(w_1 - w_3)}{w_1} \cdot 100 \qquad (2)$$

Scanning Electron Microscopy (SEM)

Freeze-dried matrices obtained after WUC were cut into small pieces (~2.5 mm), gold coated and, finally, examined by SEM, in a ZEISS EVO (Oberkochen, Germany). A beam current of 18 pA and a working distance of 7.5–8.5 mm were employed in the microscope, with an acceleration voltage of 10 kV. Image analyses were obtained at 1000×, 500× and 200× magnification. Moreover, they were analyzed by the imaging software ImageJ (Bethesda, MD, USA) [29].

2.3. Statistical Analysis

All measurements were carried out at least in triplicate. The statistical analysis was carried out using the STATGRAPHICS Centurion XVIII software (The Plains, VA, USA). The standard deviation for some selected parameters was included. Significant differences ($p < 0.05$) were indicated by superscript letters.

3. Results and Discussion

3.1. Influence of the Rugulopteryx okamurae/Glycerol (RO/GLY) Ratio

3.1.1. Blends

Mixing

After freeze drying and milling, RO seaweed was thoroughly mixed with glycerol (GLY) to obtain a homogeneous blend that was eventually injected. A picture of the resulting blends can be found in the Supplementary Material (Figure S1. Visual appearance of RO/GLY blends at different ratios (from left to right: 50/50, 60/40, 70/30)). Figure 2 shows the evolution of torque (A) and temperature (B) with mixing time for blends obtained at different RO/GLY ratios (50/50, 60/40 and 70/30). An apparent increase in both torque and temperature values was observed during the whole mixing stage as the ratio of seaweed in the blends increased, and was especially noticeable for the 70/30 system. The greater the RO/GLY ratio (i.e., the lower amount of plasticizer in the blend) resulted in a lowering of the free volume among the polymeric chains present in the seaweed biomass. As a consequence, there was greater friction which led to thermal energy dissipation that was detected as a temperature rise [22]. Moreover, 50/50 and 60/40 systems achieve a steady value for both torque (2.2, 2.8 N·m, respectively) and temperature (28.7 and 33.4 °C, respectively) after around 10 min of mixing time. At that time, both torque and temperature of the system with the greatest RO content (70/30) displayed much greater values (13.2 N·m, 43.4 °C, respectively) than the rest. Moreover, this system evolves into a sudden growth of both parameters from around 8 min, which may be related to a certain strengthening of the sample promoted by the greater proximity of the different compounds present in the biomass.

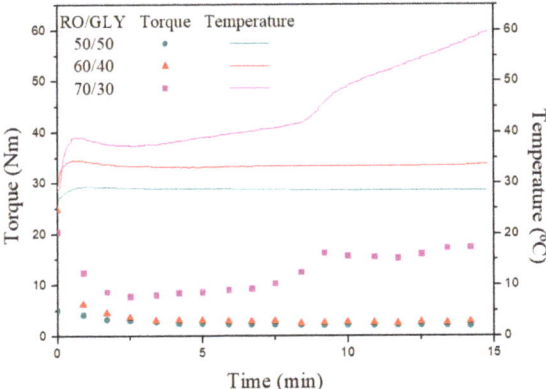

Figure 2. Torque and temperature values for different RO/Gly systems during the mixing stage.

Specific mechanical energy (SME) of every system at a specific time from the mixing curves:

$$\text{SME} = \frac{\omega}{m} \int_0^{t_{mix}} M(t)\delta(t) \tag{3}$$

where ω (in rad/s) is the mixing speed, m (in g) the sample mass, M(t) (in N·m) the torque and t_{mix} (in s) the mixing time. In order to avoid the effect of the second strengthening in 70/30, a t_{mix} of 5 min was selected for the estimation of SME shown in Table 1.

Table 1. Specific Mechanical Energy (SME) involved in the mixing of different RO/GLY systems (t_{mix}: 5 min).

RO/GLY	SME (kJ/kg)
50/50	208 ± 37
60/40	328 ± 50
70/30	805 ± 182

It is quite apparent that SME is greater as the seaweed content increases within the formulation, with 60/40 and 70/30 systems 1.6 and 3.9 times higher, respectively, than the system with the highest amount of plasticizer (50/50). It is well known that plasticizers ease the processability of polymeric samples and lower torque and temperatures are expected through the mixing due to a softening of the sample as their content increases [24].

Dynamic Mechanical Thermal Analysis (DMTA)

Figure 3A shows the effect of temperature from −50 to 150 °C on both viscoelastic moduli for the blend with a RO/GLY ratio of 60/40. Due to excessive flabbiness (50/50) or rigidity (70/30), blends with other RO/GLY rations did not lead to reliable experimental data. All blends softened as they were heated, which was reflected in a global decrease in both the elastic (E′) and viscous (E″) moduli when subjected to DMTA tests (data not shown). A marked decrease higher than two orders of magnitude was observed in both E′ and E″, leading to a minimum approximately at 75 °C. From then on, a slight increase and a tendency onto a steady value were observed. This behavior is qualitatively similar to that displayed by other blends including biomass from different sources such as soy, plasma porcine, among others [23,24,30]. Thermoplastic behavior generally implies a softening of the material, as reflected by the decrease in both viscoelastic moduli. This is typically associated with the fading of secondary interactions (e.g., hydrogen bonds) when heating a polymer (i.e., proteins).

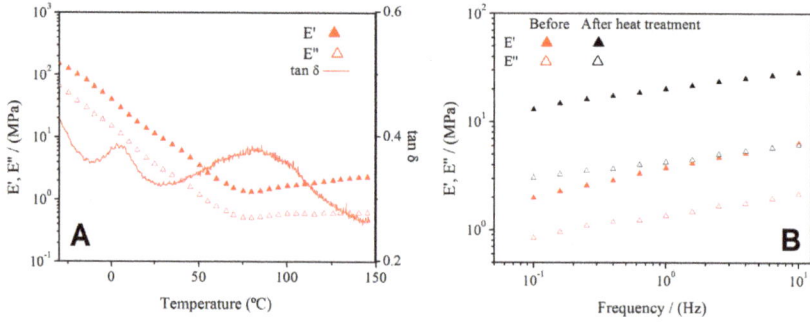

Figure 3. Evolution of elastic (E′) and viscous (E″) moduli with temperature (**A**) and with frequency before and after heat treatment (**B**) for the blend with a RO/GLY ratio of 60/40.

When observing the evolution of the loss tangent (tan δ) along with the heating (Figure 3A), it is quite clear that in spite of the softening previously commented when heating, the sample reinforced its elastic character over its viscous one. Thus, tan δ decreased from 0.58 to 0.26. Moreover, two peaks were also observed for tan δ, one around 0 °C and the other above 82 °C, which indicate thermal transitions of the material and show the heterogeneity of the sample. As a matter of fact, when cooled down to room temperature, greater E' and E" were obtained due to the formation of physical interactions (e.g., hydrogen bonds), as observed in Figure 3B for frequency sweep tests performed at 20 °C before and after the whole DMTA test.

Figure 3B shows the dependence of the viscoelastic moduli on frequency from 0.1 to 10 Hz before and after the heat treatment for a 60/40 RO/GLY ratio. The frequency sweep tests confirm the marked elastic character, since the elastic moduli is above the viscous moduli within the entire range of frequency studied. However, a slight dependence of the viscoelastic moduli on frequency is manifested in the E'-frequency slope, being 0.25 before the heat treatment and 0.16 after the heat treatment (Figure 3B). Despite the predominant elastic behavior, the blends obtained are injectable, since they behave similarly to other protein blends with successful injection molding results [31]. Moreover, these results also evidence the effect of temperature on the microstructure of the blends, decreasing the dependence on frequency, which reflects a lower protein chain mobility [32]. These results evidence the suitability of thermal processing methods for these blends.

3.1.2. Bioplastics

Previous blends were injection molded obtaining different bioplastic materials. A picture of the resulting bioplastics can be found in the Supplementary Material (Figure S2. Visual appearance of RO/GLY bioplastics at different ratios (from left to right: 50/50, 60/40, 70/30) molded at 120 °C).

Dynamic Mechanical Thermal Analysis (DMTA)

Blends with different RO/GLY ratios (50/50, 60/40, 70/30) were subjected to injection molding at a mold temperature of 120 °C, a temperature much higher than the minimum detected in the DMTA tests of the blends (~75 °C). Figure 4 shows the DMTA tests carried out from −30 to 180 °C for the different bioplastics obtained. As observed for blends, a softening of all the bioplastics took place as they were heated, which can be associated with the disruption of the secondary structure of proteins caused by the increase in temperature. However, the decrease in the viscoelastic moduli observed in the bioplastic systems was much less pronounced than in the blends, due to the strengthening induced in the samples during the injection molding process at high temperatures and pressures. During this softening, a peak in tan δ was observed at around 80 °C for all samples, pointing out a thermal transition of these materials (data not shown). Moreover, contrary to blends, no minimum was detected during the thermal treatment which implied that no further strengthening would be expected through more exhaustive conditions. It seems that the greatest thermosetting potential was already achieved during the injection-molding processing of the seaweed-based blends. All bioplastics displayed a much more cohesive structure than the original blends and could be tested correctly, tending towards steady values at high temperatures (~125 °C). Thus, higher content in seaweed resulted in greater viscoelastic moduli, observing a growing sequence for both E' and E" with the RO/GLY ratio (50/50 < 60/40 < 70/30) during the whole DMTA test. A similar response when increasing the biomass content in bioplastics made from proteins and GLY has already been observed [23,24]. The observed evolution of E' and E" with RO/Gly ratio was also found in injected bioplastics from microalgae, which was attributed to the promotion of protein–protein interactions when the plasticizer content decreased [33,34]. It should be noticed that López-Rocha et al. [33] did not find the mentioned biomass content-viscoelastic properties correlation for blends, but also after the injection molding process. Samples were submitted to drastic temperature and pressure conditions when injected, which promoted the strengthening of the samples in a more efficient way than just mixing at room temperature. Therefore, when considering the value of the elastic modulus at 20 °C (E'_{20}), the 60/40 system experienced an increase from around 14 (blend) to 219 (bioplastic) MPa as a result of the injection molding process.

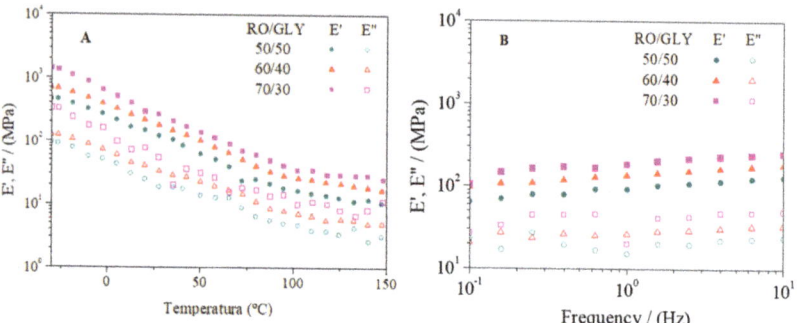

Figure 4. Evolution of viscoelastic moduli with temperature (**A**) and frequency (**B**) for bioplastics with different RO/GLY ratio (50/50, 60/40, 70/30).

Tensile Tests

Tensile tests were carried out on the bioplastics with different RO/GLY ratios at constant elongation rate of 1 mm·min^{-1} to assess the effect of the biomass content on their mechanical properties. Figure 5 shows the stress–strain curves produced by the three systems studied, showing an initial linear elastic region at lower strains in which the slope corresponds to the Young's modulus, E. When certain yield stress was exceeded, the slope began to descend as plastic deformation took place. Once the stress reached a maximum value (σ_{max}), the slope decayed abruptly, especially for the 70/30 system, which implied the breakage of the probe at a maximum deformation ε_{max}. In the systems studied, it is observed that as the content of seaweed in the sample increased, and the rigidity of the materials also increased, requiring higher stresses to achieve the same degree of deformation, which implies greater resistance to breakage. Moreover, this plot also showed the higher tenacity obtained for the ratio 70/30.

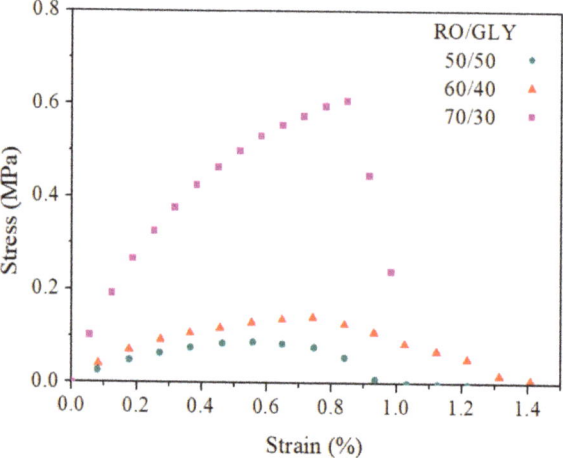

Figure 5. Stress–strain curves for bioplastics with different RO/GLY ratios (50/50, 60/40 and 70/30).

Mechanical properties estimated from tensile tests for bioplastics with different RO/GLY ratios (50/50, 60/40 and 70/30) are shown in Table 2. It is observed that both E and σ_{max} increased with the content of seaweed in the formulation of the bioplastic, displaying an exponential increase that led to a remarkable increase for the 70/30 system. This evolution is in line with that observed for the viscoelastic properties of the systems obtained previously from the DMTA tests, although the evolution of E' and E" with RO/GLY was more gradual than for the tensile properties. Regarding the ε_{max}, no clear evolution could be perceived, and the 60/40 system displayed the greatest deformability. Other authors have found a similar evolution for E and σ_{max} when increasing biomass content in injection molded bioplastics [24], although they also detected an increase in ε_{max}. The lower ε_{max} obtained for 70/30 should be related to its greater rigidity due to the higher strengthening achieved during injection molding, as commented before. This should be explained on basis of the lack of plasticizer, which allowed greater interaction between the polymeric chains (e.g., proteins, carbohydrates) of the material, leading to a more rigid and resistant structure. This phenomenon of ε_{max} variation has been reproduced by other authors for starch-based bioplastics. The plasticizer replaced existing intermolecular bonds in the material's structure with hydrogen bonds, which reduces stiffness and allows flexibility, so the higher the amount of plasticizer, the greater the deformation capacity [35]. However, when the plasticizer concentration exceeds a critical value, the anti-plasticizer phenomenon can occur. In these cases, too much plasticizer in the mix can excessively weaken the cohesion between the polymer chains making it too brittle and resulting in less elongation [36].

Table 2. Mechanical parameters (Young's modulus (E), maximum stress (σ_{max}) and maximum strain (ε_{max})) for bioplastics with different RO/GLY ratios (50/50,60/40,70/30). Different letters within a column indicate significant differences ($p < 0.05$).

RO/GLY	E (MPa)	ε_{max} (%)	σ_{max} (MPa)
50/50	0.40 ± 0.14 [a]	0.88 ± 0.07 [a]	0.089 ± 0.021 [a]
60/40	0.70 ± 0.10 [b]	1.49 ± 0.21 [b]	0.15 ± 0.03 [b]
70/30	2.61 ± 0.51 [c]	0.74 ± 0.12 [a]	0.68 ± 0.06 [c]

Similar profiles are observed in other bioplastics made with proteins such as soy protein or porcine plasma protein [23,24].

Water Uptake Capacity (WUC)

Figure 6 shows the dependence of water uptake capacity (WUC) and soluble matter loss (SML) on the RO/GLY ratio of bioplastics studied. There was a significant and progressive decrease in WUC values when increasing the seaweed content in the bioplastic, going from around 260% to 180% when the RO/GLY ratio increased from 50/50 to 70/30. Similar values and evolution with biomass content were observed in bioplastics from a microalgae consortium by López Rocha et al. [33]. The greater WUC displayed by the 50/50 system can be associated with its higher glycerol content, which promotes the formation of a porous structure that may improve the absorption process, as hydrophilic GLY tends to pass onto the aqueous immersion media [23,24]. The greater rigidity and resistance measured for the 70/30 system inhibited the swelling of the bioplastic, limiting—as a consequence—its ability to absorb water.

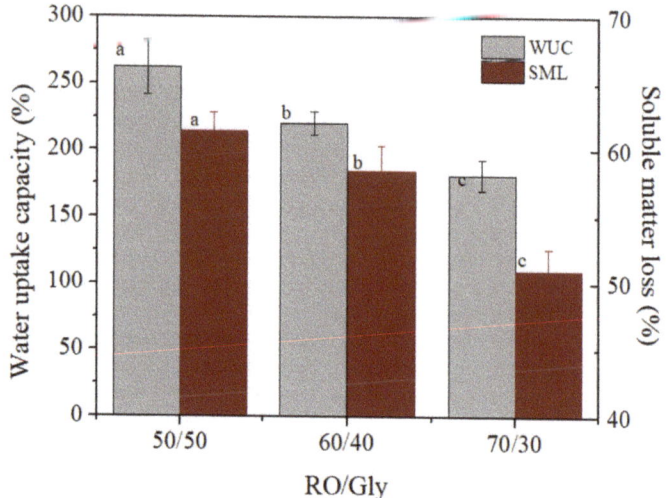

Figure 6. Water uptake capacity and soluble matter loss for the RO/Gly systems: 50/50, 60/40 and 70/30. Different letters (a, b, c) within the same column parameter (WUC or SML) indicate significant differences ($p < 0.05$).

Regarding the SML, a tendency to decrease as the RO/GLY ratio increased was detected, being significant for the 70/30 system: both 50/50 and 60/40 bioplastics lost around 60% of mass when immersed for 24 h, while the 70/30 bioplastic lost around 50%, which is coherent with a greater number of interactions strengthening the material. Most of the SML is due to the loss of glycerol, which has a highly hydrophilic behavior [37]. However, some of the biomass (~20–30%) was also lost during immersion as the SML was always higher than the original percentage of GLY in the formulation.

Scanning Electron Microscopy (SEM)

Figure 7 shows the images obtained by SEM for each of the bioplastics generated after a freeze drying process after studying the water uptake capacity. By increasing the percentage of seaweed in the formulation, the surface became less porous, which was much more noticeable in the case of the 70/30 system. Furthermore, the lower presence of pores in the structure as the RO/GLY ratio increased was strongly affected by the WUC results (Figure 6): the structure of the 50/50 system was more heterogeneous and showed many structural irregularities and pores through which water can enter, while the 70/30 system showed a lower amount of pores and interstices. Moreover, this also agrees with the viscoelastic and mechanical properties already reported, as the 70/30 system possessed higher viscoelastic moduli, rigidity and resistance supported by a more compact structure, with a lower porosity.

3.2. Influence of Molding Temperature in Bioplastic Properties

Bioplastics with a RO/GLY ratio equal to 70/30 were processed at different molding temperatures (T_m: 90, 120 and 150 °C) to assess their effect on their properties. Previously, some authors have already indicated that T_m is the processing parameter that commonly affects the properties of bioplastics to a greater extent when injection molded [23].

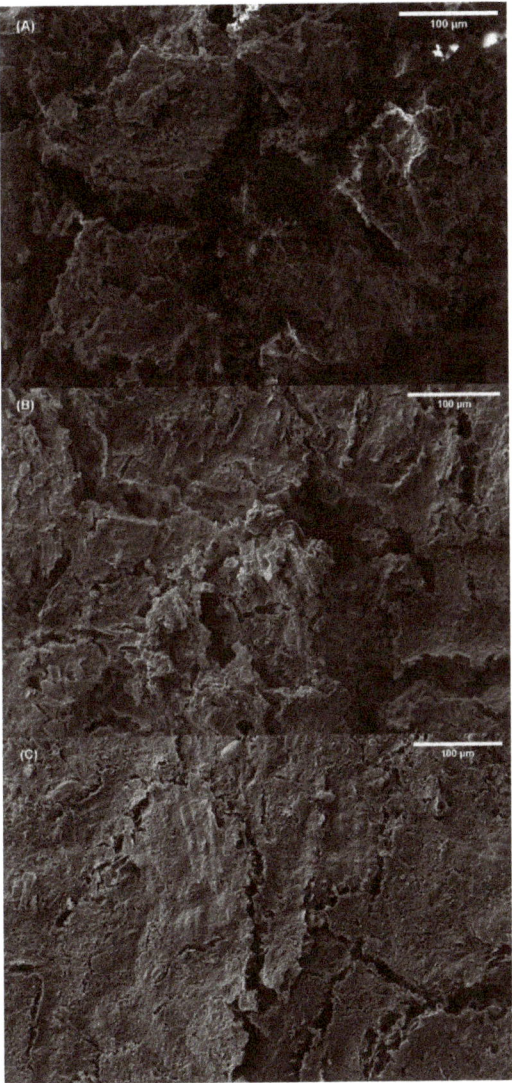

Figure 7. Scanning electron microscopy images of RO/GLY 50/50 (**A**), 60/40 (**B**) and 70/30 (**C**) systems prepared at 120 °C with a magnification of 200×.

3.2.1. Dynamic Mechanical Thermal Analysis (DMTA)

Figure 8 shows the dependence of viscoelastic moduli on temperature (A) and frequency (B) for 70/30 bioplastics molded at different temperatures (90, 120, 150 °C).

As observed in Figure 4A, all samples processed at different molding temperatures displayed the same qualitative evolution: E' and E" values, since they decreased when heating the samples and tended to a plateau value around 120 °C. Other authors have previously studied the effect of molding temperature in other bioplastic systems. Thus, Fernández-Espada et al. studied bioplastics based on soy protein molded at 40, 80 and 120 °C, not obtaining significant differences in the evolution of E' or E" moduli with temperature [23]. However, no great quantitative differences were observed for the 70/30 sample when molded at different temperatures, which would indicate that all molding temperatures used during the holding stage in the injection molding process were high enough to achieve the greatest thermosetting potential of seaweed bioplastics. It should be noted that the sample molded at the highest temperature displayed slightly lower values for both viscoelastic moduli during the whole temperature test, which could be related to the possible degradation of the biomass at 150 °C.

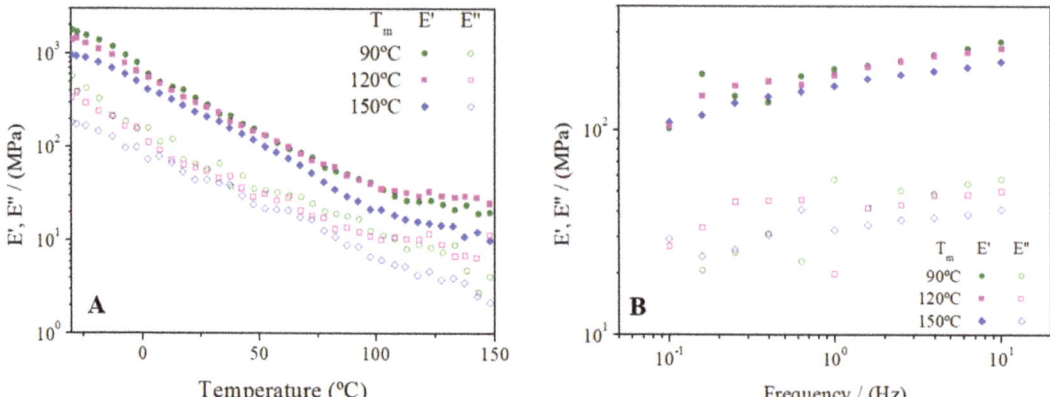

Figure 8. Dependence of viscoelastic moduli on temperature (**A**) and frequency (**B**) for RO/GLY 70/30 bioplastics processed at different molding temperature (90, 120 and 150 °C).

The frequency sweep tests performed on the 70/30 samples molded at different T_m were quite similar (Figure 4B), although the system molded at the highest temperature showed slightly lower E' and E" values. Moreover, a significant increase was observed in the loss tangents values (tan δ) when increasing T_m for the 70/30 system (0.22 ± 0.04, 0.26 ± 0.01, and 0.32 ± 0.01 for 90, 120 and 150 °C, respectively), which implied a loss of the elasticity of the material when molded at higher temperatures. The effects of molding temperature in soy protein-based bioplastics have already been studied by Paetau et al. [38], obtaining similar results. Increasing the temperature favors the crosslinking of the polymer fibers, which makes the material stiffer to some extent. After a certain temperature, proteins are degraded and a decrease in mechanical properties begins to be observed.

3.2.2. Tensile Properties

Figure 9 and Table 3 show the stress–strain curves and the main mechanical parameters, respectively, (E, σ_{max}, ε_{max}) for bioplastics with a RO/GLY ratio of 70/30 processed at different molding temperature (90, 120 and 150 °C). The stress–strain curves of the three systems are similar in shape, with an initial linear elastic behavior, the slope of which decayed slightly until reaching the maximum stress when the material ruptured. Greater stresses are required to deform the sample processed at 120 °C when compared to the rest of mold temperatures, which can resist stresses up to 0.7 MPa; twice the value of that can be applied to the rest of the samples. The initial increase from 90 to 120 °C should be related to a certain strengthening of the sample due to thermally induced interactions (e.g., disulfide bonds), while the decrease in the stress–strain curve from 120 to 150 °C should be associated with the previously commented degradation that negatively affects the mechanical properties [23].

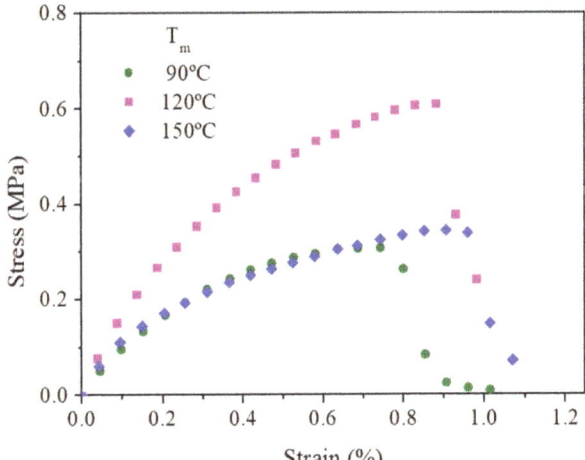

Figure 9. Tensile properties of bioplastics (RO/GLY: 70/30) processed at different molding temperature (90, 120 and 150 °C).

Table 3. Mechanical parameters (Young's modulus (E), maximum stress (σ_{max}) and maximum strain (ε_{max})) of bioplastics (RO/GLY: 70/30): for different molding temperature (90, 120 and 150 °C). Different letters within the same column parameter (WUC or SML) indicates significant differences ($p < 0.05$).

Molding Temperature (°C)	E (MPa)	σ_{max} (%)	ε_{max} (MPa)
90	1.60 ± 0.14 [a]	0.73 ± 0.002 [a]	0.33 ± 0.022 [a]
120	2.61 ± 0.51 [b]	0.74 ± 0.12 [a]	0.68 ± 0.057 [b]
150	1.47 ± 0.17 [a]	0.69 ± 0.32 [a]	0.36 ± 0.045 [a]

3.2.3. Water Uptake Capacity

Regarding the WUC, the WUC slightly decreases (Figure 10) the molding temperature increases, being 189.1 ± 6.2% for the system molded at 90 °C, 181.5 ± 11.6% for the 120 °C and 166.8 ± 2.0% for the 150 °C molded system. This effect of temperature on WUC was observed in soy-based bioplastics [23] and was related to the formation of thermally induced interactions that take place during the injection molding. The lowering in WUC is observed from 90 to 150 °C, in spite of the lower viscoelastic moduli and rigidity measured for the sample molded at 150 °C.

No significant differences were observed in the SML values of the systems with T_m, as all values were around 51.5%, which implies that molding temperature does not exert much influence on soluble matter loss. Different letters within the same column parameter (WUC or SML) indicate significant differences ($p < 0.05$).

3.2.4. Scanning Electron Spectroscopy

Figure 11 shows the SEM images for bioplastics with a RO/GLY ratio of 70/30 processed at different molding temperatures. Some differences were found between the systems, with the structure of the bioplastic molded at 90 and 150 °C showing greater irregularities and pores than the one molded at 120 °C. These imperfections may be related to a lower degree of cross-linking, resulting in a lower dimensional stability. The smoother and more compact structure observed for the system molded at 120 °C should be related to its lower water uptake capacity and higher viscoelastic properties.

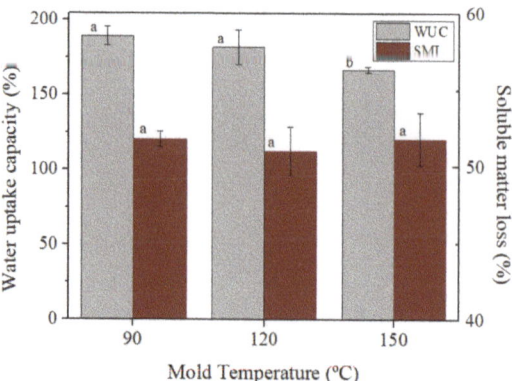

Figure 10. Water uptake capacity and soluble matter loss of the RO/Gly systems 70/30 at different molding temperature (90, 120 and 150 °C). Different letters (a, b) within the same column parameter (WUC or SML) indicate significant differences ($p < 0.05$).

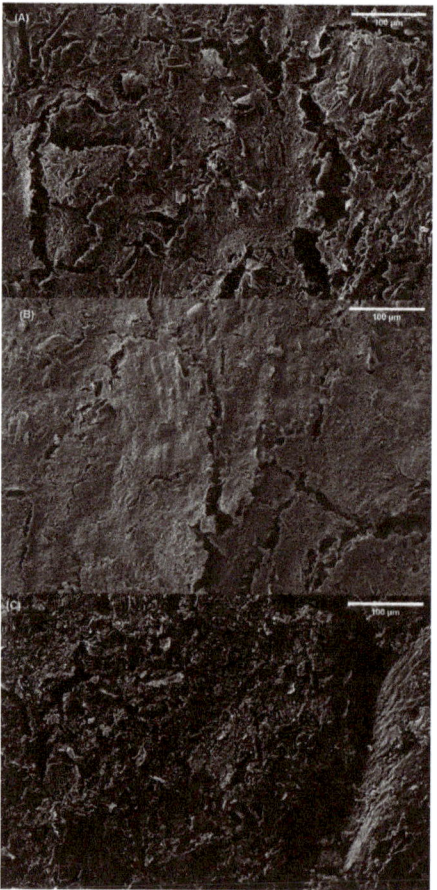

Figure 11. Scanning electron spectroscopy images of RO/GLY 70/30 bioplastics molded at 90 °C (**A**), 120 °C (**B**) and 150 °C (**C**) with a magnification of 200×.

4. Conclusions

The invasive seaweed *Rugulopteryx okamurae* can be used to produce bioplastics through injection molding employing glycerol as plasticizer. The higher the content of seaweed, the greater the viscoelastic properties, rigidity and resistance the bioplastics display. Thus, the sample with RO/Gly ratio of 70/30 displayed the higher values of those properties when compared to 50/50 or 60/40 ratios.

Regarding the DMA response, typical thermoplastic behavior was shown, as samples softened when heated until a plateau was achieved at high temperatures. The elastic behavior always prevailed over the viscous one since the storage modulus of every system remained above the loss modulus. The WUC was lowered as the biomass content increased, due to the lower presence of the hydrophilic plasticizer employed. However, some biomass was lost during the immersion of the bioplastic samples, which can be explained by the relatively low mechanical properties of all systems when compared to those obtained from other sources. In relation to this, the greater the mechanical stability of the material, the lower WUC, because the polymers are expected to be more cross-linked and there are greater impediments for water to be in the pores of the structure and bond. This also resulted in a lower SML. The impact of the mold temperature was more noticeable in the tensile test, where the system molded at 120 °C stood out slightly. A slight lowering of water absorption capacity was detected when increasing the mold temperature, although its effect was not very significant when compared to the effect of the formulation (RO/GLY ratio).

The results obtained in the present work are very promising, although further research should be performed to improve the mechanical properties of *Rugulopteryx okamurae* seaweed-based bioplastics. The formulation of composites with other biodegradable polymers and the studies of effects on different processing conditions may be a way to achieve that purpose and to find new utilities.

Supplementary Materials: The following are available online at https://www.mdpi.com/article/10.3390/polym14020355/s1, Figure S1: Visual appearance of RO/GLY blends at different ratios (from left to right: 50/50, 60/40, 70/30). Figure S2: Visual appearance of RO/GLY bioplastics at different ratios (from left to right: 50/50, 60/40, 70/30) molded at 120 °C.

Author Contributions: Conceptualization, M.F. and C.B.; methodology, M.F. and C.B.; software, I.S.; validation, M.F. and C.B.; formal analysis, M.F. and C.B.; investigation, I.S.; resources, C.B.; data curation, I.S.; writing—original draft preparation, I.S.; writing—review and editing, M.F. and C.B.; visualization, C.B.; supervision, M.F. and C.B.; project administration, A.G., C.B.; funding acquisition, A.G. and C.B. All authors have read and agreed to the published version of the manuscript.

Funding: This study was financially supported by the "FEDER/Ministerio de Ciencia e Innovación-Agencia Estatal de Investigación", through the project RTI2018-097100-B-C21, and through the PhD Grant: PRE2019-089815.

Institutional Review Board Statement: Not applicable.

Informed Consent Statement: Not applicable.

Data Availability Statement: The data presented in this study are available on request from the corresponding author.

Acknowledgments: The authors acknowledge the University of Seville for the VPPI-US grant (Ref.-II.5) to Manuel Felix, and also the Andalusian goberment and the European commission for funding the contract of Ismael Santana through the Youth Employment Initiative (EJ5-13-1). The authors also thank CITIUS for granting access to and their assistance with the Microscopy service. Finally, the authors also kindly thank Ismael Hachero (from IFAPA El Toruño) for providing the raw material used in this study. Moreover, the authors would also like to thank the Spanish "Ministerio de Ciencia e Innovación (MCI)/Agencia Estatal de Investigación (AEI)/Fondo Europeo de Desarrollo Regional (FEDER, UE)" for the financial support provided through the funding of the RTI2018-097100-B-C21 (MCI/AEI/FEDER, UE) project.

Conflicts of Interest: The authors declare no conflict of interest. The funders had no role in the design of the study; in the collection, analyses, or interpretation of data; in the writing of the manuscript, or in the decision to publish the results.

References

1. Hwang, I.-K.; Lee, W.J.; Kim, H.-S.; De Clerck, O. Taxonomic reappraisal of Dilophus okamurae (Dictyotales, Phaeophyta) from the western Pacific Ocean. *Phycologia* **2009**, *48*, 1–12. [CrossRef]
2. Dawson, E.Y. Notes on Some Pacific Mexican Dictyotaceae. *Bull. Torrey Bot. Club* **1950**, *77*, 83. [CrossRef]
3. Agatsuma, Y.; Kuwahara, Y.; Taniguchi, K. Life cycle of Dilophus okamurae (Phaeophyceae) and its associated invertebrate fauna in Onagawa Bay, Japan. *Fish. Sci.* **2005**, *71*, 1107–1114. [CrossRef]
4. García-Gómez, J.C.; Florido, M.; Olaya-Ponzone, L.; Rey Díaz de Rada, J.; Donázar-Aramendía, I.; Chacón, M.; Quintero, J.J.; Magariño, S.; Megina, C. Monitoring Extreme Impacts of Rugulopteryx okamurae (Dictyotales, Ochrophyta) in El Estrecho Natural Park (Biosphere Reserve). Showing Radical Changes in the Underwater Seascape. *Front. Ecol. Evol.* **2021**, *9*, 1–18. [CrossRef]
5. García-Gómez, J.C.; Sempere-Valverde, J.; González, A.R.; Martínez-Chacón, M.; Olaya-Ponzone, L.; Sánchez-Moyano, E.; Ostalé-Valriberas, E.; Megina, C. From exotic to invasive in record time: The extreme impact of Rugulopteryx okamurae (Dictyotales, Ochrophyta) in the strait of Gibraltar. *Sci. Total Environ.* **2020**, *704*, 135408. [CrossRef]
6. El Ayuntamiento de Tarifa ya ha Retirado Unas 400 Toneladas de Lagas Invasoras en el Último Mes. Available online: https://www.elestrechodigital.com/2020/08/19/el-ayuntamiento-de-tarifa-ya-ha-retirado-unas-400-toneladas-de-algas-invasoras-en-el-ultimo-mes/ (accessed on 30 September 2021).
7. Shrivastava, A. *Plastic Properties and Testing*; Elsevier: Oxford, UK, 2018; ISBN 9780323395007.
8. Kalia, V.C.; Raizada, N.; Sonakya, V.; Kalia, V.C. Neena Raizada and V Sonakya. *J. Sci. Ind. Res.* **2000**, *59*, 433–445.
9. Gahlawat, S.K.; Salar, R.K.; Siwach, P.; Duhan, J.S.; Kumar, S.; Kaur, P. Plant biotechnology: Recent advancements and developments. In *Plant Biotechnology: Recent Advancements and Developments*; Springer: Singapore, 2017; pp. 1–390. [CrossRef]
10. Schmaltz, E.; Melvin, E.C.; Diana, Z.; Gunady, E.F.; Rittschof, D.; Somarelli, J.A.; Virdin, J.; Dunphy-Daly, M.M. Plastic pollution solutions: Emerging technologies to prevent and collect marine plastic pollution. *Environ. Int.* **2020**, *144*, 106067. [CrossRef] [PubMed]
11. Cox, K.D.; Covernton, G.A.; Davies, H.L.; Dower, J.F.; Juanes, F.; Dudas, S.E. Human Consumption of Microplastics. *Environ. Sci. Technol.* **2019**, *53*, 7068–7074. [CrossRef]
12. Ashter, S.A. 5—Types of Biodegradable Polymers. In *Introduction to Bioplastics Engineering*; Elsevier: Oxford, UK, 2016; pp. 81–151. ISBN 9780323393966.
13. European Bioplastics Bioplastics Glossary. Available online: https://www.european-bioplastics.org/glossary/ (accessed on 30 September 2021).
14. Thiruchelvi, R.; Das, A.; Sikdar, E. Materials Today: Proceedings Bioplastics as better alternative to petro plastic. *Mater. Today Proc.* **2021**, *37*, 1634–1639. [CrossRef]
15. Tauer, K. Fundamental principles of polymeric materials, second edition. Stephen L. Rosen. A volume in the SPE monograph series. A wiley-interscience publication. John Wiley & Sons, Inc. 1993, 420 pages. *Acta Polym.* **1993**, *44*, 210. [CrossRef]
16. Wilson, A. *Plasticisers: Principles and Practice*; Institute of Materials: London, UK, 1995; ISBN 9780901716767.
17. Wypych, G. *Handbook of Plasticizers*, 3rd ed.; Elsevier Inc.: Amsterdam, The Netherlands, 2017; ISBN 9781895198973.
18. Vieira, M.G.A.; da Silva, M.A.; dos Santos, L.O.; Beppu, M.M.; Adeodato Vieira, M.G.; da Silva, M.A.; dos Santos, L.O.; Beppu, M.M.; Vieira, M.G.A.; da Silva, M.A.; et al. Natural-based plasticizers and biopolymer films: A review. *Eur. Polym. J.* **2011**, *47*, 254–263. [CrossRef]
19. Sikora, J.; Majewski, Ł.; Puszka, A. Modern Biodegradable Plastics-Processing and Properties: Part I. *Materials* **2020**, *13*, 1986. [CrossRef]
20. Ashter, S.A. 7—Processing Biodegradable Polymers. In *Introduction to Bioplastics Engineering*; Elsevier: Oxford, UK, 2016; pp. 211–225. ISBN 9780323393966.
21. Ashter, S.A. 8—Extrusion of Biopolymers. In *Introduction to Bioplastics Engineering*; Elsevier: Oxford, UK, 2016; pp. 211–225. ISBN 9780323393966.
22. Jerez, A.; Partal, P.; Martínez, I.; Gallegos, C.; Guerrero, A. Rheology and processing of gluten based bioplastics. *Biochem. Eng. J.* **2005**, *26*, 131–138. [CrossRef]
23. Fernández-Espada, L.; Bengoechea, C.; Cordobés, F.; Guerrero, A. Thermomechanical properties and water uptake capacity of soy protein-based bioplastics processed by injection molding. *J. Appl. Polym. Sci.* **2016**, *133*, 1–10. [CrossRef]
24. Álvarez-Castillo, E.; Bengoechea, C.; Rodríguez, N.; Guerrero, A. Development of green superabsorbent materials from a by-product of the meat industry. *J. Clean. Prod.* **2019**, *223*, 651–661. [CrossRef]
25. Beacham, W. *Algae-Based Bioplastics a Fast-Growing Market*; ICIS: Maastricht, The Netherlands, 2010.
26. Mæhre, H.K.; Malde, M.K.; Eilertsen, K.E.; Elvevoll, E.O. Characterization of protein, lipid and mineral contents in common Norwegian seaweeds and evaluation of their potential as food and feed. *J. Sci. Food Agric.* **2014**, *94*, 3281–3290. [CrossRef] [PubMed]

27. *ISO-527-2:2012*; Plastics—Determination of Tensile Properties—Part 2: Test Conditions for Moulding and Extrusion Plastics. ISO Central Secretariat: Geneva, Switzerland, 2012.
28. Cuadri, A.A.A.; Romero, A.; Bengoechea, C.; Guerrero, A. The Effect of Carboxyl Group Content on Water Uptake Capacity and Tensile Properties of Functionalized Soy Protein-Based Superabsorbent Plastics. *J. Polym. Environ.* **2018**, *26*, 2934–2944. [CrossRef]
29. Schneider, C.A.; Rasband, W.S.; Eliceiri, K.W. NIH Image to ImageJ: 25 years of image analysis. *Nat. Methods* **2012**, *9*, 671–675. [CrossRef] [PubMed]
30. Bier, J.M.; Verbeek, C.J.R.; Lay, M.C. Thermal transitions and structural relaxations in protein-based thermoplastics. *Macromol. Mater. Eng.* **2014**, *299*, 524–539. [CrossRef]
31. Felix, M.; Perez-Puyana, V.; Romero, A.; Guerrero, A. Development of protein-based bioplastics modified with different additives. *J. Appl. Polym. Sci.* **2017**, *134*, 45430. [CrossRef]
32. Ferry, J.D. *Viscoelastic Properties of Polymers*; Wiley: Hoboken, NJ, USA, 1980; ISBN 9780471048947.
33. López Rocha, C.J.; Álvarez-Castillo, E.; Estrada Yáñez, M.R.; Bengoechea, C.; Guerrero, A.; Orta Ledesma, M.T. Development of bioplastics from a microalgae consortium from wastewater. *J. Environ. Manag.* **2020**, *263*, 110353. [CrossRef]
34. Zeller, M.A.; Hunt, R.; Jones, A.; Sharma, S. Bioplastics and their thermoplastic blends from Spirulina and Chlorella microalgae. *J. Appl. Polym. Sci.* **2013**, *130*, 3263–3275. [CrossRef]
35. Sanyang, M.L.; Sapuan, S.M.; Jawaid, M.; Ishak, M.R.; Sahari, J. Effect of plasticizer type and concentration on tensile, thermal and barrier properties of biodegradable films based on sugar palm (Arenga pinnata) starch. *Polymers* **2015**, *7*, 1106–1124. [CrossRef]
36. Zhang, Y.; Rempel, C.; Liu, Q. Thermoplastic Starch Processing and Characteristics-A Review. *Crit. Rev. Food Sci. Nutr.* **2014**, *54*, 1353–1370. [CrossRef] [PubMed]
37. Félix, M.; Martín-Alfonso, J.E.; Romero, A.; Guerrero, A. Development of albumen/soy biobased plastic materials processed by injection molding. *J. Food Eng.* **2014**, *125*, 7–16. [CrossRef]
38. Paetau, I.; Chen, C.Z.; Jane, J.L. Biodegradable plastic made from soybean products. 1. Effect of preparation and processing on mechanical-properties and water-absorption. *Ind. Eng. Chem. Res.* **1994**, *33*, 1821–1827. [CrossRef]

Article

Cost Function Analysis Applied to Different Kinetic Release Models of *Arrabidaea chica* Verlot Extract from Chitosan/Alginate Membranes

Luis Concha [1], Ana Luiza Resende Pires [2], Angela Maria Moraes [2], Elizabeth Mas-Hernández [3,4], Stefan Berres [5,*] and Jacobo Hernandez-Montelongo [1,4,*]

1. Department of Physical and Mathematical Sciences, Catholic University of Temuco, Temuco 4813302, Chile; luisconcha.c@gmail.com
2. School of Chemical Engineering, University of Campinas, Campinas 13083-852, Brazil; analurespi@gmail.com (A.L.R.P.); ammoraes@unicamp.br (A.M.M.)
3. Department of Mathematical Engineering, University of La Frontera, Temuco 4811230, Chile; elimher@gmail.com
4. Bioproducts and Advanced Materials Research Nucleus (BioMA), Catholic University of Temuco, Temuco 4813302, Chile
5. Department of Information Systems, University of Bio-Bio, Concepcion 4051381, Chile
* Correspondence: stefan.berres@gmail.com (S.B.); jacobo.hernandez@uct.cl (J.H-M.)

Abstract: This work focuses on the mathematical analysis of the controlled release of a standardized extract of *A. chica* from chitosan/alginate (C/A) membranes, which can be used for the treatment of skin lesions. Four different types of C/A membranes were tested: a dense membrane (CA), a dense and flexible membrane (CAS), a porous membrane (CAP) and a porous and flexible membrane (CAPS). The *Arrabidae chica* extract release profiles were obtained experimentally in vitro using PBS at 37 °C and pH 7. Experimental data of release kinetics were analyzed using five classical models from the literature: Zero Order, First Order, Higuchi, Korsmeyer–Peppas and Weibull functions. Results for the Korsmeyer–Peppas model showed that the release of *A. chica* extract from four membrane formulations was by a diffusion through a partially swollen matrix and through a water filled network mesh; however, the Weibull model suggested that non-porous membranes (CA and CAS) had fractal geometry and that porous membranes (CAP and CAPS) have highly disorganized structures. Nevertheless, by applying an explicit optimization method that employs a cost function to determine the model parameters that best fit to experimental data, the results indicated that the Weibull model showed the best simulation for the release profiles from the four membranes: CA, CAS and CAP presented Fickian diffusion through a polymeric matrix of fractal geometry, and only the CAPS membrane showed a highly disordered matrix. The use of this cost function optimization had the significant advantage of higher fitting sensitivity.

Keywords: cost function; controlled release; *Arrabidae chica* Verlot; chitosan/alginate membranes

1. Introduction

Arrabidaea chica Verlot is a type of shrub found in tropical America, from the south of Mexico to Brazil, being very common in the Amazon rainforest [1]. As the extract of *A. chica* is an important source of tannins, flavonoids and anthocyanins, it presents different medicinal properties, such as antioxidant, antiseptic, anti-inflammatory and antifungal activities [2]. One of its main components is the anthocyanin 'carajurina', which can be used as a marker for the detection and quantification of the extract of *A. chica*.

One of the strategies to deliver drugs and other medicinal substances is through controlled release systems, which maintain the drug concentration in the blood or in target tissues as long as possible at a desired value, being able to control the drug release rate and duration [3]. Different biopolymers have been used to deliver drugs and other compounds for the treatment of skin lesions in a local, targeted and controlled manner [4]. One of the most versatile biopolymers is chitosan, which is obtained mainly from shells of crustaceans, such as shrimp [5]. In this regard, Servat-Medina et al. (2015) [6] synthesized chitosan nanoparticles loaded with different concentrations of *A. chica* extract (10 to 25% relative to chitosan mass) for the treatment of skin ulcers. Recently, in 2020, our group reported synthesizing different types of chitosan/alginate (C/A) membranes to be used as controlled release systems of *A. chica* extract, as an alternative for the treatment of skin lesions [7]. C/A membranes are attractive because they are insoluble in water, stable to pH variations and capable of incorporating different bioactive agents into their matrix.

Mathematical modeling of drug delivery and predictability of drug release has been a field of academic and industrial importance for several decades [8]. In 1961, Higuchi published his famous equation allowing for a surprisingly simple description of the drug release mechanism from an ointment base. Numerous models have been proposed since then, including empirical/semi-empirical as well as mechanistic realistic ones [9]. Consequently, many different mathematical approaches have been proposed to assess the similarity between drug dissolution and mass transfer profiles [10].

A versatile mathematical tool is the cost function analysis, which is an optimization process that consists of measuring the difference between real experimental data with the prediction of a model. The objective of this analysis is to minimize this difference, that is, to find the parameters that allow the model to fit the data as closely as possible [11]. The cost function can be performed for any type of mathematical model and compared with various experimental data; regarding numerical optimization methods, local gradient methods such as conjugate gradients [12,13] or global parameter population methods such as genetic algorithms [11,14] or deep learning strategies [15] can be applied.

The method in this work focuses on the quantitative assessment of the controlled release of a standardized extract of *A. chica* from C/A membranes. First, in vitro experimental *A. chica* kinetic release profiles were obtained from four different formulations of C/A membranes: a dense membrane (CA), a dense and flexible membrane (CAS), a porous membrane (CAP), and a porous and flexible membrane (CAPS). Later, the mechanisms of the extract release kinetics were determined comparing the r^2 coefficient obtained using five classic models from the literature [16]: Zero Order, First Order, Higuchi, Korsmeyer–Peppas and Weibull model functions. Finally, as the main novelty of this work, the release profiles were analyzed using an optimization method implemented by our group that employs the cost function approach to determine the model parameters that best fit to the experimental data.

2. Materials and Methods

2.1. Materials

Chitosan (C, from shrimp shells 96% deacetylated and $M_W = 1.26 \times 10^6$ g/mol), medium viscosity sodium alginate (A, from *Macrocystis pyrifera* $M_W = 9.11 \times 10^4$ g/mol), Kolliphor P188 (a pore-forming surfactant) and phosphate buffer solution (PBS) 0.01 M (0.138 M NaCl, 0.0027 M KCl, pH = 7.4) were acquired from Sigma-Aldrich (Sao Paulo, Brazil). Silpuran 2130 A/B (a silicone polymer) was obtained from Wacker Chemie AG (Munich, Germany), and glacial acetic acid, calcium chloride dihydrate and sodium salt from Merck KGaA (Sao Paulo, Brazil). The standardized *Arrabidaea chica* Verlot extract was supplied by the Division of Chemistry of Natural Products of the Center for Chemical, Biological and Agricultural Research (CPQBA) at the University of Campinas (Campinas, Brazil). The used water was deionized in a Milli-Q System from Millipore.

2.2. Membranes synthesis

Four formulations of C/A membranes containing *A. chica* extract (10% in weight) were obtained according to the protocol previously described by Pires et al. (2020) [7]. The main synthesis differences are summarized as follows: (A) CA membrane: prepared with chitosan 1% (*m/v*) in acetic acid 2% and alginate 0.5% (C:A = 1:2 *v/v*); (B) CAS membrane:

synthesized with the same CA formulation, but including 10% of Silpuran 2130 A/B; (C) CAP membrane: obtained by formulating CA membrane plus 10% in weight of the Kolliphor P188; (D) CAPS membrane: It was synthesized by formulating CAS plus 10% in weight of Kolliphor P188. In all cases, the membranes were cross-linked with calcium ions.

2.3. Morphology of the Membranes

Membrane samples containing the *A. chica* extract were microscopically observed and photographed using a Nikon digital camera (COOLPIX model S3300). The cross section morphology of the membranes was analyzed using a scanning electron microscope (model Leo 440i, Leica). Samples of 2 cm × 1 cm were fixed on a suitable support and metalized (mini-Sputter coater, SC 7620) by depositing a thin layer of gold (92 Å) on their surfaces.

2.4. Release Experiments

To obtain the release profiles, membrane samples (2 cm × 2 cm) containing the *A. chica* extract were previously weighed and immersed in a 10 mL of PBS solution containing 20% ethanol at pH 7.4, 37 °C and 100 rpm. At predetermined time intervals up to 48 h, 1 mL of the solution was withdrawn for absorbance analysis using a spectrophotometer (Thermo Scientific Evolution–220, Thermo Fisher Scientific, Waltham, MA, USA) at 470 nm and then returned to the vial. The analytical curve was prepared using *A. chica* extract also dissolved in a PBS solution containing 20% ethanol. All experiments were performed in triplicate, and mean values were used. Sink conditions were maintained during the drug release experiments and at predetermined time intervals, the supernatant solution was completely renewed.

2.5. Mathematical Models

The mechanism of drug release was determined by fitting the mathematical models to the experimental data using OriginPro 8.5 software. Five models were studied, the main characteristics of which are outlined below [17–20].

In the zero-order model, the drug is released at a constant rate independent of concentration, and dissolution from dosage forms that do not disaggregate and release the drug slowly. This model is represented by the equation

$$Q(t) = Q_0 + k_0 t, \qquad (1)$$

where $Q(t)$ is the amount of drug released at time t, $Q_0 = Q(t=0)$ is the initial amount of the extract in the solution and k_0 is the zero-order proportional constant.

In the first-order model, the release is a concentration-dependent process, and the equation that gives the release behavior is

$$Q(t) = Q_0 \left(1 - \exp(-k_1 t)\right), \qquad (2)$$

where $Q(t)$ and Q_0 are again the amount of drug released at time t and the initial amount of the extract in the solution, respectively, and k_1 is the first-order release kinetic constant.

The Higuchi model implies more assumptions, such as that the initial drug concentration in the matrix is higher than drug solubility; drug diffusion takes place only in one dimension (edge effect must be negligible); drug particles are smaller than system thickness; matrix swelling and dissolution are negligible; drug diffusivity is constant; and perfect sink conditions are always attained in the release environment. The general release equation is given by

$$Q(t) = k_H t^{1/2}, \qquad (3)$$

where $Q(t)$ is the amount of drug released in time t per unit area, and k_H is the Higuchi dissolution constant. It represents a Fickian diffusion of drugs without the matrix dissolution taken into account.

The Korsmeyer–Peppas model is a simple relationship to describe drug release from a polymeric system. This semi-empirical model analyzes both Fickian and non-Fickian release of drug from swelling as well as non-swelling materials; however, it is applied just up to 60% of the drug amount released. The Korsmeyer–Peppas release equation is given by

$$\frac{Q(t)}{Q_\infty} = k_{KP} t^n, \qquad (4)$$

where the ratio $Q(t)/Q_\infty$ is the fraction of drug released at time t, k_{KP} is the Korsmeyer–Peppas kinetic constant, which characterizes the drug–matrix system and n is the exponent that indicates the drug release mechanism.

The Weibull model is an alternative description for the dissolution and release processes, and can be applied for most types of dissolution curves. The Weibull equation expresses the cumulative fraction of the drug, $Q(t)$, in a solution at time t, by the following expression:

$$Q(t) = 1 - \exp\left(-(t - T_i)^\beta / \alpha\right), \qquad (5)$$

where α is related to the specific surface of the dosage matrix form, β is mainly related to the mass transport characteristics of the device and T_i represents the delay time before starting the dissolution or release process, which in most cases is 0.

3. Cost function Analysis

3.1. Definition of Cost Function

We introduce the notation $Q(\mathbf{e}; t)$ for a general model that depends on the parameter vector $\mathbf{e}$. For example, for the Korsmeyer–Peppas model, the parameter set is specified as

$$\mathbf{e} = (e_1, e_2) = (k_{KP}, n).$$

A standard technique that enables the interpretation of experimental data with a quantitative model is the formulation as an inverse problem, where the direct problem is formulated as a mathematical model, such that the distance of the model $Q(\mathbf{e}; t_i)$ to the data

$$(t_i, \hat{Q}_i), \qquad i = 1, \ldots, N$$

is described by a cost function [11,21]

$$F(\mathbf{e}) = \sum_{i=1}^{n} \mu_i \left| Q(\mathbf{e}; t_i) - \hat{Q}_i \right|^p, \qquad (6)$$

where $p \in \{1, 2, \ldots\}$ accounts for different metrics of the distance between model and data, and $\mu_i = \mu(t_i)$ contains weights of the data points. For $p = 2$ (in comparison to $p = 1$), we have an underestimation respective overestimation of the measurement errors (under the assumption that the model is correct) for small respective high distances between data and model. Though, $p = 2$ applies the concept of least squares, which is more common, in spite of the bias.

Regarding the weights, there are various choices, as the following prototypes, among which there are multiple possible variants:

1. Equal weights $\mu_i \equiv 1$ for all i.
2. Switch off at a threshold time t^*,

$$\mu_i = \begin{cases} 1 & \text{for } t_i < t^* \\ 0 & \text{for } t_i > t^*, \end{cases}$$

3. Adaptive weights, that gives higher emphasis to smaller times, such as, for example,

$$\mu(t_i) = \frac{1}{c + t_i}, \quad c > 0.$$

The switch off in case (2) at a threshold time t^* applies, i.e., should apply, for model approximations that do not satisfy asymptotic upper limits, such as the models of Higuchi or Korsmeyer–Peppas, which are designed to be valid for lower times only. As their unlimited asymptotic behavior is qualitatively wrong, above a threshold value the models are quantitatively wrong. In favor of the Higuchi and Korsmeyer–Peppas models, the weighting choice was case (2) with data switch off at $t^* = 8$ h. For the Weibull model we choose case (1).

The inverse problem is solved by minimizing the cost function,

$$\min_{\mathbf{e} \in \mathbb{R}^n} F(\mathbf{e}), \tag{7}$$

that gives the optimal set of parameters $\mathbf{e}$. The mathematical model can have the form of differential equations that in turn include parametric functions, or the mathematical model is expressed directly in terms of parametric functions. For the minimization of the cost function, an issue might be (1) high correlations between the parameters, or (2) its non-convexity. In the case of high parameter correlation that touch the scale of computational errors induced by the machine error, even the optimal numerical method cannot compensate a wrong model choice; it is an issue of model choice. In the case of non-convexity of the cost function, the optimization method needs to be chosen carefully, or the experimentation with different optimization methods is subject of research, e.g., global optimization methods, where populations of parameter sets are optimized, instead of local methods with only one single parameter set.

3.2. Cost Function Minimization: Optimality Conditions

The goal is to minimize the cost function, that is, to find the parameters of each model that fits the experimental data as closely as possible.

As a criterion of optimality [22], there is a necessary condition and a sufficient condition. For two-variable models, the necessary condition for optimality is that the gradient is equal to zero in each of its components,

$$\nabla f(e_1, e_2) = \begin{pmatrix} \frac{\partial}{\partial e_1} f(e_1, e_2) \\ \frac{\partial}{\partial e_2} f(e_1, e_2) \end{pmatrix} = \begin{pmatrix} 0 \\ 0 \end{pmatrix}. \tag{8}$$

The sufficient condition of optimality is that the Hessian matrix

$$H(e_1, e_2) = \begin{pmatrix} \partial_{e_1 e_1} f(e_1, e_1) & \partial_{e_1 e_2} f(e_1, e_2) \\ \partial_{e_2 e_1} f(e_1, e_2) & \partial_{e_2 e_2} f(e_2, e_2) \end{pmatrix} \tag{9}$$

is positive definite for a minimum, i.e., according to the definition

$$\begin{pmatrix} q_1 & q_2 \end{pmatrix} H(e_1, e_2) \begin{pmatrix} q_1 \\ q_2 \end{pmatrix} > 0, \quad \forall q_1, q_2 \in \mathbb{R}^2. \tag{10}$$

This definition can be written as

$$q_1^2 f_{e_1 e_1} + 2 q_1 q_2 f_{e_1 e_2} + q_2^2 f_{e_2 e_2} > 0, \quad \forall q_1, q_2. \tag{11}$$

A general criterion of a square matrix of any size to be positive definite is that all eigenvalues are positive.

3.3. Equivalence of Optimization Methods

The comparison of model predictions with data,

$$f_i(\mathbf{e}) = Q(\mathbf{e}, t_i) - \hat{Q}_i, \quad i = 1, \ldots, N, \tag{12}$$

defines a residual function $f : \mathbb{R}^n \to \mathbb{R}^N$, where we want to find its zeros, which can only be calculated approximately if the system is overdetermined. An approximation criterion is the sum of squares

$$F(\mathbf{e}) = \sum_{i=1}^{N} f_i^2(\mathbf{e}), \tag{13}$$

which is equivalent to the cost function (6), but for the specific choice of $p = 2$ and with equal weights, i.e. case (1).

There are two different paths to obtain optimal parameters:

- Approximate the zeros of the overdetermined system (12).
- Calculate the zeros of the gradient of the cost function (13), defined as the sum of squares of the residual.

An iterative procedure that solves an overdetermined system of nonlinear equations, such as (6) is the Gauss–Newton algorithm. This means that an initial estimate of the vector parameter must be provided.

4. Methodology: Implementation

The five mathematical models of Section 2.5 were tested in parallel. The generated process was automated by subroutines (see Supplementary Materials).

Prior to the calculations, the derivatives and the gradients were obtained for one- and two-variable models, respectively, in order to verify the fulfillment of the necessary condition. The sufficient condition was satisfied, as for one-variable models the second derivative, and for two-variable models the elements in the Hessian matrix were different from zero.

In the implementation, the corresponding equations and the experimental data were consistent enough to be called by the common cost function subroutine.

The algorithm for the cost function is given in the Algorithm S1 (see Supplementary Materials for Algorithms). The following criterion has to be satisfied:

$$c = \sum_{i=1}^{N} \sum_{j=1}^{J} |u_i(t_j) - \hat{u}_{ij}|^2, \tag{14}$$

where c is the cost function, N is the number of experiments for each membrane in a given time t (in our case $N = 3$) and J is the number of data in the experiment. A syntax was chosen from the model subroutine (Algorithm S2):

$$\mathbf{u} = \mathrm{uModel}(\hat{\mathbf{t}}),$$

where the vector of variables consisted of the observation times, to calculate the solution of the model at given times. The description of steps to implement this methodology is presented in Algorithm S4.

It is important to consider that a better fit is expected with the cost function used in models with several parameters. However, that does not necessarily mean that the type of model as such is better or that it effectively better explains the phenomenon. Moreover, a higher number of parameters causes optimization algorithms to have problems when approaching the 'optimal' data, given the higher correlation within the same parameters.

5. Results and Discussion

The obtained membranes presented the geometrical form of a thin slab. Figure 1 shows photographs and SEM images of the membrane samples loaded with *A. chica* extract. The observed intense red color in the photographs (Figure 1A–D) was due to the insoluble anthocyanin pigments carajurin and carajurone included in the *A. chica* extract [23]. The membranes produced without Kolliphor P188 surfactant turned out to be dense and compact (CA and CAS). However, for the formulations synthesized using the surfactant, the membranes were porous and thicker (CAP and CAPS). When the silicone Silpuran 2130 was included in the formulation, samples were soft and flexible to the touch (CAS and CAPS). In summary, the general characteristics of the obtained membranes were as follows: CA was dense, thin and rigid; CAS was dense, thin but flexible due to the silicone; CAP was rigid but thick and porous due to air bubbles in its matrix generated by the surfactant; and CAPS was thick and porous due to the surfactant, but flexible because of the silicone.

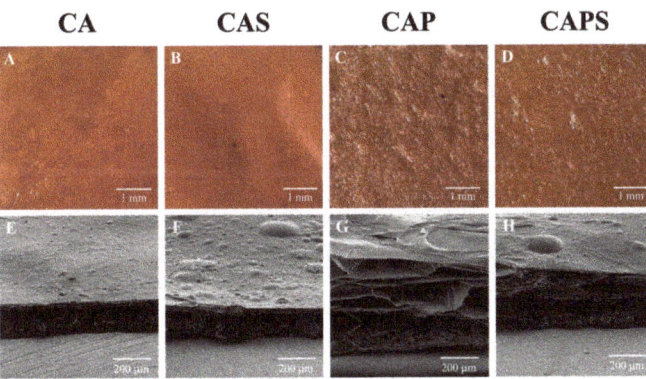

Figure 1. Photographs and SEM images of membrane samples loaded with *A. chica* extract: CA (**A**,**E**), CAS (**B**,**F**), CAP (**C**,**G**) and CAPS (**D**,**H**).

A. chica extract release profiles are shown in Figure 2. Although all samples were loaded with the same percentage of extract (10% in weight), CA and CAS membranes released higher amounts of extract per mass of polymer than CAP and CAPS. The burst release observed in the first minutes of the process can be attributed to the fraction of the drug which is adsorbed or weakly bound to the surface area of the polymer rather than to the drug incorporated into the polymer matrix [24]. Moreover, the lower amounts in porous formulations can be explained by the higher dispersion of the mixture due to the presence of the surfactant with a consequent increase in the interphases of the extract with the polymers. For all cases, the maximum amount of extract released was reached at 24 h. Re-absorption of the extract by the membranes was not observed because the maximum released values of extract were maintained constant for up to 48 h.

To gain a deeper insight into the mechanisms that govern the release of *A. chica* extract from the membranes, five mathematical models were fitted to the experimental data: Zero-order, First order, Higuchi, Korsmeyer–Peppas and Weibull. Experimental data were normalized and only evaluated up to 8 h because these models are semi-empiric and better valid for the first stages of release; as mentioned previously in Section 3.1, where the weights on the cost function switch off (7) at a threshold value of time t^*.

The switch off is an artificial fix for a model that is valid for short time intervals and loses pertinence at bigger times. In order to avoid this artificial switch off, one can select admissible functions for a semi-empirical model that satisfy physically reasonable criteria. For the drug release model, these characteristics should address at least the following components:

1. Zero release at zero time.
2. Increasing cumulative release for advancing time.
3. Existence of an upper bound for the total release at sufficiently long period.

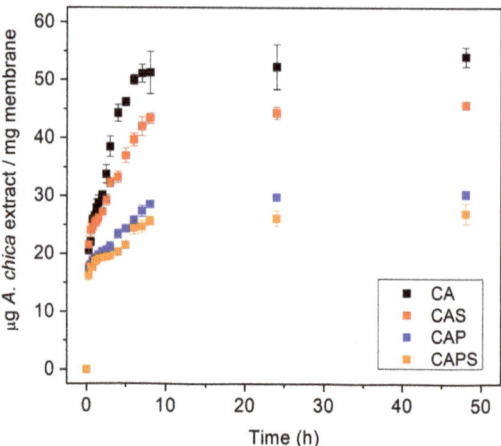

Figure 2. Experimental *A. chica* extract release profiles from the C/A membranes. Results represent mean ± SD of three measurements.

These characteristics can be expressed as

$$Q(0) = 0, \quad \frac{\partial Q(t)}{\partial t} > 0, \quad \lim_{t \to \infty} Q(t) < \infty.$$

With help of these model design indications, we can assess the various models (see Table 1 for a qualitative model comparison).

Table 1. Model evaluation: Satisfaction status of criteria by considered models.

Model	Function	$Q(0) = 0$	$\lim_{t \to \infty} Q(t) < \infty$
Zero-order equation	$Q(t) = Q_0 + k_0 t$	NO	NO
First order model	$Q(t) = \left(1 - \exp(-k_1 t)\right)$	YES	YES
Higuchi model	$Q(t) = k_H t^{1/2}$	YES	NO
Korsmeyer–Peppas model	$\dfrac{Q(t)}{Q_\infty} = k_{KP} t^n$	YES	NO
The Weibull model	$Q(t) = 1 - \exp\left(-(t - T_i)^\beta / \alpha\right)$	YES	YES

Simulations of the release profiles from each membrane using the indicated models are shown in Figure 3. From these simulations, the kinetic parameters were extracted and presented in Table 2. The r^2 coefficient (coefficient of determination) was used to compare the results between models. The value of r^2 is usually between 0 and 1, and generally a higher value means that the model fits the data better; some r^2 equal to or higher than 0.95 is considered as a good fitting by linear regression. Therefore, according to the r^2 values generated by each model (Zero-order, First-order, Higuchi, Korsmeyer–Peppas and Weibull) for each membrane (CA, CAP, CAPS and CAPS), the models that best fit the release profiles were Korsmeyer–Peppas and Weibull. In the case of the Korsmeyer–Peppas model, r^2 values of all samples were higher than 0.95, but in the case of the Weibull model, the fitting for the CA and CAP membranes returned values of almost 0.95 and for the CAS and CAPS membranes, values were higher than 0.95.

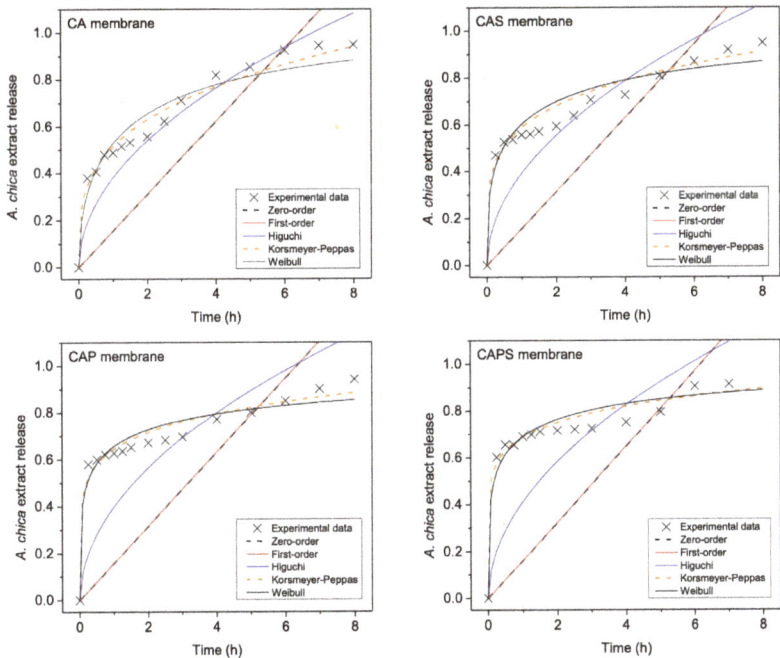

Figure 3. Simulations of the release profiles from each membrane (CA, CAS, CAP and CAPS) using the Zero-order, First-order, Higuchi and Korsmeyer–Peppas and Weibull models.

Table 2. In vitro release kinetics of *A. chica* from C/A membranes.

Sample	Zero-Order		First-Order		Higuchi		Korsmeyer-Peppas			Weibull			
	k_0 (h^{-1})	r^2	k_1 (h^{-1})	r^2	k_H (h$^{-1/2}$)	r^2	k_{KP} (h^{-n})	n	r^2	α (-)	β (-)	T_i (h)	r^2
CA	0.1556	−0.1982	0.0015	−0.1924	0.3827	0.7979	0.5170	0.2872	0.9818	1.7645	0.5014	0.0000	0.9476
CAS	0.1574	−0.9428	0.0015	−0.9351	0.3937	0.5193	0.5897	0.2095	0.9792	1.2335	0.3793	0.0000	0.9507
CAP	0.1580	−1.7429	0.0015	−1.7154	0.4005	0.1652	0.6455	0.1519	0.9720	1.2636	0.2843	0.0000	0.9487
CAPS	0.1602	−2.1602	0.0016	−2.1497	0.4139	−0.0494	0.6889	0.1264	0.9701	1.7013	0.3009	0.0000	0.9531

Korsmeyer–Peppas is a model determined on the basis of experimental data and it is a shortened version of the solution of the diffusion equation by Crank (1975) [25]. In this model, the release mechanisms are dependent on the sample geometry (thin films, cylinders, spheres). In that sense, when the Korsmeyer–Peppas model is applied to thin films, such as the membranes of this study, if the release parameter n is less than 0.5, it means that drug diffusion is occurring through a partially swollen matrix and through a solution-filled network mesh [26]. If $n = 0.5$, the Fickian diffusion is controlling the release and the solvent penetration is the rate-limiting step. On the other hand, if $0.5 < n < 1$, then this is related to a non-Fickian release, which is the release of the extract controlled by both diffusion and erosion mechanisms. If $n = 1$, then the release corresponds to the zero-order model, where the release of the extract is independent of time [9]. In that sense, according to the results from the Korsmeyer–Peppas model, the n values for the four types of membranes were lower than 0.5; this indicates that the release of the extract was controlled by a diffusion occurring through a partially swollen matrix, which could be produced by the polymer chains relaxation. Moreover, the k_{KP} parameter is related to the interaction between the extract and the constituent polymers of the membrane, and higher values mean higher rates of release. According to this, the k_{KP} values obtained for the samples presented the following tendency:

$$CAPS > CAP > CAS > CA.$$

This means that the porous membranes (CAPS and CAP) generated a faster release probably due to the pores of the polymeric matrix, which facilitated the absorption of water from the release medium (PBS). On the contrary, for membranes without pores (CAS and CA), which are more compact and dense, the absorption of water would be limited, and as a consequence, the release of the extract, too. As CAS and CA samples obtained higher amounts of extract release than CAPS and CAP (Figure 2), this means that faster release of CAPS and CAP was mainly controlled by the porous membrane cross-linked structure rather than the concentration differential. On the other hand, although the Weibull model has been criticized for the lack of a kinetic basis for its use and for the non-physical nature of its parameters, according to different works, the Weibull function demonstrated that the exponent β, for polymeric matrices, is an indicator of the mechanism of transport of the drug through the matrix related to the exponent n of the power law model [17,27–29]: a value of β less than or equal to 0.75 was associated with Fick diffusion in either fractal or Euclidean space, while a combined mechanism (Fick diffusion and swelling controlled transport) was associated with β values in the range $0.75 < \beta < 1$. For $\beta > 1$, drug release involves complex mechanisms, which imply that the release rate does not change monotonically. In fact, the release rate initially increases nonlinearly up to an inflection point and then decreases asymptotically.

According to the above, the results obtained for the Weibull β parameter (Table 2) indicated that the mechanism of release was Fickian diffusion (β less than or equal to 0.75). However, it is also important to identify the polymeric matrix geometry in this β range [27]: for $\beta < 0.35$, diffusion occurs in highly disordered spaces, differently than the percolation cluster; for $0.35 < \beta < 0.69$, diffusion occurs in a fractal substrate; for $0.69 < \beta < 0.75$, diffusion takes place in a normal Euclidean space. These results suggest that the non-porous membranes (CA and CAS) present a fractal structure, and the porous formulations (CAP and CAPS) show highly disordered structures. Fractal organization of CA and CAS samples could be originated during the association of alginate carboxylic groups ($-COO^-$) with chitosan amino groups ($-NH3^+$) [7,30]; meanwhile, the polymeric matrices of CAP and CAPS were highly disorganized by the pores formation.

As a summary, the release of *A. chica* extract from all membrane formulations occurred by a Fickian difussion, what was confirmed by both Korsmeyer–Peppas and Weibull models. Moreover, the Weibull model suggested that non-porous membranes (CA and CAS) had fractal geometry and that porous membranes (CAP and CAPS) were highly disorganized structures.

Furthermore, in this work, the cost function is presented as a tool to analyze different mathematical models that simulate experimental data of release profiles of *A. chica* extract from C/A membranes to a greater extent. Accordingly, Figure 4 shows the cost function simulations applied to each model (Zero-order, First-order, Higuchi, Korsmeyer–Peppas and Weibull) in each membrane type (CA, CAS, CAP and CAPS), and Table 3 presents the parameters obtained. The lower the fitted cost function value (F), the more efficient the model. Conversely, when the cost function value is higher, the model is less effective.

According to the results obtained by the cost function, F values in Table 3, the best simulations were performed by

$$\text{Weibull} > \text{Korsmeyer–Peppas} > \text{Higuchi} > \text{First-order} > \text{Zero-order,}$$

which is similar to the previous results. However, it is important to highlight that using the cost function, the Weibull model fits better to the release profiles than the Korsmeyer-Peppas model for all the membranes. Note that in the case of the Weibull model e_1 and e_2 correspond to α and β, respectively, and the e_3 value, which is T_i, was not reported because it is zero for all cases (Table 2). According to the cost function results, CA, CAS and CAP formulations would present a Fickian diffusion through a polymeric matrix of fractal geometry, and only the CAPS membrane would have a highly disordered matrix. This partially different result compared to the case when the cost function is not applied can be due to the fact that F values of the Weibull model are around 10 times lower—i.e., 10 times mores precise—than the Korsmeyer–Peppas model. Without the use of the cost function, the r^2 of both models are very similar. In that sense, the cost function presents a significant advantage, which is a higher fitting sensitivity.

Table 3. Summary of results of the cost function fitting.

Results	CA	CAS	CAP	CAPS
Zero-order				
$e_1(k_0)$	0.153 ± 0.006	0.143 ± 0.005	0.162 ± 0.008	0.159 ± 0.004
F	$89,011 \pm 6396$	$77,665 \pm 5433$	$99,925 \pm 9832$	$94,746 \pm 6789$
First-order				
$e_1(k_1)$	0.340 ± 0.013	0.360 ± 0.003	0.512 ± 0.007	0.561 ± 0.009
F	$15,192 \pm 3755$	$17,626 \pm 311$	$36,954 \pm 2754$	$31,566 \pm 1140$
Higuchi				
$e_1(k_H)$	0.392 ± 0.700	0.390 ± 0.400	0.411 ± 0.100	0.381 ± 0.800
F	8076 ± 302	7995 ± 165	8901 ± 44	7615 ± 319
Korsmeyer-Peppas				
$e_1(k_{KP})$	0.514 ± 0.004	0.549 ± 0.010	0.61 ± 0.008	0.671 ± 0.003
$e_2(n)$	0.321 ± 0.005	0.241 ± 0.006	0.192 ± 0.001	0.152 ± 0.003
F	2388 ± 102	2767 ± 73	3570 ± 97	4550 ± 67
Weibull				
$e_1(\alpha)$	0.133 ± 0.003	0.13 ± 0.002	0.11 ± 0.003	0.09 ± 0.012
$e_2(\beta)$	0.521 ± 0.005	0.51 ± 0.003	0.402 ± 0.005	0.32 ± 0.006
F	292 ± 2	271 ± 6	298 ± 5	316 ± 5

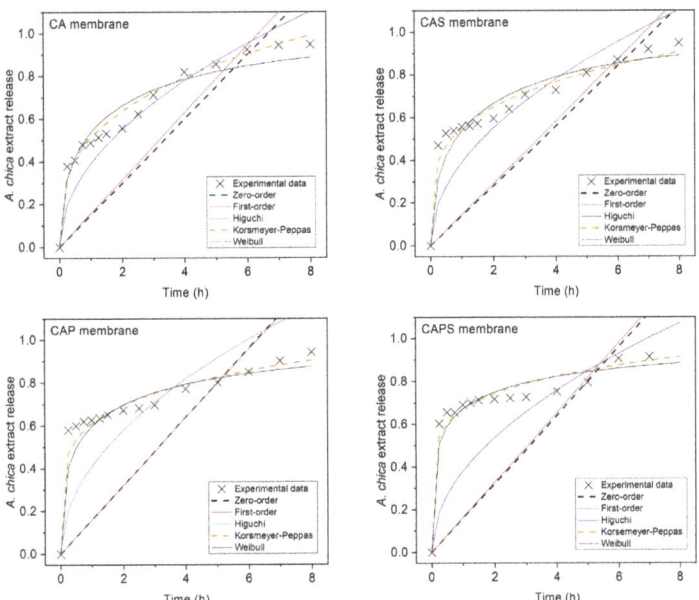

Figure 4. Cost function simulations of the mathematical models (Zero-order, First-order, Higuchi and Korsmeyer–Peppas and Weibull) used in each membrane: CA, CAS, CAP and CAPS.

6. Conclusions

This work consisted on studying the performance of *in vitro* experimental *A. chica* extract release profiles obtained from four different formulations of chitosan/alginate membranes: a dense membrane (CA), a dense and flexible membrane (CAS), a porous membrane (CAP), and a porous and flexible membrane (CAPS). The mechanism of the extract release kinetics was determined comparing classic models from the literature: Zero Order, First Order, Higuchi, Korsmeyer–Peppas and Weibull. Furthermore, in order to improve the mathematical analysis between the models, and as the main novelty of this work, the cost function was presented as a tool to analyze these different mathematical models that simulated the release profiles and were compared to experimental data. Our method explored how some metrics and weights of the cost function impact on the results of the release models that describe experimental information for drug delivery release processes. Our results indicated that the use of the proposed model parameter optimization by the cost function, which better fits the experimental data, had the significant advantage of showing a higher fitting sensitivity.

Supplementary Materials: The following supporting information can be downloaded at: https://www.mdpi.com/article/10.3390/polym14061109/s1. Algorithm S1: Cost function; Algorithm S2: Pseudo-code for the implementation of the model function; Algorithm S3: Residual function; Algorithm S4: Central routine that calls the optimization algorithms; Algorithm S5: Gauss-Newton method.

Author Contributions: Conceptualization, A.M.M., S.B. and J.H.-M.; membranes fabrication, A.L.R.P. and A.M.M.; release profiles, A.L.R.P. and A.M.M.; mathematical analysis, L.C., E.M.-H., S.B. and J.H.-M.; writing—original draft preparation, L.C. and J.H-M. All authors have read and agreed to the published version of the manuscript.

Funding: This research was funded by FONDECYT, Chile grant number 11180395, the National Council for Scientific and Technological Development, Brazil grants 307829 /2018-9 and 152053/2014-0, the Coordination for the Improvement of Higher Education Personnel, Brazil finance code 001.

Institutional Review Board Statement: Not applicable.

Informed Consent Statement: Not applicable.

Data Availability Statement: The data that support the findings of this study are available from the corresponding author upon reasonable request.

Acknowledgments: L.C. thanks to the Master Program in Applied Mathematics from Catholic University of Temuco (UCT, Chile). E.M-H. acknowledges support by ANID Fondecyt Postdoctorate Project 3200839.

Conflicts of Interest: The authors declare no conflict of interest. The funders had no role in the design of the study; in the collection, analyses, or interpretation of data; in the writing of the manuscript; or in the decision to publish the results.

References

1. Barbosa, W.L.R.; Pinto, L.D.N.; Quignard, E.; Vieira, J.M.D.S.; Silva Jr, J.O.C.; Albuquerque, S. *Arrabidaea chica* (HBK) Verlot: Phytochemical approach, antifungal and trypanocidal activities. *Braz. J. Pharmacogn.* **2008**, *18*, 544–548. [CrossRef]
2. de Sá, J.C.; Almeida-Souza, F.; Mondêgo-Oliveira, R.; da Silva Oliveira, I.D.S.; Lamarck, L.; Magalhães, I.D.F.B.; Ataídes-Lima, A.F.; Ferreira, H.S.; Abreu-Silva, A.L. Leishmanicidal, cytotoxicity and wound healing potential of *Arrabidaea chica* Verlot. *BMC Complement. Altern. Med.* **2016**, *16*, 1–11. [CrossRef] [PubMed]
3. Dash, S.; Murthy, P.N.; Nath, L.; Chowdhury, P. Kinetic modeling on drug release from controlled drug delivery systems. *Acta Pol. Pharm.—Drug Res.* **2010**, *67*, 217–223.
4. Holzapfel, B.M.; Reichert, J.C.; Schantz, J.T.; Gbureck, U.; Rackwitz, L.; Nöth, U.; Jakob, F.; Rudert, M.; Groll, J.; Hutmacher, D.W. How smart do biomaterials need to be? A translational science and clinical point of view. *Adv. Drug Deliv. Rev.* **2013**, *65*, 581–603. [CrossRef] [PubMed]
5. De Los Ríos Escalante, P.; Arancibia, E. A checklist of marine crustaceans known from Easter Island. *Crustaceana* **2016**, *89*, 63–84. [CrossRef]

6. Servat-Medina, L.; González-Gómez, A.; Reyes-Ortega, F.; Sousa, I.M.O.; Queiroz, N.D.C.A.; Zago, P.M.W.; Jorge, M.P.; Monteiro, K.M.; de Carvalho, J.E.; San Román, J.; et al. Chitosan–tripolyphosphate nanoparticles as *Arrabidaea Chica* Stand. Extr. Carrier: Synth. Charact. Biocompat. Antiulcerogenic Act. *Int. J. Nanomed.* **2015**, *10*, 3897–3909. [CrossRef] [PubMed]
7. Pires, A.L.R.; Westin, C.B.; Hernandez-Montelongo, J.; Sousa, I.M.O.; Foglio, M.A.; Moraes, A.M. Flexible, dense and porous chitosan and alginate membranes containing the standardized extract of *Arrabidaea chica* Verlot for the treatment of skin lesions. *Mater. Sci. Eng. C* **2020**, *112*, 110849. [CrossRef]
8. Parmar, A.; Sharman, S. Engineering design and mechanistic mathematical models: Standpoint on cutting edge drug delivery. *Trends Anal. Chem.* **2018**, *100*, 15–35. [CrossRef]
9. Siepmann, J.; Siepmann, F. Mathematical modeling of drug delivery. *Int. J. Pharm.* **2008**, *364*, 328–343. [CrossRef]
10. Zhang, Y.; Huo, M.; Zhou, J.; Zou, A.; Li, W.; Yao, C.; Xie, S. DDSolver: An add-in program for modeling and comparison of drug dissolution profiles. *AAPS J.* **2010**, *12*, 263–271. [CrossRef]
11. Coronel, A.; Berres, S.; Lagos, R. Calibration of a sedimentation model through a continuous genetic algorithm. *Inverse Probl. Sci. Eng.* **2019**, *27*, 1263–1278. [CrossRef]
12. Berres, S.; Bürger, R.; Coronel, A.; Sepúlveda, M. Numerical identification of parameters for a flocculated suspension from concentration measurements during batch centrifugation. *Chem. Eng. J.* **2005**, *111*, 91–103. [CrossRef]
13. Berres, S.; Bürger, R.; Coronel, A.; Sepúlveda, M. Numerical identification of parameters for a strongly degenerate convection-diffusion problem modelling centrifugation of flocculated suspensions. *Appl. Numer. Math.* **2005**, *52*, 311–337. [CrossRef]
14. Tang, H.W.; Xin, X.K.; Dai, W.H.; Xiao, Y. Parameter identification for modeling river network using a genetic algorithm. *J. Hydrodyn.* **2010**, *22*, 246–253. [CrossRef]
15. Peralta, B.; Soria, R.; Berres, S.; Caro, L.; Mellado, A.; Schiappacasse, N. Detection of Anomalous Pollution Sensors Using Deep Learning Strategies. In *IOP Conference Series: Earth and Environmental Science*; IOP Publishing: Bristol, UK, 2020; Volume 503.
16. Caccavo, D. An overview on the mathematical modeling of hydrogels' behavior for drug delivery systems. *Int. J. Pharm.* **2019**, *560*, 175–190. [CrossRef] [PubMed]
17. Weibull, W. A statistical distribution function of wide applicability. *J. Appl. Mech.* **1951**, *18*, 293–297. [CrossRef]
18. Higuchi, T. Mechanism of sustained-action medication. Theoretical analysis of rate of release of solid drugs dispersed in solid matrices. *J. Pharm. Sci.* **1963**, *52*, 1145–1149. [CrossRef]
19. Korsmeyer, R.W.; Gurny, R.; Doelker, E.; Buri, P.; Peppas, N.A. Mechanisms of potassium chloride release from compressed, hydrophilic, polymeric matrices: Effect of entrapped air. *J. Pharm. Sci.* **1983**, *72*, 1189–1191. [CrossRef]
20. Peppas, N.A. Analysis of Fickian and non-Fickian drug release from polymers. *Pharm. Acta Helv.* **1985**, *60*, 110–111.
21. Berres, S.; Bürger, R.; Garcés, R. Centrifugal settling of flocculated suspensions: A sensitivity analysis of parametric model functions. *Dry. Technol.* **2010**, *28*, 858–870. [CrossRef]
22. Ravindran, A.; Reklaitis, G.V.; Ragsdell, K.M. *Engineering Optimization: Methods and Applications*; John Wiley & Sons: Hoboken, NJ, USA, 2006.
23. Aro, A.A.; Simões, G.F.; Esquisatto, M.A.M.; Foglio, M.A.; Carvalho, J.E.; Oliveira, A.L.R.; Gomes, L.; Pimentel, E.R. *Arrabidaea Chica* extract improves gait recovery and changes collagen content during healing of the Achilles tendon. *Injury* **2013**, *44*, 884–892. [CrossRef] [PubMed]
24. Hoffman, A.S. The origins and evolution of "controlled" drug delivery systems. *J. Control. Release* **2008**, *132*, 153–163. [CrossRef] [PubMed]
25. Crank, J. *The Mathematics of Diffusion*; Clarendon: New York, NY, USA, 1975; 414p.
26. Bacaita, E.S.; Ciobanu, B.C.; Popa, M.; Agop, M.; Desbrieres, J. Phases in the temporal multiscale evolution of the drug release mechanism in IPN-type chitosan based hydrogels. *Phys. Chem. Chem. Physics.* **2014**, *16*, 25896–25905. [CrossRef] [PubMed]
27. Papadopoulou, V.; Kosmidis, K.; Vlachou, M.; Macheras, P. On the use of the Weibull function for the discernment of drug release mechanisms. *Int. J. Pharm.* **2006**, *309*, 44–50. [CrossRef] [PubMed]
28. Kobryn, J.; Sowa, S.; Gasztych, M.; Drys, A.; Musial, W. Influence of Hydrophilic Polymers on the β Factor in Weibull Equation Applied to the Release Kinetics of a Biologically Active Complex of Aesculus hippocastanum. *Int. J. Polym. Sci.* **2017**, *3486384*, 1–8. [CrossRef]
29. Paolino, D.; Tudose, A.; Celia, C.; Di Marzio, L.; Cilurzo, F.; Mircioiu, C. Mathematical Models as Tools to Predict the Release Kinetic of Fluorescein from Lyotropic Colloidal Liquid Crystals. *Materials* **2019**, *12*, 1–23. [CrossRef]
30. Hernandez-Montelongo, J.; Nascimento, V.; Hernández-Montelongo, R.; Beppu, M.; Cotta, M. Fractal analysis of the formation process and morphologies of hyaluronan/chitosan nanofilms in layer-by-layer assembly. *Polymer* **2020**, *191*, 1–8. [CrossRef]

Article

Optimization of Multiple $W_1/O/W_2$ Emulsions Processing for Suitable Stability and Encapsulation Efficiency

Manuel Felix *, Antonio Guerrero and Cecilio Carrera-Sánchez

Departamento de Ingeniería Química, Escuela Politécnica Superior, Universidad de Sevilla, 41011 Sevilla, Spain; aguerrero@us.es (A.G.); cecilio@us.es (C.C.-S.)
* Correspondence: mfelix@us.es

Abstract: Double emulsions are a type of multiple emulsions, which can be defined as a multi-compartmentalized system where the droplets are dispersed into the continuous phase containing other emulsions. Although double food-grade emulsions have been manufactured, there is a lack of scientific background related to the influence of different processing conditions. This work analyses the influence of processing variables in ($W_1/O/W_2$) double emulsions: passes through the valve homogenizer, pressure applied, lipophilic emulsifier concentration, the ratio between the continuous phase (W_2) and the primary emulsion (W_1/O), and the incorporation of xanthan gum (XG) as a stabilizer. The results obtained show that these emulsions can be obtained after selecting suitable processing conditions, making them easily scalable in industrial processes. In terms of droplet size distribution, the input of higher energy to the system (20 MPa) during emulsification processing led to emulsions with smaller droplet sizes ($D_{3,2}$). However, more monodispersed emulsions were achieved when the lowest pressure (5 MPa) was used. As for the number of passes, the optimal (emulsions more monodispersed and smaller droplet sizes) was found around 2–3 passes, regardless of the valve homogenizer pressure. However, emulsions processed at 20 MPa involved lower encapsulation efficiency (E_E) than emulsions processed at 5 MPa (87.3 ± 2.3 vs. 96.1 ± 1.8, respectively). The addition of XG led to more structured emulsions, and consequently, their kinetic stability increased. The results obtained indicated that a correct formulation of these $W_1/O/W_2$ double emulsions allowed the optimal encapsulation of both hydrophilic and lipophilic bioactive compounds. Thus, the development of food matrices, in the form of multiple emulsions, would allow the encapsulation of bioactive compounds, which would result in the development of novelty food products.

Keywords: droplet size distribution; emulsification; multiple emulsions; polyglycerol polyricinoleate (PGPR)

1. Introduction

Functional food products are based on the presence of functional ingredients (bioactive compounds) in the food matrix; the possibility of developing this new generation of food products requires the employment of strategies able to preserve the desired components during their storage and after their intake thorough the intestinal tract [1]. One approach for the development of these functional foods includes the use of current technologies applied to emulsion-based products, making them a reliable alternative to the current food market [2]. In this way, the most versatile path to modify the composition of food arises from the possibility of introducing bioactive ingredients during its preparation [3]. Based on the solubility of these ingredients, they could be classified into two big families: lipophilic and hydrophilic compounds [4]. The former ingredients (e.g., carotenoids, E-vitamin, etc.) are soluble in the oil phase of conventional emulsions, although their encapsulation in nanoemulsion-based systems has been demonstrated to improve not only the physical stability but also their absorption after intake [3]. However, hydrophilic components (e.g., polyphenols, peptides, etc.) cannot be included in the continuous phase of conventional emulsions since their biodisponibility is reduced after digestion [5]. An alternative for introducing these bioactive components in food matrices is to include them in the primary emulsion of double emulsions, protecting them from chemical changes during digestion [6,7].

In this context, the development of double emulsions as food matrices opens up new possibilities in the design and development of functional foods [8]. These food matrices could be used as carriers of water-soluble bioactive components. However, technological strategies must be applied to adapt current food processing methods in the development of this type of emulsion, optimizing the presence of bioactive compounds [8].

Double emulsions are a type of multiple emulsion, which can be defined as a multicompartmentalized system where the droplets disperse into the continuous phase containing other emulsions [9]. These systems are characterized by the coexistence of oil-in-water (O/W) and water-in-oil (W/O) emulsions, in which the dispersed phase (in the form of droplets) is the continuous phase of the primary emulsion (which means that it has smaller droplets of the other phase disperse into it) [10]. Thus, the most common case is water-in-oil-in-water emulsions ($W_1/O/W_2$), although those of oil-in-water-in-oil ($O_1/W/O_2$) can also be used in specific applications (i.e., they exhibit a fine texture, showing a smooth touch upon application) [6,11]. Once produced, double emulsions are subjected to several coalescence and diffusion phenomena that consequently impact product properties, such as texture or encapsulation performance [12].

The use of multiple (double) emulsions in food matrices brings unexplored opportunities to introduce bioactive compounds in food products and, consequently, in human consumption habits. Although double food-grade emulsions have been already manufactured, there is a lack of scientific background related to the influence of different processing conditions.

The general objective of this work is to analyze the influence of processing variables that affect the development of double emulsion ($W_1/O/W_2$). The encapsulation efficiency (E_E, parameter to be optimized) was assessed by tartrazine, whose use as a marker was initially proven. The different variables studied were: passes through the valve homogenizer, pressure applied, lipophilic emulsifier (polyglycerol polyricinoleate, PGPR), the ratio between the continuous phase (W_2) and the primary emulsion (W_1/O), and the incorporation of a hydrocolloid (xanthan gum) as a stabilizer of the double emulsions generated.

2. Materials and Methods

2.1. Materials

The lipophilic surfactant polyglycerol polyricinoleate (PGPR) was supplied by Campobetica (Malaga, Spain). This emulsifier was used for the stabilization of primary emulsions (W_1/O). BiPro® whey protein isolate (WPI), supplied by AGROPUR (Longueuil, QC, Canada), was the emulsifier used for the stabilization of secondary emulsions ($W_1/O/W_2$). Xanthan from Xanthomonas campestris was provided by Merck (Branchburg, NJ, USA). Chemical reagents (i.e., HCl, NaOH, NaCl, Tartrazine and Trizma) were purchased from Sigma–Aldrich company (St. Louis, MO, USA). The oil used (sunflower oil) was purchased from a local market. Emulsions were prepared using deionized water.

2.2. Methods

2.2.1. Preparation of $W_1/O/W_2$ Double Emulsions

Firstly, the primary W_1/O emulsion was prepared. For these emulsions, the aqueous phase was stabilized at pH 7.0 using Trizma (50 mM). NaCl was added for ionic-strength control (50 mM), and tartrazine was used as a marker (0.06 wt.%). This aqueous phase (W_1) was dispersed into the continuous oily phase, which contained PGPR (1, 3 or 5 wt.%). The W_1/O ratio of these primary emulsions was, in all cases, 20/80. This primary emulsion was prepared by applying high energy to obtain small-sized droplets. Thus, the oil and the W_1 phase were blended and mixed using a rotor-stator homogenizer (Utraturrax® T-25, IKA, Staufen, Germany) at 8000 rpm. Subsequently, the emulsion obtained was passed through a high-pressure homogenizer at 50 MPa (Emulsiflex 2000-CS, Avestin, Ottawa, ON, Canada). Once primary emulsions were prepared, secondary emulsions were generated by mixing the primary emulsion with an aqueous WPI solution (0.5 wt.%) at pH 7, W_2. In order to adjust the osmotic pressure balance between the aqueous phases W_1 and W_2, NaCl was also added to this outer aqueous phase. Soft processing conditions (i.e., low energy input) were used in this case to avoid the breakdown of primary emulsions. The W_1/O primary emulsion was mixed with the outer phase (W_2) using the same rotor-stator mixer, but at 1000 rpm. These pre-emulsions were also passed through the Emulsiflex valve homogenizer, selecting, in this case, 5 and 20 MPa. The $(W_1/O)/W_2$ ratios of multiple emulsions were 10/90, 25/75 and 40/60. These double emulsions were prepared according to the method proposed by Lynch [13]. Moreover, the influence of the addition of xanthan gum (XG) as a stabilizer was also tested. These emulsions were prepared following the same procedure described above with a $(W_1/O)/W_2$ ratio of 40/60 (processed in absence of XG). Once emulsions were generated (using 5 wt.% PGPR according to the highest EE value), XG dissolved into W_2 was added up to reach the $(W_1/O)/W_2$ ratio of 25/75. XG solution was dispersed into $(W_1/O)/W_2$ emulsions by soft magnetic stirring. The final XG concentration studied was 0.125 and 0.25 wt.%. Supplementary Table S1 summarizes the systems performed in this research.

2.3. Characterisation of Emulsions

2.3.1. Droplet Size Distribution

Droplet size distribution (DSD) of the double emulsions $(W_1/O/W_2)$ was determined with the particle size analyzer based on laser diffraction Mastersizer 2000 (Malvern, UK). To facilitate the rupture of possible flocs, 0.5 mL of emulsion samples were diluted in a 1% SDS (sodium dodecyl sulfate) solution. Immediately after this, the droplet size distribution of the deflocculated systems was determined [14]. The mean Sauter diameter ($D_{3,2}$) and volumetric diameter ($D_{4,3}$) were calculated according to Equations (1) and (2):

$$D_{3,2} = \frac{\sum n_i d_i^3}{\sum n_i d_i^2} \quad (1)$$

$$D_{4,3} = \frac{\sum n_i d_i^4}{\sum n_i d_i^3} \quad (2)$$

where n_i is the number of droplets that have d_i diameter.

The span parameter (Equation (3)) was calculated to analyze the DSD profiles:

$$\text{Span} = \frac{d(90) - d(10)}{d(50)} \quad (3)$$

where $d(90)$, $d(10)$ and $d(50)$ refer to the 10, 50 and 90 volume percentile of droplets with diameters smaller or equal to these values.

The Flocculation Index (FI, Equation (4)) was calculated to study the flocculation phenomena occurring during emulsion storage:

$$\text{FI (\%)} = \frac{|D_{4,3} - D_{4,3\,(\text{SDS})}|}{D_{4,3}} \times 100 \quad (4)$$

where $D_{4,3}$ and $D_{4,3\,(\text{SDS})}$ are the volumetric diameter without and with SDS solution.

2.3.2. Rheological Properties

Linear viscoelastic properties were analyzed by means of small amplitude oscillatory shear (SAOS) tests using the AR-2000 rheometer (TA Instruments, New Castle, DE, USA). All measurements were carried out within the linear viscoelastic range (LVR), which was determined by means of strain sweep tests. These tests were performed at 1 Hz and from

0.01 to 10% strain. Mechanical spectra were obtained from frequency sweep tests (from 0.1 to 30 Hz) at a constant strain (1%). These tests were carried out using a low inertia aluminum smooth plate (60 mm) at a gap of 1 mm. The temperature was fixed at 25 °C by a peltier system

2.3.3. Emulsion Stability

The overall stability of emulsions was assessed using a vertical scan analyzer Turbiscan MA 2000 (Formulaction, Toulouse, France) through backscattering measurements of a pulsed light source (λ = 850 nm) as a function of the height of the cylindrical glass tube containing the sample [15]. Emulsions were stored at 5 °C and measured at room temperature. Then, the profiles of backscattering of light from emulsions (ΔBS/100%) versus the height were plotted as a function of storage time.

2.4. Encapsulation Capacity

2.4.1. Measurement of Marker Concentration in the Outer Aqueous Phase (W_2)

One of the most important steps in the determination of the marker concentration is to ensure that the concentration determined for the marked is not affected by the presence of the protein used, the processing conditions or the storage time. In this section, experiments were carried out to ensure the correct measurement of the marker in the final double emulsions [16].

The first determination consisted of checking the correct determination of the marker (tartrazine) in the presence of a WPI solution. The marker was dissolved in Trizma buffer (50 mM, pH 7) at the same concentration as it was added to the inner aqueous phase (W_1) in the double emulsion (0.06 wt.%). Subsequently, this solution was added to the outer aqueous phase W_2 at the same ratio as in the final double emulsions. The W_2 phase contained 0.5 wt.% WPI protein since this is the protein concentration used in the W_2 of the final double emulsions. $W_1 + W_2$ were mixed at room temperature for 30 min using a magnetic stirrer. The concentration of the marker in the overall solution was calculated by measuring the absorbance at 435 nm of the corresponding solution using a spectrophotometer (Thermo Scientific, Waltham, MA, USA), using the WPI in Trizma buffer as the blank reference. Tartrazine solution was also added to a Trizma buffer solution (without WPI) and treated, but in this case, using the buffer solution as the blank reference. The value of absorbance obtained was interpolated in a calibration curve obtained after dissolving different known concentrations of the marker in Trizma buffer at pH 7 (W_1).

Measurements were carried out on the same day of $W_1 + W_2$ mixing and after 7 storage days in an oven at 45 °C. The marker measurable concentration ($MC_m(\%)$) in the WPI solution was calculated according to the Equation (5):

$$MC_m(\%) = \frac{C_{m_{WPI}}}{C_{m_b}} \times 100 \qquad (5)$$

where $C_{m_{WPI}}$ is the concentration of the marker measured in the $W_1 + W_2$ (with WPI) mixtures and C_{m_b} is the concentration of the marker according to the dilutions carried out (without WPI).

2.4.2. Measurement of the Recovery Yield (Ry) of Marker

The recovery yield is defined as the concentration of marker used, which is present in the aqueous phase recovered after separating the cream phase and an aqueous phase by centrifugation. The percentage of recovery yield is relative to the concentration of marker in the outer aqueous phase after emulsion preparation [16].

For these measurements, a conventional oil-in-water emulsion (O/W_2) was prepared using the same oil phase (O) and the same outer aqueous phase (W_2) as in the W_1/O/W_2 double emulsion (ratio 10/90). The oil phase (O) was gradually added to the aqueous phase (W_2) while they were mixing using an Ultraturrax® homogenizer (IKA, Staufen, Germany) at 9500 rpm and 1 min. Subsequently, the pre-emulsion was passed through a high-pressure homogenizer (Emulsiflex 2000-CS, Avestin, Ottawa, ON, Canada). These O/W_2 conventional emulsions were diluted in the W_1 phase (reaching the ratio eventually used in W_1/O/W_2 double emulsions). The W_1 contained either (i) the marker at 0.06 wt.% for the determination of marker recovery or (ii) without the marker for the sample blank.

According to Dickinson et al. [17], this procedure ensured that the marker was completely present in the outer aqueous phase. Consequently, this emulsion can be used as the standard where 100% of the marker was present in the outer aqueous phase (W_2). The determination of aqueous phase color requires a transparent medium. Thus, an aliquot of the emulsions was diluted with the W_1 phase (1:1), the mixture was homogenized by a soft stirring and, subsequently, it was separated by centrifugation (15,000× g at 4 °C for 30 min) in a cream phase (which contained the oil droplets) and an aqueous phase. The aqueous phase was filtered (0.2 μm syringe microfilter), and the concentration of the marker presented in the resulting filtered aqueous phase was calculated by the measurement of the absorbance at 435 nm in a spectrophotometer (Thermo Scientific) using the subnatant obtained from O/W_2 emulsion with added Trizma buffer without the added marker as a reference blank.

The value of the absorbance obtained was compared with the concentration of the marker in the outer aqueous phase (which depends on the ratio (W_1/O)/W_2). The recovery yield can be calculated according to Equation (6) [16]:

$$\text{Ry (\%)} = \frac{C_{m_{OAP(1,7)}}}{C_{m_b}} \times 100 \qquad (6)$$

where $C_{m_{OAP(1,7)}}$ was the concentration of the marker used present in the outer aqueous phase (recovered after centrifugation) after 1 and 7 days of emulsion aging. C_{m_b} was the concentration of the marker used in the outer aqueous phase. This latter concentration was determined by an equivalent dilution of the aqueous marker solution with Trizma buffer as occurred in the emulsion.

The concentration of the marker present in the outer aqueous phase was calculated on the same day of emulsion preparation and 7 days later to determine the influence of aging on the Ry parameter.

2.4.3. Encapsulation Efficiency (E_E) Measurement

The E_E of the double W_1/O/W_2 emulsion was calculated by the determination of the concentration of the marker in the outer aqueous phase (W_2) just after the preparation of the emulsions.

The E_E of double W_1/O/W_2 emulsions was prepared according to the experimental procedure indicated in Section 2.2.1. Once double emulsions were prepared, they were subjected to the phase separation explained in Section 2.4.2. Double emulsions were prepared with the marker (tartrazine) diluted in the W_1 and in the absence of the marker to obtain the blank for the measurement of the absorbance of the aqueous phase. The concentration of the marker in the outer aqueous phase (W_2) was measured by the same method as described above for measuring the recovery yield but replacing the O/W_2 emulsion with the W_1/O/W_2 emulsion. The E_E (%) was calculated with respect to the concentration of the marker present in the aqueous phase obtained after emulsion separation ($W_1 + W_2$) according to Equation (7):

$$E_E(\%) = 100 - \frac{C_{m_{OAP}}}{C_{m_b}} \times 100 \qquad (7)$$

where $C_{m_{OAP}}$ is the marker concentration measured in the aqueous phase (which has been recovered from the double emulsions), C_{m_b} is the marker concentration initially added to the inner aqueous phase.

The E_E was calculated on days 0 and 15 to determine the influence of storage time on this parameter.

2.5. Statistical Analysis

Emulsions were prepared in duplicate. Three replicates were carried out for each measurement. Significant differences were determined by performing ANOVA tests using STATGRAPHICS Centurion XVIII software version number (Statgraphics Tecnologies Inc., The Plains, VA, USA).

3. Results and Discussion

3.1. Evaluation of Tartrazine as a Marker for Double Emulsions Stabilized by WPI

The effectiveness of tartrazine as a marker for encapsulating hydrophilic compounds in $W_1/O/W_2$ double emulsions must be addressed before any experiment. To this end, it is essential to identify a marker added to the internal aqueous phase (W_1), whose concentration in the outer aqueous phase (W_2) can be determined after double emulsion preparation. Moreover, changes in the concentration of the marker should not be elucidated in the outer aqueous phase during double emulsion storage [16].

Table 1 shows the marker measurable concentration (MC_m) and recovery yield (Ry) for the marker used in this work (tartrazine) in the aqueous phase recovered when 100% of the marker is in the outer aqueous phase (W_2). Table 1 indicates that the marker used can be accurately measured in the external aqueous phase (W_2) of the double $W_1/O/W_2$ emulsion by measuring the absorbance of the outer aqueous phase following the procedure indicated in Section 2.4 since the marker concentration measurable (MC_m) reached $100 \pm 0.5\%$. Moreover, there are no changes in the concentration of tartrazine after 7 days of external phase (W_2) storage containing tartrazine at 45 °C ($MC_m = 98.8 \pm 0.2\%$).

Table 1. Marker concentration measurable (MC_m) and recovery yield (Ry) for the marker (tartrazine) in the aqueous phase recovered standard where 100% of the marker is in the outer aqueous phase (W_2). Different letters within a row indicate significant differences ($p < 0.05$).

Day	MC_m (%)	Ry (%)
0	100.0 ± 0.5 [a]	95.5 ± 1.5 [a]
7	98.8 ± 0.2 [b]	98.9 ± 0.2 [a]

As for the recovery yield (Ry), this table also reveals that the marker selected (tartrazine) can be used for the evaluation of this parameter since the $95.5 \pm 1.5\%$ tatrazine remains in the inner aqueous phase (W_1) the same day emulsion preparation, and $98.9 \pm 0.2\%$ after 7 days, where no significant differences were found ($p < 0.05$). These results also show that the tartrazine marker does not preferentially migrate with the cream phase formed after emulsion centrifugation. The high level of marker present in the recovered aqueous phase after centrifugation suggests that tartrazine is equally distributed throughout the aqueous phase of the entire emulsion.

These results indicate that tartrazine does not associate with WPI and that this marker can be used for the evaluation of emulsion efficiency (E_E) for the double $W_1/O/W_2$ emulsion generated hereinafter. These parameters were more suitable than the values found when methylene blue and B_{12} vitamin were used as markers in double emulsions stabilized by sodium caseinate [16].

3.2. Processing Conditions, Optimization Passes and Valve Homogenizer Pressure
Droplet Size Distribution Measurements

The ratio 25/75 was used for double emulsions (($W_1/O)/W_2$) during the optimization of the number of passes through the high-pressure homogenizer (from 1 to 5). The homogenization pressures evaluated were 5 and 20 MPa. The PGPR concentration used for these emulsions was 5 wt.%.

Figure 1 shows the droplet size distribution (DSD) of the $W_1/O/W_2$ emulsions obtained after processing the secondary emulsion at 5 MPa (A) and 20 MPa (B) as a function of emulsion passes through the valve homogenizer. All measurements plotted were carried out in the presence of 1% SDS (deflocculating agent) to avoid DSD influenced by the bridging flocculation phenomena [14]. When the homogenization pressure was 5 MPa (Figure 1A), unimodal DSD profiles were obtained in all cases; however, the distribution becomes narrower as emulsion passes increased. Thus, the distributions always started around 2 μm; however, the highest droplets moved from c.a. 40 to 20 μm, where the reduction of droplet sizes most important took place after passes 1 and 2. This reduction in droplet size can be expected since there is an increase in the energy input during emulsion processing as the number the passes increases [18].

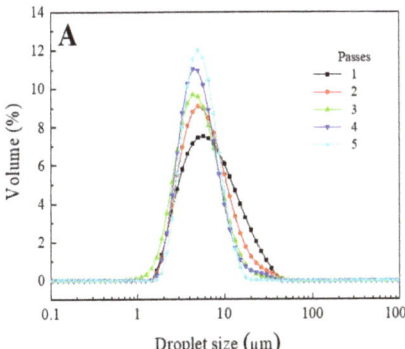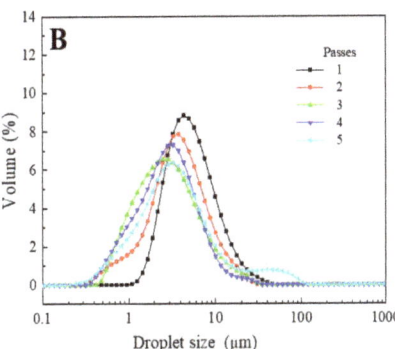

Figure 1. Droplet size of $W_1/O/W_2$ multiple emulsions obtained after processing the secondary emulsion at 5 MPa (**A**) and 20 MPa (**B**), as a function emulsion passes through the high-pressure homogenizer.

On the other hand, when the homogenization pressure increased up to 20 MPa (Figure 1B), the unimodal DSD was obtained only for one emulsion passing through the valve homogenizer. A higher number of passes involved the emergence of other tenuous droplet populations at the ends or the beginning of the main peak. Thus, for passes 2 and 3, shorter droplets were obtained due to the above-mentioned increase in energy input. However, this tendency changed for passes 4 and 5, where higher droplet populations were found. This behavior has been related to excessive processing energy, leading to a higher coalescence rate after passing the emulsion through the valve homogenizer [19]. In this sense, a comparison of Figure 1A,B show that an increase in pressure leads to a shift of the droplet size distribution profiles towards smaller sizes, although giving rise to apparently wider distributions. Moreover, an excess of energy (caused by the number of passes) led to an increase in droplet sizes due to coalescence. A similar influence of valve homogenizer pressure was found by Floury et al. [18] for conventional emulsions stabilized by whey protein isolate.

In order to quantify the tendencies observed by droplet size distributions after processing the emulsions at two different homogenizer pressures (5 and 20 MPa) and passes (up to 5). Table 2 summarizes the Sauter mean diameter, the span parameter and the flocculation index ($D_{3,2}$, span and FI, respectively) for all emulsions studied in this section. The values observed in this Table confirm the results previously observed in the drop size distribution curves since an increase in the number of steps led to a significant decrease in Span (higher droplet polydispersity). As for the $D_{3,2}$, the increase in passes involved a decrease in this parameter, especially for the multiple emulsions processed at 5 MPa. However, more than three passes did not involve significant differences ($p < 0.05$).

Table 2. Parameters ($D_{3,2}$, Span and FI) from droplet size distribution as a function of the number of emulsion passes through the high-pressure homogenizer and the two pressures studied (5 MPa and 20 MPa). Different superscript letters within a column indicate significant differences ($p < 0.05$).

Passes	5 MPa			20 MPa		
	$D_{3,2}$ (µm)	Span	FI (%)	$D_{3,2}$ (µm)	Span	FI (%)
1	5.2 ± 0.3 [a]	2.2 ± 0.1 [a]	5.2 ± 3.2 [a]	4.1 ± 0.3 [a]	1.9 ± 0.4 [a]	20.0 ± 8.0 [a]
2	4.7 ± 0.2 [b]	1.8 ± 0.1 [b]	9.5 ± 2.1 [b]	2.3 ± 0.2 [b]	2.2 ± 0.3 [a]	13.2 ± 6.8 [b]
3	4.0 ± 0.1 [c]	1.6 ± 0.2 [c]	3.6 ± 2.2 [a]	1.9 ± 0.2 [b]	2.6 ± 0.2 [b]	6.4 ± 4.2 [c]
4	4.3 ± 0.2 [c]	1.4 ± 0.1 [c]	14.0 ± 4.1 [c]	1.9 ± 0.1 [b]	2.2 ± 0.3 [a]	6.0 ± 4.0 [c]
5	4.5 ± 0.2 [c,d]	1.2 ± 0.1 [d]	9.0 ± 2.9 [b]	2.0 ± 0.3 [b]	4.0 ± 0.2 [c]	28.0 ± 8.2 [a]

The FI parameter accounts for the electrostatic interactions of oil droplets. Flocculation can modify the rheological behavior of emulsions, creating a network that prevents phase separation when highly concentrated emulsions are prepared. However, it can also promote droplet coalescence. In general, low flocculated emulsions were obtained. The standard deviation obtained in this parameter suggests that this flocculation was not too strong, and the system was deflocculated/flocculated by the simple mechanical agitation when the measurement was performed [20].

Thus, in terms of droplet size distribution, the input of higher energy to the system (20 MPa) during emulsification processing led to emulsions with smaller droplet sizes ($D_{3,2}$). However, more monodispersed emulsions were achieved when the lowest pressure (5 MPa) was used. As for the number of passes, the optimal (emulsions more monodispersed and smaller droplet sizes) was found around 2–3 passes, regardless of the valve homogenizer pressure.

The processing parameters selected were also based on the E_E of the multiple emulsions generated. In fact, the objective of the present work is to maximize this key parameter. Thus, Table 3 shows the E_E for emulsions as a function of the number of emulsion passes through the high-pressure homogenizer and the two pressures studied (5 MPa and 20 MPa). This table evidences a clear reduction in the E_E when the homogenization pressure increases. In this sense, the encapsulation pressure was higher when emulsions were processed at 5 MPa than when they were processed at 20 MPa (i.e., 96.1 ± 1.8 vs. 87.3 ± 2.3 for three passes).

Table 3. Encapsulation efficiency (E_E) for the emulsions as a function of the number of emulsion passes through the high-pressure homogenizer (steps) and the two pressures studied (5 MPa and 20 MPa). Different superscript letters within a column indicate significant differences ($p < 0.05$).

Passes	E_E (%) 5 MPa	E_E (%) 20 MPa
1	97.0 ± 2.1 [a]	86.1 ± 3.1 [a]
2	97.6 ± 1.2 [a]	54.1 ± 6.1 [b]
3	96.1 ± 1.8 [a]	87.3 ± 2.3 [a]
4	95.4 ± 1.5 [a]	74.3 ± 5.6 [c]
5	97.0 ± 1.1 [a]	87.3 ± 2.9 [a]

No significant differences ($p < 0.05$) were found when the emulsions were processed at the lower pressure (5 MPa). On the other hand, when emulsions were processed at the highest pressure (20 MPa), a significant decrease in E_E with the number of passes was found. However, this tendency was not clear for the emulsions processed at this pressure. Schuch et al. [21] evaluated the influence of processing conditions on the E_E in double emulsions. They found that the E_E directly depended on the size of W_1/O (primary emulsions) since the bigger the primary emulsion droplets, the higher the encapsulation efficiency. However, an excess of energy in secondary emulsions led to a decrease in this parameter.

From these results, and their comparison with the results obtained from droplet size distributions, hereinafter, the emulsion will be processed at 5 MPa and three passes through the high-pressure homogenizer. These conditions generated emulsions with a high encapsulation efficiency, at the same time as the droplet size profiles obtained were unimodal, which has been related to higher emulsion stability than polydispersed droplet size distributions [22].

3.3. Processing Conditions Optimization, Influence of PGPR

Droplet Size Distribution Measurements

The optimization of PGPR concentration in double emulsions was carried out using emulsions containing 25/75 of the (W_1/O)/W_2 ratio. Moreover, according to the previous results obtained, the pressure in the valve homogenizer was 5 MPa, and emulsions were passed through it three times.

Figure 2 shows the droplet size distribution of W_1/O/W_2 multiple emulsions obtained after using 1, 3 and 5 wt.% PGPR for the primary emulsion. All measurements were carried out in the presence of 1% SDS. The DSD profiles reveal practically single-mode curves, exhibiting a maximum value around 5 μm. In the case of 3% and 5% of PGPR, the curves obtained were practically the same. However, the system containing 1% of PGPR showed a widening in the 1 μm range, resulting in a greater droplet dispersion.

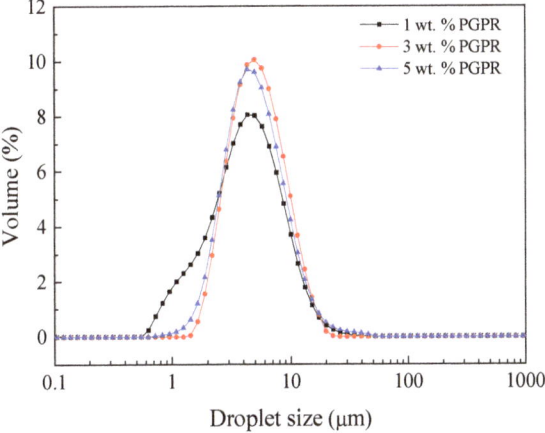

Figure 2. Droplet size distribution of W_1/O/W_2 multiple emulsions obtained after using 1, 3 and 5 wt.% PGPR for the primary emulsion.

Narrower distributions are desired in emulsions since it determines the potential destabilization processes (such as coalescence and further creaming) [23,24]. This change in DSD profile was also obtained by Altuntas [25] for double emulsions stabilized by PGPR–lecithin mixtures, which evidences the importance of PGPR in double emulsions.

Table 4 summarizes parameters from droplet size distribution ($D_{3,2}$, Span and FI) as well as E_E to evaluate the influence of PGPR in these emulsions. This table indicates that the lowest $D_{3,2}$ was obtained for the system containing 1 wt.% PGPR. However, this system also led to the highest Span value (higher polydispersity). Eisinaite et al. [7] found similar results; they observed that the initial droplet size of the double emulsions first increased with the increase in the amount of PGPR and then decreased. They suggest that this behavior could be a result of a combined effect: at higher concentrations of PGPR, the water transfer capacity was higher and swelling was promoted, while the interfacial tension of the droplets was lower, which leads to less swelling.

Table 4. Parameters from droplet size distribution ($D_{3,2}$, Span and FI) as well as encapsulation efficiency (E_E) at three different PGPR concentrations (1, 3 and 5 wt.%). Different superscript letters within a column indicate significant differences ($p < 0.05$).

% PGPR	$D_{3,2}$ (µm)	Span	FI (%)	E_E (%)
1	2.9 ± 0.1 [a]	1.9 ± 0.2 [a]	15.2 ± 8.2 [a]	69.1 ± 4.1 [a]
3	4.3 ± 0.1 [b]	1.5 ± 0.2 [b]	16.8 ± 9.3 [a]	84.1 ± 3.8 [b]
5	4.0 ± 0.1 [b]	1.6 ± 0.2 [b]	3.6 ± 2.2 [b]	96.1 ± 1.8 [c]

Moreover, the flocculation index (FI) calculated for these freshly prepared emulsions indicates that it decreased as the PGPR concentration increased. This result is not desired since flocculation phenomena can increase the viscosity of concentrated emulsions; however, in diluted emulsions, it can conduct droplet coalescence [26].

This Table also indicates that a decrease in PGPR led to a significant decrease in E_E, finding the best results for the emulsion containing 5 wt.% PGPR (96.1 ± 1.8%). These results agree with other studies where double $W_1/O/W_2$ emulsions were prepared with various concentrations of PGPR (0.5–5% wt.%) in the oil phase. Eisinaite et al. [7] found that the emulsions showed good physical stability, with encapsulation efficiency close to 100% only at high concentrations of PGPR (> 2 wt.%). According to these results, the PGPR selected for the further emulsions manufactured was 5.0 wt.%

3.4. Processing Conditions Optimization, Influence of (W1/O)/W2 Ratio

Hereinafter, double emulsions were operated under the following operating conditions: low pressure for the valve homogenizer in the final double emulsion (5 MPa), three passes through it and 5 wt.% PGPR in the oil phase for the primary emulsion (W_1/O). This section analyzes the ratios 10/90, 25/75 and 40/60 between the primary emulsion W_1/O and the aqueous phase of the secondary emulsion, W_2, as a function of the aging time.

3.4.1. Droplet Size Distribution and Backscattering Measurements

Figure 3 shows the DSD profiles obtained for the $W_1/O/W_2$ multiple emulsions for three different O/W ratios when the secondary emulsion is prepared: 10/90 (Figure 3A), 25/75 (Figure 3B) and 40/60 (Figure 3C). Moreover, the droplet stability of the emulsions prepared over storage time was also analyzed. All measurements plotted were carried out in the presence of 1% SDS. The DSD profiles obtained for all double emulsions generated were fairly similar, obtaining a unimodal distribution and the droplet sizes were within 2 and 20–30 µm. Moreover, these DSD profiles remained unaltered as a function of storage time, indicating high stability of the three ratios analyzed (10/90, 25/75 and 40/60) over storage time.

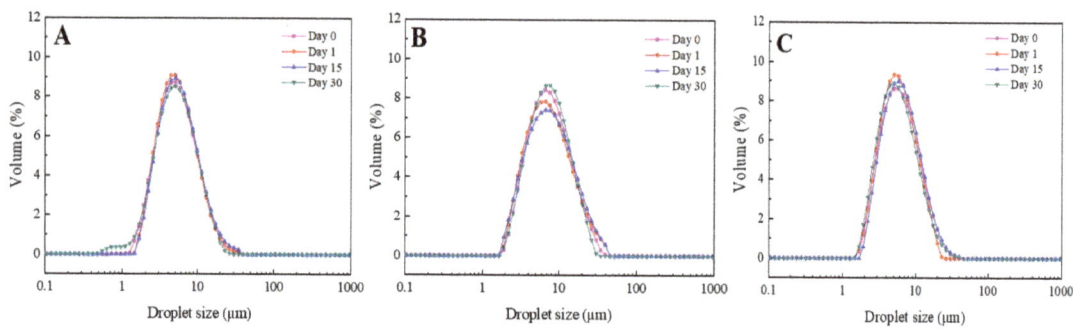

Figure 3. Droplet size distribution of $W_1/O/W_2$ multiple emulsions obtained after three different O/W ratios for the secondary emulsion: 10/90 (**A**), 25/75 (**B**) and 40/60 (**C**).

In order to evaluate the evolution of the droplet size of the emulsions, Table 5 shows parameters from DSD measurements ($D_{3,2}$, span and FI) of $W_1/O/W_2$ multiple emulsions obtained after evaluating three different O/W ratios for the secondary emulsion (10/90, 25/75 and 40/60) over 30 days storage time. These parameters did not vary significantly over time, where the FI obtained indicates poorly flocculated emulsions. These results suggest that suitable emulsions with the desired stability were obtained regardless of the $(W_1/O)/W_2$ ratio, indicating that the WPI concentration used in the W_2 phase (0.5 wt.%) was appropriate for the stabilization of the double emulsions [27].

Table 5. Parameters from droplet size distribution ($D_{3,2}$, span and FI) of $W_1/O/W_2$ multiple emulsions obtained after evaluating three different O/W ratios (10/90, 25/75 and 40/60) for the secondary emulsion over 30 days storage time. Different superscript letters within a column indicate significant differences ($p < 0.05$).

Day	10/90			25/75			40/60		
	$D_{3,2}$ (μm)	Span	FI (%)	$D_{3,2}$ (μm)	Span	FI (%)	$D_{3,2}$ (μm)	Span	FI (%)
0	4.1 ± 0.2 [a]	1.8 ± 0.2 [a]	10.2 ± 3.8 [a]	4.8 ± 0.2 [a]	1.8 ± 0.3 [a]	7.8 ± 3.1 [a]	5.8 ± 0.2 [a]	1.9 ± 0.2 [a]	13.7 ± 5.0 [a]
1	4.2 ± 0.1 [a]	1.7 ± 0.2 [a]	12.1 ± 3.1 [a]	4.6 ± 0.1 [a]	1.6 ± 0.2 [a]	12.2 ± 4.0 [b]	5.6 ± 0.1 [a]	2.1 ± 0.2 [a]	10.3 ± 4.1 [a]
15	4.4 ± 0.1 [a]	1.8 ± 0.1 [a]	6.9 ± 4.1 [b]	5.0 ± 0.2 [a]	1.7 ± 0.3 [a]	11.1 ± 4.2 [b]	5.6 ± 0.2 [a]	2.1 ± 0.1 [a]	6.1 ± 3.2 [b]
30	3.8 ± 0.3 [a]	1.8 ± 0.2 [a]	7.4 ± 2.7 [b]	4.9 ± 0.2 [a]	1.8 ± 0.1 [a]	12.6 ± 5.2 [b]	5.8 ± 0.1 [a]	1.7 ± 0.2 [b]	9.7 ± 3.7 [b]

Moreover, comparing the three systems between them, it can be concluded that the emulsion with the lowest proportion of primary emulsion (10/90) has the lowest Sauter mean diameter ($D_{3,2}$). Thus, as the proportion of primary emulsion increased, the droplet sizes increased slightly, finding the highest $D_{3,2}$ values and the wider DSD for the 40/60 system. With regard to these results, it can be found in the literature that an increase in the dispersed phase comes hand-in-hand with increasing the droplet size [28–30], as long as the emulsifier concentration and emulsification energy remain constant.

In order to analyze the stability of emulsions against creaming, Figure 4 shows the differential profiles of the backscattering variations (ΔBS%) when samples destabilize as a function of time.

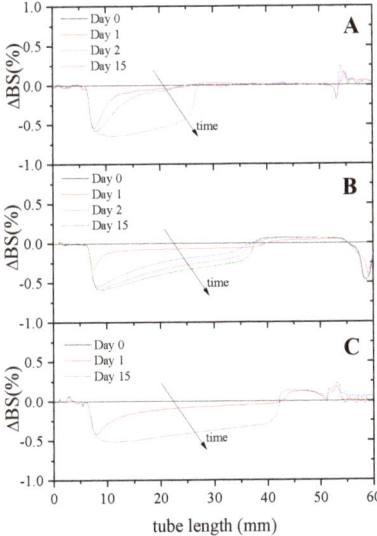

Figure 4. Changes in backscattering profiles (DBS%) as a function of sample height with storage time of $W_1/O/W_2$ multiple emulsions obtained after three different O/W ratios for the secondary emulsion: 40/60 (**A**), 25/75 (**B**) and 10/90 (**C**). The arrows indicate the storage time.

At the beginning of the experiment, the backscattering of light was rather constant along with the entire height, as there was an even distribution of droplets throughout the system. Over time, droplets move upwards due to gravitational forces, which causes a decrease in the backscattering at the bottom of the emulsions (clarification) since the droplet concentration decreases. Conversely, an increase in the backscattering was detected at the top of the tube due to an increase in the droplet concentration (creaming). Clarification phenomenon is observed for all emulsions; however, creaming was only observed for the 10/90 and 25/75 systems.

3.4.2. Rheological Characterization

Figure 5 shows complex moduli (G*) of multiple emulsions as a function of frequency obtained for three different (W_1/O)/W_2 ratios for the secondary emulsion: 10/90, 25/75 and 40/60.

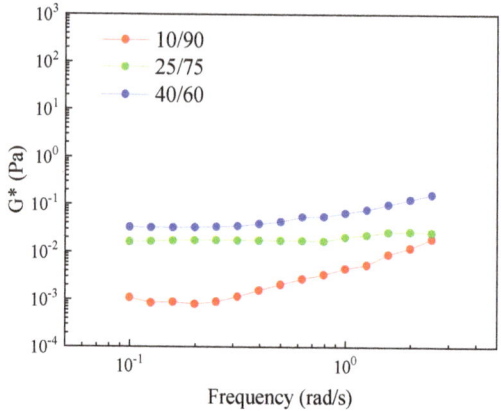

Figure 5. Complex moduli (G*) of multiple emulsions as a function of frequency obtained for three different (W_1/O)/W_2 ratios for the secondary emulsion: 10/90, 25/75 and 40/60.

This Figure shows an increase in the values of the complex modulus (G*) values when the proportion of the primary emulsion (W_1/O) increases (from 10/90 to 40/60), which eventually results in an increase in the viscosity of the system. This same behavior has been previously found for conventional emulsions stabilized by proteins and was attributed to the formation of a network as a consequence of the closeness of the protein-stabilized oil droplets [31]. Moreover, the dependence of the complex moduli (G*) can also be analyzed, where an ideal gel-like response was related to a low dependence of G* with frequency [32]. Thus, the highest dependence of G* on frequency was observed for the system with the lowest W_1/O content, indicating that this system not only had lower values of complex modulus but also that the system had a poorer rheological response.

Table 6 shows complex moduli at 0.2 Hz ($G^*_{0.2}$) as a function of time (1, 15 and 30 days) obtained for $W_1/O/W_2$ multiple emulsions at three different O/W ratios for the secondary emulsion (10/90, 25/75 and 40/60) to quantify the dependence of the $G^*_{0.2}$ on the concentration of the dispersed phase and the storage time.

This Table confirms a decrease in G* when the W_1/O decreases, as well as a decrease in the value of G* over storage time for the three systems studied. The decrease in rheological functions with storage time has been previously observed for systems that suffered from destabilization, either by phase separation (creaming) or by coalescence [33]. In the systems studied in this work, the phenomenon of coalescence must be ruled out, as shown by the independence of the droplet size over time (Table 5). Thus, these results suggest that the structural stability of the systems could be increased by the addition of a hydrocolloid that structures the emulsions generated and prevent them from changing in their structure over storage time.

Table 6. Complex modulus at 0.2 Hz ($G^*_{0.2}$) as a function of time (1, 15 and 30 days) obtained for $W_1/O/W_2$ multiple emulsions at three different O/W ratios for the secondary emulsion (10/90, 25/75 and 40/60). Different superscript letters within a column indicate significant differences ($p < 0.05$).

Day	E_E (%)		
	10/90	25/75	60/40
1	93.3 ± 2.0 [a]	99.8 ± 0.2 [a]	97.2 ± 1.1 [a]
15	95.2 ± 2.5 [a]	95.0 ± 1.2 [b]	94.7 ± 1.5 [b]
30	93.0 ± 3.1 [a]	91.0 ± 2.0 [c]	95.0 ± 1.2 [b]

3.4.3. Encapsulation Efficiency

Table 7 shows E_E as a function of time (1, 15 and 30 days) obtained for $W_1/O/W_2$ multiple emulsions at three different O/W ratios for the secondary emulsion (10/90, 25/75 and 40/60). This table shows suitable results in all cases where E_E was higher than 90% in all cases, regardless of the storage time. Only a slight release of the marker was seen in all systems.

Table 7. Encapsulation efficiency (E_E) as a function of time (1, 15 and 30 days) obtained for $W_1/O/W_2$ multiple emulsions at three different O/W ratios for the secondary emulsion (10/90, 25/75 and 40/60). Different superscript letters within a column indicate significant differences ($p < 0.05$).

Day	$G^*_{0.2}$ (Pa)		
	10/90	25/75	40/60
1	0.023 ± 0.002 [a]	0.055 ± 0.002 [a]	0.079 ± 0.001 [a]
15	0.015 ± 0.003 [b]	0.038 ± 0.004 [b]	0.026 ± 0.003 [b]
30	0.014 ± 0.002 [c]	0.015 [c] ± 0.002 [c]	0.025 ± 0.002 [b]

These results would confirm the suitability of the processing conditions used for these emulsions since time emulsions did not suffer a significant decrease in this parameter after one-month storage. These results indicate that the formulation of these $W_1/O/W_2$ double emulsion systems in the three ratios studied allows the optimal encapsulation of the marker. Thus, the development of food matrices, in the form of multiple emulsions, would allow the encapsulation of bioactive compounds, which could result in the development of novelty food products.

3.5. Processing Conditions Optimization, Addition of Xanthan Gum

Up until now, the results indicate that the double emulsions generated had moderate stability together with an excellent E_E. However, the structural stability of these systems could be increased by the addition of a hydrocolloid that would structure them and avoid destabilization phenomena such as emulsion creaming and further flocculation that could lead to unintended coalescence over storage time.

This section addresses the analysis of the microstructural characteristics of emulsions after the addition of xanthan gum (XG). It can be noticed that the E_E (%) of these systems was not determined since the addition of XG led to turbid emulsions after the centrifugation stage was carried out, which prevented reliable absorbance measurements. However, it is reasonable to assume that the E_E measures will not be affected by the addition of XG because it was dispersed by soft magnetic stirring after the double emulsion preparation.

3.5.1. Droplet Size Distribution and Backscattering Measurements

Table 8 shows parameters from the DSD measurements ($D_{3,2}$ and span) of $W_1/O/W_2$ multiple emulsions obtained at two different xanthan gum (XG) concentrations (0, 0.125 and

0.25 wt.%). These results agree with the results previously obtained for 40/60 (W_1/O)/W_2, confirming that the processing conditions followed did not influence the emulsions generated. Thus, the DSD parameters in Table 8 for the final 25/75 emulsions are practically the same as the parameters obtained for the 40/60 system (shown in Table 5) since these 25/75 double emulsions come from the dilution of 40/60 double emulsions. As for the influence of aging on these emulsions, this table shows that neither $D_{3,2}$ nor Span parameters have undergone significant changes after one-month emulsion storage. The results indicated that the emulsions did not seem to be affected by coalescence phenomena over the storage time analyzed (which agrees with previous results) [26].

Table 8. Parameters from Droplet size distribution ($D_{3,2}$ and span) of $W_1/O/W_2$ multiple emulsions obtained at two different xanthan gum (XG) concentrations: 0, 0.125 and 0.25 wt.%.

Day	0 wt.% XG		0.125 wt.% XG		0.25 wt.% XG	
	$D_{3,2}$ (μm)	Span	$D_{3,2}$ (μm)	Span	$D_{3,2}$ (μm)	Span
0	4.3 ± 0.2 [a]	1.6 ± 0.2 [a]	4.3 ± 0.2 [a]	1.5 ± 0.3 [a]	4.1 ± 0.2 [a]	1.7 ± 0.3 [a]
1	4.6 ± 0.1 [a]	1.7 ± 0.3 [a]	4.0 ± 0.3 [a]	1.6 ± 0.3 [a]	4.1 ± 0.3 [a]	1.8 ± 0.2 [a]
15	4.0 ± 0.3 [a]	1.5 ± 0.3 [a]	3.9 ± 0.3 [a]	1.6 ± 0.2 [a]	3.8 ± 0.4 [a]	1.7 ± 0.3 [a]
30	3.9 ± 0.3 [a]	1.8 ± 0.3 [a]	3.8 ± 0.4 [a]	2.0 ± 0.2 [b]	3.8 ± 0.3 [a]	2.0 ± 0.3 [a]

In order to analyze the stability of these emulsions against creaming, Figure 6 shows the changes in backscattering profiles (ΔBS%) at two different XG concentrations (0, 0.125 and 0.25 wt.%). These graphs show that an increase in XG concentration gives rise to an increase in stability against creaming; this phenomenon disappeared completely at the highest concentration studied (0.25% XG). These results indicate that the addition of the hydrocolloid causes an increase in stability as a consequence of the structuring of the system [34].

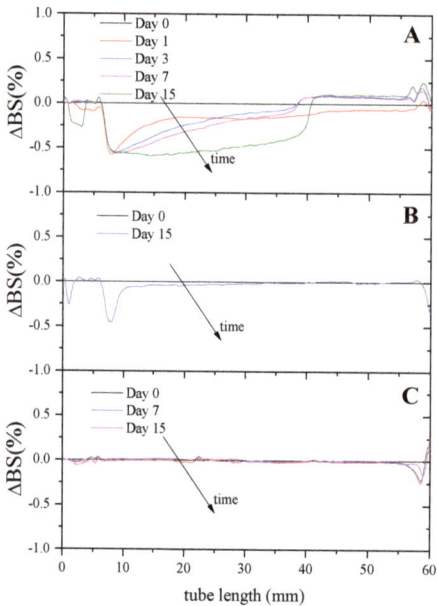

Figure 6. Backscattering profiles of $W_1/O/W_2$ multiple emulsions obtained at two different xanthan gum (XG) concentrations: 0 (**A**), 0.125 (**B**) and 0.25 (**C**) wt.%.

Previous results indicated the occurrence of microstructural changes in these emulsions. Consequently, rheological tests were carried out to investigate if the addition of

XG reduces it. Bulk rheology helps the understanding of the physical origin of emulsion creaming and flocculation [35].

3.5.2. Rheological Characterization

Figure 7 shows viscoelastic moduli (G' and G'') of $W_1/O/W_2$ multiple emulsions as a function of frequency obtained at three different XG concentrations (0, 0.125 and 0.25 wt.%) in the same day emulsion preparation. The rheological response obtained corresponds to a weak gel-like structure, where the elastic component is predominant (G' > G''), even for the lowest XG concentration. This response has been previously observed for diluted emulsions (which is the case) due to the lack of emulsion flocculation [28].

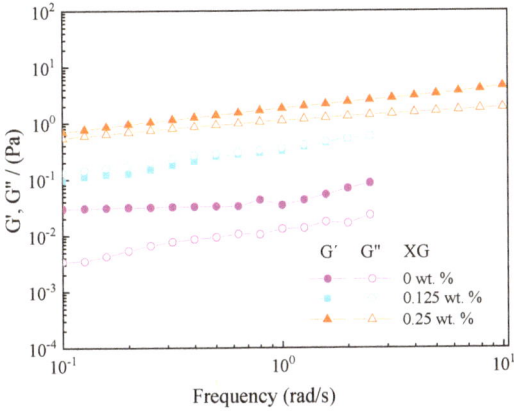

Figure 7. Viscoelastic moduli (G' and G'') as a function of frequency of $W_1/O/W_2$ multiple emulsions obtained at three different xanthan gum (XG) concentrations (0, 0.125 and 0.25 wt.%) obtained the same day emulsion preparation.

The influence of XG on structuring the $W_1/O/W_2$ multiple emulsions generated is corroborated since an increase in XG concentration involved an increase of the viscoelastic moduli (G' and G''). Thus, the emulsions with the lowest XG concentration exhibited the weakest microstructure (lower viscoelastic moduli), and the emulsion with the highest XG concentration exhibited the highest G' and G'' values. The gel-like behavior of emulsions has been attributed either to the packing effect of droplets surrounded by proteins and/or the formation of a biopolymer network [36].

4. Conclusions

The results show that multiple $W_1/O/W_2$ emulsions can be obtained after selecting suitable processing conditions, which in turn, makes them easily scalable for industrial processes. This has great importance since these emulsions could be used for encapsulation of both hydrophilic and lipophilic components. The analysis of the processing conditions showed that several passes of the emulsion through the valve homogenizer led to narrower droplet size distributions. However, excessive processing resulted in droplet coalescence due to an excess of energy input, which involved larger droplet sizes. Thus, in terms of droplet size distribution, the input of higher energy to the system (20 MPa) during emulsification processing led to emulsions with smaller droplet sizes ($D_{3,2}$). However, more monodispersed emulsions were achieved when the lowest pressure (5 MPa) was used. As for the number of passes, the optimal (emulsions more monodispersed and smaller droplet sizes) was found around 2–3 passes regardless of the valve homogenizer pressure. However, emulsions processed at 20 MPa involved lower EE values than emulsions processed at 5 MPa (87.3 ± 2.3 vs. 96.1 ± 1.8, respectively).

Although microstructural properties of emulsions can be modulated by the ratio between the primary and secondary emulsions ((W_1/O)/W_2 ratio), all the emulsions generated without XG presented similar stability and EE (~95%). This result suggests that different products exhibiting different textures (i.e., mayonnaise, salad dressings, milkshake, etc.) could be developed based on these emulsions. However, emulsions without XG resulted in a higher destabilization over storage time.

These results indicate that the formulation of these W1/O/W2 double emulsion systems allows the optimal encapsulation of the marker. Thus, the development of food matrices, in the form of multiple emulsions, would allow the encapsulation of bioactive compounds, which could result in the development of novelty food products.

Supplementary Materials: The following supporting information can be downloaded at: https://www.mdpi.com/article/10.3390/foods11091367/s1, Table S1. Correlation of systems analyzed in this research, evaluating two parameters: the concentration of the oil phase ratio in the secondary emulsion as well as the effect of the thickening agent. Figure S1. Images of double emulsion droplets (emulsion containing 25/75 of (W_1/O)/W_2) obtained from optical microscopy.

Author Contributions: Conceptualization, M.F., A.G. and C.C.-S.; methodology, C.C.-S.; software, M.F. and C.C.-S.; validation, C.C.-S.; formal analysis, M.F., A.G. and C.C.-S.; investigation, M.F.; resources, A.G. and C.C.-S.; data curation, M.F. and C.C.-S.; writing—original draft preparation, M.F.; writing—review and editing, C.C.-S. and A.G.; visualization, M.F.; supervision, C.C.-S. and A.G.; project administration, A.G.; funding acquisition, A.G. All authors have read and agreed to the published version of the manuscript.

Funding: This research was funded by Consejería de Transformación económica, Industria, Conocimiento y Universidades de la Junta de Andalucía-Agencia Andaluza del Conocimiento by the grant Ref. PY20_01046, Junta de Andalucía/FEDER, UE.

Institutional Review Board Statement: Not applicable.

Informed Consent Statement: Not applicable.

Data Availability Statement: The datasets generated for this study are available on request to the corresponding author.

Acknowledgments: The authors acknowledge Cristina Caro for her support in the lab. Authors also acknowledge the financial support of the Consejería de Transformación económica, Industria, Conocimiento y Universidades de la Junta de Andalucía-Agencia Andaluza del Conocimiento by the grant Ref. PY20_01046, Junta de Andalucía/FEDER, UE.

Conflicts of Interest: The authors declare no conflict of interest.

References

1. Araiza-Calahorra, A.; Akhtar, M.; Sarkar, A. Recent advances in emulsion-based delivery approaches for curcumin: From encapsulation to bioaccessibility. *Trends Food Sci. Technol.* **2018**, *71*, 155–169. [CrossRef]
2. McClements, D.J. Encapsulation, protection, and delivery of bioactive proteins and peptides using nanoparticle and microparticle systems: A review. *Adv. Colloid Interface Sci.* **2018**, *253*, 1–22. [CrossRef] [PubMed]
3. Fathi, M.; Mozafari, M.R.; Mohebbi, M. Nanoencapsulation of food ingredients using lipid based delivery systems. *Trends Food Sci. Technol.* **2012**, *23*, 13–27. [CrossRef]
4. Rein, M.J.; Renouf, M.; Cruz-Hernandez, C.; Actis-Goretta, L.; Thakkar, S.K.; da Silva Pinto, M. Bioavailability of bioactive food compounds: A challenging journey to bioefficacy. *Br. J. Clin. Pharmacol.* **2013**, *75*, 588–602. [CrossRef] [PubMed]
5. Jang, D.J.; Jeong, E.J.; Lee, H.M.; Kim, B.C.; Lim, S.J.; Kim, C.K. Improvement of bioavailability and photostability of amlodipine using redispersible dry emulsion. *Eur. J. Pharm. Sci.* **2006**, *28*, 405–411. [CrossRef] [PubMed]
6. Muschiolik, G.; Dickinson, E. Double Emulsions Relevant to Food Systems: Preparation, Stability, and Applications. *Compr. Rev. Food Sci. Food Saf.* **2017**, *16*, 532–555. [CrossRef]
7. Eisinaite, V.; Estrada, P.D.; Schroën, K.; Berton-Carabin, C.; Leskauskaite, D. Tayloring W/O/W emulsion composition for effective encapsulation: The role of PGPR in water transfer-induced swelling. *Food Res. Int.* **2018**, *106*, 722–728. [CrossRef]
8. Ding, S.; Serra, C.A.; Vandamme, T.F.; Yu, W.; Anton, N. Double emulsions prepared by two—Step emulsification: History, state-of-the-art and perspective. *J. Control. Release* **2019**, *295*, 31–49. [CrossRef]
9. Diep, T.T.; Dao, T.P.; Vu, H.T.; Phan, B.Q.; Dao, D.N.; Bui, T.H.; Truong, V.; Nguyen, V. Double emulsion oil-in water-in oil (O/W/O) stabilized by sodium caseinate and k-carrageenan. *J. Dispers. Sci. Technol.* **2018**, *39*, 1752–1757. [CrossRef]

10. Iancu, M.-N.; Chevalier, Y.; Popa, M.; Hamaide, T. Internally Gelled W/O and W/O/W Double Emulsions. *E-Polymers* **2009**, *9*, 1184. [CrossRef]
11. Nafisi, S.; Maibach, H.I. Chapter 22—Nanotechnology in Cosmetics. In *Cosmetic Science and Technology*; Sakamoto, K., Lochhead, R.Y., Maibach, H.I., Yamashita, Y., Eds.; Elsevier: Amsterdam, The Netherlands, 2017; pp. 337–369, ISBN 978-0-12-802005-0.
12. Leister, N.; Karbstein, H.P. Evaluating the Stability of Double Emulsions—A Review of the Measurement Techniques for the Systematic Investigation of Instability Mechanisms. *Colloids Interfaces* **2020**, *4*, 8. [CrossRef]
13. Lynch, A.G.; Mulvihill, D.M. Effect of sodium caseinate on the stability of cream liqueurs. *Int. J. Dairy Technol.* **1997**, *50*, 1–7. [CrossRef]
14. Puppo, M.C.; Speroni, F.; Chapleau, N.; de Lamballerie, M.; Añón, M.C.; Anton, M. Effect of high-pressure treatment on emulsifying properties of soybean proteins. *Food Hydrocoll.* **2005**, *19*, 289–296. [CrossRef]
15. Mengual, O.; Meunier, G.; Cayré, I.; Puech, K.; Snabre, P. TURBISCAN MA 2000: Multiple light scattering measurement for concentrated emulsion and suspension instability analysis. *Talanta* **1999**, *50*, 445–456. [CrossRef]
16. Regan, J.O.; Mulvihill, D.M. Water soluble inner aqueous phase markers as indicators of the encapsulation properties of water-in-oil-in-water emulsions stabilized with sodium caseinate. *Food Hydrocoll.* **2009**, *23*, 2339–2345. [CrossRef]
17. Dickinson, E.; Evison, J.; Gramshaw, J.W.; Schwope, D. Flavour release from a protein-stabilized water-in-oil-in-water emulsion. *Food Hydrocoll.* **1994**, *8*, 63–67. [CrossRef]
18. Floury, J.; Desrumaux, A.; Lardières, J. Effect of high-pressure homogenization on droplet size distributions and rheological properties of model oil-in-water emulsions. *Innov. Food Sci. Emerg. Technol.* **2000**, *1*, 127–134. [CrossRef]
19. Ciron, C.I.E.; Gee, V.L.; Kelly, A.L.; Auty, M.A.E. Comparison of the effects of high-pressure microfluidization and conventional homogenization of milk on particle size, water retention and texture of non-fat and low-fat yoghurts. *Int. Dairy J.* **2010**, *20*, 314–320. [CrossRef]
20. Wang, S.; Yang, J.; Shao, G.; Qu, D.; Zhao, H.; Yang, L.; Zhu, L.; He, Y.; Liu, H.; Zhu, D. Soy protein isolated-soy hull polysaccharides stabilized O/W emulsion: Effect of polysaccharides concentration on the storage stability and interfacial rheological properties. *Food Hydrocoll.* **2020**, *101*, 105490. [CrossRef]
21. Schuch, A.; Wrenger, J.; Schuchmann, H.P. Production of W/O/W double emulsions. Part II: Influence of emulsification device on release of water by coalescence. *Colloid. Surface A* **2014**, *461*, 344–351. [CrossRef]
22. McClements, D.J. *Food Emulsions: Principles, Practices, and Techniques*; CRC Press: Boca Raton, FL, USA, 2015; ISBN 9781498726689.
23. Ooi, Z.Y.; Othman, N.; Choo, C.L. The Role of Internal Droplet Size on Emulsion Stability and the Extraction Performance of Kraft Lignin Removal from Pulping Wastewater in Emulsion Liquid Membrane Process. *J. Dispers. Sci. Technol.* **2016**, *37*, 544–554. [CrossRef]
24. Goodarzi, F.; Zendehboudi, S. A Comprehensive Review on Emulsions and Emulsion Stability in Chemical and Energy Industries. *Can. J. Chem. Eng.* **2015**, *37*, 544–554. [CrossRef]
25. Altuntas, O.Y.; Sumnu, G.; Sahin, S. Preparation and characterization of W/O/W type double emulsion containing PGPR–lecithin mixture as lipophilic surfactant. *J. Dispers. Sci. Technol.* **2016**, *38*, 486–493. [CrossRef]
26. Tadros, T. *Emulsion Formation and Stability*; Wiley: Hoboken, NJ, USA, 2013; ISBN 9783527647965.
27. Dickinson, E.; Golding, M.; Povey, M.J.W. Creaming and Flocculation of Oil-in-Water Emulsions Containing Sodium Caseinate. *J. Colloid Interface Sci.* **1997**, *185*, 515–529. [CrossRef] [PubMed]
28. Taherian, A.R.; Fustier, P.; Ramaswamy, H.S. Effect of added oil and modified starch on rheological properties, droplet size distribution, opacity and stability of beverage cloud emulsions. *J. Food Eng.* **2006**, *77*, 687–696. [CrossRef]
29. Dłużewska, E.; Stabiecka, A.; Maszewska, M. Effect of oil phase concentration on rheological properties and stability of beverage emulsion. *Acta Sci. Pol. Technol. Aliment.* **2006**, *5*, 147–156.
30. Gallegos, C.; Partal, P.; Franco, J.M. Droplet-size distribution and stability of lipid injectable emulsions. *Am. J. Heal. Pharm.* **2009**, *66*, 162–166. [CrossRef]
31. Pal, R.; Rhodes, E. Viscosity/Concentration Relationships for Emulsions. *J. Rheol.* **1989**, *33*, 1021–1045. [CrossRef]
32. Tadros, T.F. Correlation of viscoelastic properties of stable and flocculated suspensions with their interparticle interactions. *Adv. Colloid Interface Sci.* **1996**, *68*, 97–200. [CrossRef]
33. Felix, M.; Romero, A.; Guerrero, A. Influence of pH and Xanthan Gum on long-term stability of crayfish-based emulsions. *Food Hydrocoll.* **2017**, *72*, 372–380. [CrossRef]
34. Huang, X.; Kakuda, Y.; Cui, W. Hydrocolloids in emulsions: Particle size distribution and interfacial activity. *Food Hydrocoll.* **2001**, *15*, 533–542. [CrossRef]
35. Dickinson, E. Structure, stability and rheology of flocculated emulsions. *Curr. Opin. Colloid Interface Sci.* **1998**, *3*, 633–638. [CrossRef]
36. Calero, N.; Munoz, J.; Cox, P.W.; Heuer, A.; Guerrero, A. Influence of chitosan concentration on the stability, microstructure and rheological properties of O/W emulsions formulated with high-oleic sunflower oil and potato protein. *Food Hydrocoll.* **2013**, *30*, 152–162. [CrossRef]

Article

Occurrence of Linear Alkylbenzene Sulfonates, Nonylphenol Ethoxylates and Di(2-ethylhexyl)phthalate in Composting Processes: Environmental Risks

Julia Martín *, Carmen Mejías, Marina Arenas, Juan Luis Santos, Irene Aparicio and Esteban Alonso

Departamento de Química Analítica, Escuela Politécnica Superior, Universidad de Sevilla, C/Virgen de África 7, E-41011 Sevilla, Spain; cmpadilla@us.es (C.M.); mamolina@us.es (M.A.); jlsantos@us.es (J.L.S.); iaparicio@us.es (I.A.); ealonso@us.es (E.A.)
* Correspondence: jbueno@us.es

Abstract: Composting is an important waste management strategy, providing an economical and environment-friendly approach to sanitizing and stabilizing biosolids for land soil amendment. However, the resulting product can contain a large number of organic pollutants that may have adverse effects on the ecosystem. This paper presents the occurrence of eight widely used organic pollutants (four linear alkylbenzene sulfonates (LAS C10-C13), nonylphenol and its mono- and di-ethoxylates (NPE) and a di(2-ethylhexyl)phthalate (DEHP)) in full-scale composting processes. LAS homologues were detected at the highest concentrations (range of $\sum$LAS: 2068–9375 mg kg^{-1} dm), exceeding the limit fixed in the EU Directive draft. The concentration levels of the NPE and DEHP were significantly lower (up to 27.5 and 156.8 mg kg^{-1} dm, respectively) and did not exceed their fixed limits in the EU Directive draft. Ecotoxicological risk assessment for when compost is amended onto soils has also been evaluated. The concentrations measured represented a medium-low risk for most compounds, although it was not enough in the case of LAS C11 and C13 and NP.

Keywords: linear alkylbenzene sulfonate; nonylphenol ethoxylates; di(2-ethylhexyl)phthalate; occurrence; composting processes; risk assessment

1. Introduction

Every year, a large amount of sewage sludge (≈15 million tons of sludge (dry matter [dm]) in the European Union (EU) 28) requiring disposal is produced as a result of wastewater treatments [1]. Approximately 40% of the produced sludge is used for agricultural purposes as a source of organic matter and nutrients. However, this percentage has large variations between the EU member states (from 0% in Malta or the Netherlands to more than 50% in countries like Spain or France) [1,2]. In Spain, particularly in the Andalusia region, this practice is even more pronounced (64%), since the decline in soil quality occurs due to loss of the organic fraction [1–3].

Sewage sludge is rich in nutrients such as nitrogen and phosphorus and contains valuable organic matter. Composting provides an economical and environment-friendly approach to sanitizing and stabilizing biosolids for land soil amendment and enrichment, since its agronomic value is increased as a result of fermentation and maturation [4]. However, one of the issues with this practice is the presence of a large number of organic pollutants that, even when present at low concentrations, may have adverse effects on the ecosystem and which may even be concentrated during the process. In the European Union, the application of sludge onto soil is regulated by Directive 86/278/CEE, which establishes a limit of 10 tons (Tn) of dry matter (dm) of treated sewage sludge per hectare and year. This directive sets limit values for seven heavy metals in sewage sludge used in agriculture. Currently, it is considered out of date and has been earmarked by the commission as a candidate for revision for around 10 years [3,5]. In 2000, the elaboration of the "3rd draft of the working document on sludge" set stricter limits on heavy metals and also studied the possibility of analyzing some priority organic compounds such as linear alkyl sulfonates (LAS), nonylphenol (n)ethoxylates (NPE) and phthalates, among others, in sludge [6–9]. These compounds can enter the sewage system through both industrial and domestic sources. One of the most controversial aspects of the document is the question of which organic contaminants should be monitored and if limit values for these compounds in sludge should be set for land applications of sludge. There are no unanimous criteria for the classes of pollutants that must be controlled in sewage sludge or for limiting the concentrations that determine the applicability or lack thereof for these sludges with crop soils. Some limit values for the concentrations of these compounds in sewage sludge applied to the soil have been fixed in some particular countries, such as Denmark (DEHP, LAS, NPE and PAH), Sweden (NPE, PAH and PCBs) or Austria and Germany (PCBs and PCDD/F) [2].

The presence of these organic compounds at detectable amounts (from $\mu g\ kg^{-1}$ to $mg\ kg^{-1}$ dm) in treated sewage sludge has led to concerns that land applications of biosolids may result in the accumulation of contaminants in the soil and their subsequent translocation through the food chain [10]. With this work, we wanted to reveal the necessity of introducing programs to monitor the presence of priority organic pollutants in sewage sludge and reduce the potential sources that may enter the environment. The aim of this work was to investigate the occurrence, removal and environmental risk assessment of eight priority organic pollutants (four LAS homologues (C10-C13), nonylphenol (NP) and its mono- and di-ethoxylates (NP1EO and NP2EO) and a di(2-ethylhexyl)phthalate (DEHP)) in two piles of sewage sludge along the composting process in a full-scale study.

2. Materials and Methods

2.1. Chemicals and Reagents

HPLC-grade acetone, acetonitrile, methanol and water were supplied by Romil Ltd. (Barcelona, Spain). Analytical grade formic acid and ammonium formate were obtained from Panreac (Barcelona, Spain).

A commercial LAS mixture containing C10 (12.3%), C11 (32.1%), C12 (30.8%) and C13 (23.4%) was obtained from Petroquímica Española (PETRESA). Technical-grade NP and DEHP were supplied by Riedel-de Haën (Seelze, Germany). The solutions, individual stock (1000 mg L^{-1}) and mixture (10 mg L^{-1}) were prepared in methanol and stored at 4 °C.

Oasis MCX cartridges (60 mg) for solid phase extraction (SPE) were acquired from Waters (Milford, MA, USA).

2.2. Biosolid Samples

The composting process for two piles of biosolids (pile A and B) was carried out in dynamic full-scale batteries that were thermally controlled. Aeration was facilitated by mechanical turning. This composting system is the most widely used system. The material is piled up, mixed and turned periodically, thus avoiding compaction and delivering oxygen to the system. Each pile of biosolid was composted for 127 days. The two piles were equal and were measured to ensure the representativeness of the results. Eight samples of each composting pile were taken during the composting process: 3 during the dryness phase on days 0 (initial product), 10 and 21, 3 during the fermentation phase on days 37, 65 and 85 and 2 during the maturation phase on days 99 and 127 (final product). Each sample (1 kg) was composed of 5 aliquots of biosolids (using a gripping device) from different parts of the pile. Once in the lab, the samples were lyophilized (0.01-mbar vacuum after being frozen at −18 °C for 24 h) and sieved (<0.1 mm).

2.3. Determination of Sludge Characterization Parameters

The parameters measured to characterize the sludge samples were the pH, conductivity, soluble salts, nitrogen Kjeldahl (NK), crude protein, carbon (C), organic matter, C/N, Ca, Mg, total phosphorus (TP), extracted phosphorus (EP), total potassium (TK), extracted potassium (EK), moisture, temperature and heavy metals Pb, Fe, Mn, Ni, Cu, Cr, Cd, Zn, Co,

Al and Hg. Standard methods for wastewater analysis and standard techniques compiled by APHA-AWWA-WPCF were used [11].

2.4. Analysis of LAS, NPE and DEHP

LAS, NPE and DEHP were extracted by ultrasonic solvent extraction (USE) following a previously published method [12]. Aliquots of 2 g of biosolids were put in contact with two aliquots of acetone (3 mL) and one aliquot of methanol (3 mL) for each extraction step (10 min). The supernatants were separated by centrifugation (4000 rpm, 20 min), combined and cleaned up by SPE with Oasis MCX cartridges. The samples were passed through the SPE cartridges (previously conditioned) at a flow rate of 1 mL min^{-1}. Next, an aliquot of 5 mL of the cleaned extract was evaporated to dryness and reconstituted in 1 mL of a methanol:water mixture (1:1, v/v).

Analytical determination was performed on an Agilent 1200 series HPLC coupled to a 6410 triple quadrupole (QqQ) mass spectrometer (MS) (Agilent, USA) using a previously developed method [13]. For compound separation, a Zorbax Eclipse XDB–C18 Rapid Resolution HT (50 × 4.6 mm i.d.; 1.8 µm) column (Agilent, USA) was used. The mobile phase was composed of an aqueous 15 mM ammonium formate solution (0.1% v/v formic acid) (solvent A) and acetonitrile (0.1% v/v formic acid) (solvent B). Chromatographic elution started with a linear gradient from 75% to 100% of solvent B in 5 min at a flow-rate of 0.6 mL min^{-1} and held to 5 min. The column was thermostated at 35 °C.

The analytical method was previously validated. Table 1 shows a summary of some of the most significant parameters of the method. Method detection (MDL) and quantification (MQL) limits were calculated as the concentrations corresponding to signal-to-noise ratios of 3 and 10, respectively, by means of spiked samples at low concentration levels. External calibration was used for quantitation. Quality control was applied by analyzing the procedural blanks, standard solutions and spiked samples in duplicate every 10 samples to evaluate possible contaminations and instrumental analysis variability.

Table 1. Recovery (R), precision (% expressed as relative standard deviation (RSD)), method detection limits (MDLs) and method quantification limit (MQL) in compost samples.

Compound	R (%)	RSD (%)	MDL (µg kg^{-1} dm)	MQL (µg kg^{-1} dm)
LAS C$_{10}$	79	4.4	0.18	0.61
LAS C$_{11}$	97	4.5	0.51	1.69
LAS C$_{12}$	97	4.2	0.50	1.66
LAS C$_{13}$	99	5.0	0.39	1.32
NP$_1$EO	77	8.6	1.52	5.07
NP$_2$EO	94	1.5	0.32	1.05
NP	73	4.8	0.31	1.04
DEHP	76	8.3	6.94	23.1

2.5. Ecotoxicological Risk Assessment

Ecotoxicological risk assessment (ERA) was assessed using risk quotient (RQ) values in soil amended with compost. The RQ was calculated as the quotient between the predicted environmental concentration in the soil (PEC$_{soil}$) and the predicted no-effect concentration (PNEC). The PEC$_{soil}$ values provided an estimation of the concentration of a substance expected in the soil after a one-dose application of compost (Equation (1)):

$$PEC_{soil} = C_{compost} \cdot APPL_{compost} / DEPTH_{soil} \cdot RHO_{soil} \quad (1)$$

where, according to the European Union Technical Guideline Document [14], C$_{compost}$ is the concentration measured in the compost (µg kg^{-1} dm), APPL$_{compost}$ is the application rate of dry compost onto soils (0.5 kg m^{-2} year); DEPTH$_{soil}$ is the mixing depth (0.20 m) and RHO$_{soil}$ is the bulk density of wet soil (1700 kg m^{-3}) for agricultural soils.

PNECs are usually calculated using the lowest acute Lethal Concentration 50 (LC_{50}) or Effective Concentration 50 (EC_{50}) toxicity data in fish, Daphnia magna or algae and dividing them by an assessment factor of 1000 to consider the worst case scenario.

It should be noted that the data collected from the literature on the toxicity of the selected compounds in terrestrial organisms were very limited. According to the European Union Technical Guideline Document, $PNEC_{soil}$ was estimated through the $PNEC_{water}$ and partition coefficients (K_d) approach (Equation (2)) [15], assuming that the sensitivity of the terrestrial organisms was comparable:

$$PNEC_{solid} = PNEC_{water} \cdot K_d \quad (2)$$

The ecotoxicological data and K_d values from the literature are compiled in Table 2. RQ values >1 would indicate that the compound did imply a significant risk to the terrestrial microorganisms in the soil.

Table 2. Ecotoxicological data, solid water partition coefficient (K_d) and predicted no-effect concentration (PNEC) of each compound in water and soil.

	Ecotoxicological Data			$PNEC_{water}$		$PNEC_{soil}$
	Organism	Test	Toxicological Value (mg L^{-1})	(µg L^{-1})	Log K_d	(µg kg^{-1})
C10	*Daphnia magna* (invertebrate)	LC_{50} (48 h)	13.9 [1]	13.9	2.72 [1]	7294.8
C11	*Nannochloropsis gaditana* (algae)	EC_{50} (72 h)	1.38 [2]	1.38	2.60 [1]	549.4
C12	*Pimephales promelas* (fish)	EC_{50} (48 h)	3.2 [2]	3.2	3.53 [1]	10,843.0
C13	*Nannochloropsis gaditana* (algae)	EC_{50} (72 h)	0.18 [2]	0.18	2.09 [5]	22.1
NP$_2$EO	*Mysidopsis bahia* (invertebrate)	LC_{50} (48 h)	0.11 [3]	0.11	3.76 [6]	633.0
NP$_1$EO	*Mysidopsis bahia* (invertebrate)	LC_{50} (48 h)	0.11 [3]	0.11	3.64 [7]	480.2
NP	*Mysidopsis bahia* (invertebrate)	LC_{50} (96 h)	0.02 [3]	0.02	2.67 [7]	9.4
DEHP	*Selenastrum capricornutum* (algae)	LC_{50} (48 h)	0.1 [4]	0.1	3.90 [8]	794.3

[1] Ying, 2006 [16]. [2] Garrido-Perez et al., 2008 [17]. [3] Fenner et al., 2002 [18]. [4] Rhodes et al., 1995 [19]. [5] Feijtel et al., 1999 [20]. [6] During et al., 2002 [21]. [7] Yu et al., 2008 [22]. [8] http://echa.europa.eu/documents/10162/060d4981-4dfb-4e40-8c69-6320c9debb01; Accessed December 2021 [23]. LC_{50}: Lethal Concentration 50; EC_{50}: Effective Concentration 50.

3. Results

3.1. Characterization Parameters

Temperature plays an important role in composting efficiency. In both compost piles, the temperature reached its maximum at the beginning of the fermentation step in about 37 days (41 °C in pile A and 49 °C in pile B). However, while the temperature of pile B was higher than 40 °C during days 21–65, the temperature measured in pile A was higher than 40 °C only in the first days of the fermentation step. The rise in temperature during composting was mainly due to the evolution of metabolic heat and the initial temperature of the pile, so this would indicate that the transformation of readily available substrates in pile B was higher than in pile A.

Table 3 shows those values at the initial and final stages of composting of the two biosolid piles. The initial moisture content was 79 and 73% in pile A and B, respectively. These values decreased continuously over 85 days, reaching their minimum values (8% and 6% in pile A and B, respectively) and then remaining almost constant until the end of the composting process (11% moisture content in both piles). The moisture content varied as a function of the aeration rate, agitation, the bulking agent used and the metabolic production of heat. Therefore, the similar progression of the moisture content in both piles would indicate similar conditions.

Table 3. Characterization parameters of the two biosolid piles.

Parameter	Units	Pile A		Pile B	
		Day 0	Day 127	Day 0	Day 127
Kjeldahl nitrogen	% dm	5.7	2.8	4.8	2.8
Crude protein	% dm	35.8	17.4	29.8	17.4
Carbon	% dm	41	17	40	21
Organic matter	% dm	71	30	69	37
C/N	-	7.2	6.1	8.4	7.6
Ca (CaO)	% dm	1.1	1.6	1.6	1.5
Mg (MgO)	% dm	0.65	0.17	0.41	0.34
Extracted P (P_2O_5)	% dm	2.9	0.972	1.96	0.589
Total P (P_2O_5)	% dm	6	4.7	5	4.5
Extracted K (K_2O)	% dm	0.21	0.16	0.12	0.11
Total Pb	mg kg^{-1} dm	135	99	89	81
Total K (K_2O)	% dm	0.36	0.54	0.28	0.55
Conductivity	mS cm^{-1} 25 °C dm	4.7	4.2	4.2	4.8
Soluble salts	%	0.30	0.27	0.27	0.31
Total Fe	mg kg^{-1} dm	9054	16,099	14,054	18,345
Total Mn	mg kg^{-1} dm	223	418	181	324
Total Ni	mg kg^{-1} dm	<20	23	28	30
Total Cu	mg kg^{-1} dm	249	182	300	223
Total Cr	mg kg^{-1} dm	49	56	55	63
Total Cd	mg kg^{-1} dm	1.0	0.8	1.3	1.1
Total Zn	mg kg^{-1} dm	771	643	1043	859
Total Co	mg kg^{-1} dm	<10	<10	<10	25
Total Al	mg kg^{-1} dm	12,509	26,984	14,843	28,639
Total Hg	mg kg^{-1} dm	1.20	-	1.30	-
pH	Und pH dm	8.2	8.3	8.1	8.0
Moisture	%	79	11	73	11

-: no detected.

The metal content showed a different behavior. The concentrations of K, Fe, Mn, Ni, Cr and Al increased during the composting process in both piles. This could be explained by the lost weight of the studied piles due to the degradation of the organic matter (between 30 and 40%). This fact was previously explained by other authors [24,25]. The total N in pile A decreased from 5.7 to 2.8% dm, and this decrease was notably lower in pile B (from 4.8 to 2.8% dm).

The sludge to be composted had a pH between 8.1 and 8.2 and was slightly acidic (between 6) a few days later. During fermentation, the pH value dropped to 4.5–5.5 due to bacterial activity and the formation of weak acids. Subsequently, the fermentable material produced an alkaline reaction due to the formation of ammonia resulting from the degradation of proteins and amino acids. The pH reached values close to neutral or slightly alkaline at the end of the process (pH 8.0–8.3).

The C/N ratio was reduced by the loss of C as CO_2 and N during composting. This reduction was a bit more notable in pile A (15%) than in pile B (9%), which was related to the higher decrease in the C (59%) and N (51%) content in pile A. Other parameters, such as conductivity, remained almost constant during the composting process in both piles.

The appropriate characteristics of the initial product and the composting process contributed to the increase in the agronomic value of the compost obtained in both piles.

3.2. Concentrations of Organic Compounds

All of the studied priority organic compounds (LAS, NPE and DEHP) were detected in all samples analyzed (Table 4) regardless of the pile of compost, except in one sample from pile B, where the concentration of NP was lower than the MQL of the applied analytical method.

Table 4. Mean and range concentration levels of priority organic compounds.

Group	Compound	Pile A		Pile B	
		Mean mg kg^{-1}dm	Range mg kg^{-1}dm	Mean mg kg^{-1}dm	Range mg kg^{-1}dm
LAS	C_{10}	127	67.9–214	242	128–371
	C_{11}	889	502–1451	1690	888–2520
	C_{12}	1478	793–2462	2585	1395–3647
	C_{13}	1207	705–1926	2093	1159–2836
NPE	NP_1EO	1.39	0.93–2.28	1.93	1.23–2.79
	NP_2EO	13.6	11.2–16.8	14.8	8.68–20.1
	NP	6.45	4.89–9.26	5.26	<MQL–10.6
Phthalate	DEHP	7.61	4.91–12.8	12.4	6.56–15.8

LAS homologues were detected at the highest concentrations in all compost samples, being lower in pile A (range of ∑LAS: 2068–6033 mg kg^{-1} dm) than in pile B (range of ∑LAS: 3569–9375 mg kg^{-1}dm), especially in the case of the homologues C11, C12 and C13. The highest concentrations were found in the case of the homologue C12, followed by the homologues C13, C11 and C10. This distribution was similar to those found in household products such as laundry detergents, dishwashing liquids, shampoos and other personal care products, which points to the domestic contribution to the samples. These elevated concentrations could be attributed to their intense usage and the strong tendency to partition into and persist in compost, despite being largely removed to the water phase in wastewater treatment plants [26,27]. Taking into consideration the limit values fixed by the Working Document on Sludge in the case of LAS (2600 mg kg^{-1}dm), the concentrations of LAS measured in both piles of compost were higher than this limit in all steps of the composting processes.

The concentration levels of NPE and DEHP were similar in both piles of compost. Considering NPE, the higher concentrations were found in the case of NP2EO, followed by NP and NP1EO. In both cases, the concentrations measured were lower than the limit values fixed in the Working Document on Sludge (50 mg kg^{-1}dm for NPE and 100 mg kg^{-1}dm for DEHP). In the case of pile A, concentrations ranged from 18.7 mg kg^{-1}dm to 26.9 mg kg^{-1}dm for NPE, and in pile B, they ranged from 17.8 to 27.5 mg kg^{-1}dm. In the case of DEHP, these concentrations ranged from 4.91 to 12.8 mg kg^{-1}dm and from 6.56 to 15.8 mg kg^{-1}dm in pile A and B, respectively.

3.3. Occurrence of Organic Pollutants during the Composting Process

The concentrations of the studied compounds measured during the composting process in piles A and B are shown in Figure 1. The concentrations measured in pile A showed 2 punctual rises at days 10 (dryness phase) and 99 (maturation phase), especially in the case of the LAS homologues. As a result, only a slight decrease in the concentrations was observed during composting. Regarding pile B, a continuous decrease in the concentrations was observed, reaching higher global removal in pile B.

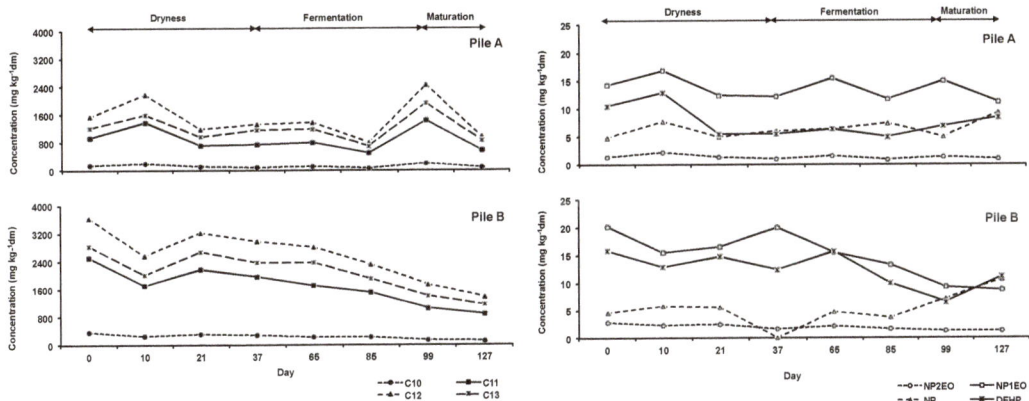

Figure 1. Evolution of the concentrations of LAS homologues, NPE and DEHP during composting in piles A and B (each sample was measured in triplicate, and the RSD was ≤1.5% in all cases).

Regarding the LAS homologues, their concentrations were reduced by between 20 and 31% during the dryness step in pile A (Figure 2). However, due to the increase in the concentrations measured at day 99, these concentrations were increased (28–36%) during the fermentation step and finally decreased during the maturation process. Globally, the reduction in the concentrations of the LAS homologues achieved during composting was from 30 to 50%. In the case of pile B, the concentrations measured for the LAS homologues decreased as follows: dryness by 6–16%; fermentation by 28–30%; maturation by 40–43%; and global by 59–66%. The higher global decrease in the concentrations of the LAS homologues was observed in pile B, probably due to the higher temperature achieved in pile B during composting (thermophilic phase), because thermophilic temperatures are more favorable to biological activity associated with the removal of these compounds. In a similar study, Pakou et al. [28] achieved removal efficiencies between 77 and 91%. The authors concluded that the initial concentration of the LAS affected the removal percentage. However, at low LAS concentration values, limited bioavailability led to a residual concentration of LAS that was difficult to reduce any further. The maturation stage's duration is important for achieving a complete removal of these compounds.

Unlike in the case of the LAS homologues, NPE showed a similar behavior during the composting process of pile A (Figure 2), in which their concentration did not vary significantly. The concentration of NP during composting increased from 4.89 to 9.26 mg kg^{-1}dm and from 4.59 to 10.6 mg kg^{-1}dm in pile A and pile B (Figure 1), respectively, probably due to the degradation of NPEO. A similar observation was published by several authors [8,28,29], and a considerable removal of NPEs (64–95%) was reached in some cases [28,29].

The mesophilic temperatures in the fermentation phase of pile B favored the removal of NP (33%), but this effect was counteracted by the generation of NP from the degradation of NPEO. This effect was also observed by Moeller and Reeh [30], in whose study the effect of the thermophilic conditions resulted in a net accumulation of NP, with transitory elevated concentrations of the degradation products. However, some authors have reported losses of NP during composting [31]. The removal (%) depends on their initial concentrations in the compost mixture [28], bulking agents and mesophilic phase [29]. Some conditions during the process such as the ventilation, the pH or the microbial community can influence the degradation and removal of NP. Zheng et al. [29] observed an increase in the removal % of NP from 19.7% to 41.6% with prolonged ventilation from 5 to 15 min during composting.

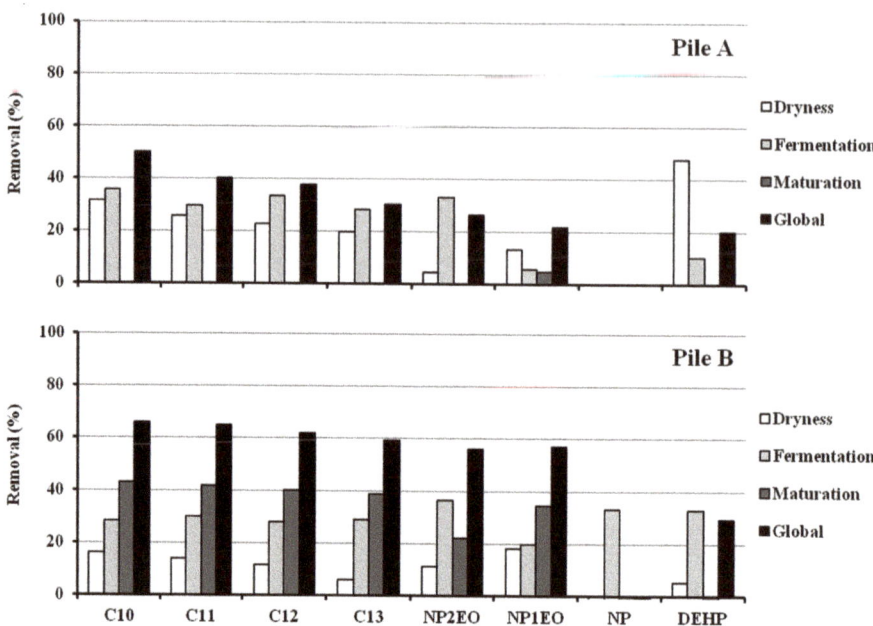

Figure 2. Removal rates of LAS homologues, NPEs and DEHP during the different phases of composting in piles A and B.

In the case of DEHP, its concentration decreased during composting from 10.5 to 8.35 mg kg^{-1}dm in pile A and from 15.8 to 11.2 mg kg^{-1}dm in pile B. Similar results were previously reported by Gibson et al. [31] and Poulsen and Bester [32]. The concentration levels of DEHP in the final product of composting were generally <100 mg kg^{-1}, with removal percentages between 50 and 97% [33].

Like what happened for the rest of the pollutants studied, the losses of this pollutant were higher in pile B (29%), especially during the fermentation step (33%), than in pile A (20%), where the higher losses were achieved in the dryness step (48%) (Figure 2). This suggests that as much microbial degradation, favored by mesophilic temperatures, as volatilization driven by water loss is possible, as DEHP is semi-volatile.

3.4. Risk Assessment

The calculated RQs in soil amended with compost from piles A and B are shown in Figure 3. The red horizontal line drawn at RQ 1 denotes the limit between medium and high risk. Overall, the risk associated with the presence of the selected compounds was medium-low for most of them. However, it was not enough to not represent a potential risk to the environment in the case of LAS C11 and C13 in both piles. These homologues had a PNEC$_{soil}$ value lower than the other homologues, mainly due to the low toxicity data reported in the literature for these compounds. Regarding NPE and DEHP, in general, their RQ values were lower than those found in the case of LAS homologues. For the NPE group, an increase in the risk was observed with the decrease in the number of ethoxy groups. The only toxicological effect expected was the one caused by NP in pile B. González et al. [8] estimated the ERA of the selected compounds after the application of compost to the soil, and their results also revealed a potential toxic effect in the case of LAS C13 and NP during the first 23 and 56 days, respectively, after the application of sewage sludge to the soil.

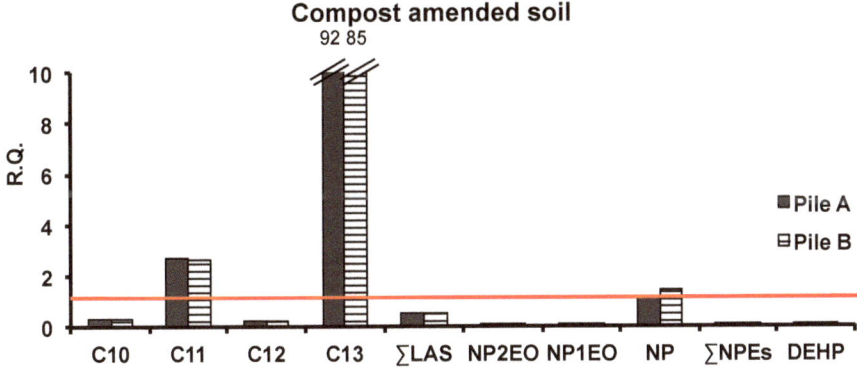

Figure 3. Risk quotients (RQs) of studied compounds in soil amended with compost.

Up to now, the lack of legislation on the content of the selected organic pollutants in sludge-amended soil made it impossible to state with certainty that the results obtained so far were sufficient to avoid medium-to-long-term damage to living microorganisms. Future advances in this area require the collection of ecotoxicological data from terrestrial microorganisms and the incorporation of additional endpoints (including chronic exposure), as well as the analysis of these compounds at lower, environmentally relevant concentrations. Additionally, other treatments can be used prior to sludge application to agriculture land in order to reduce the load of organic pollutants. Hydrothermal carbonization, biodrying or hydrothermal liquefaction have been recently investigated as potential technologies [1].

4. Conclusions

The occurrence of eight priority organic pollutants in two piles of sewage sludge along the composting process in a full-scale study was evaluated. All the studied priority organic compounds (LAS, NPE and DEHP) were detected in all samples analyzed. While LAS homologues were detected at the highest concentrations ($\sum$LAS: 2068–9375 mg kg^{-1}dm), exceeding the limit fixed in the EU Directive draft and also pointing out the domestic contribution to the samples, the concentration levels for NPE and DEHP were significantly lower and did not exceed their fixed limits in the EU Directive draft.

A decrease in the concentrations was observed along the composting process for all compounds, except for NP, probably as consequence of the degradation of NPEO. Overall, the removal was higher in the pile that reached the thermophilic phase (pile B), which probably involved specific microorganisms present at this temperature. The environmental risk assessment showed, in spite of the decrease in the concentrations of these compounds, a potential risk due to LAS homologues C11 and C13 and NP and highlighted the need to carry out a more accurate and comprehensive environmental risk assessment taking into consideration ecotoxicological data in terrestrial microorganisms for these pollutants, as well as the incorporation of additional endpoints.

Author Contributions: Conceptualization, E.A. and J.M.; methodology, J.M.; validation, I.A.; formal analysis, J.L.S.; investigation, C.M. and M.A.; resources, C.M. and M.A.; data curation, J.L.S.; writing—review and editing, J.M. and I.A.; visualization, J.L.S.; supervision, E.A.; project administration, E.A. and I.A.; funding acquisition, E.A. and I.A. All authors have read and agreed to the published version of the manuscript.

Funding: This research was funded by the Junta de Andalucía, Consejería de Economía, Conocimiento, Empresas y Universidad (Project No. P20_00556).

Institutional Review Board Statement: Not applicable.

Informed Consent Statement: Not applicable.

Data Availability Statement: Not applicable.

Conflicts of Interest: The authors declare no conflict of interest.

References

1. Mejías, C.; Martín, J.; Santos, J.L.; Aparicio, I.; Alonso, E. Occurrence of pharmaceuticals and their metabolites in sewage sludge and soil: A review on their distribution and environmental risk assessment. *Tr. Environ. Anal. Chem.* **2021**, *30*, e00125. [CrossRef]
2. Santos, J.L.; Martín, J.; Aparicio, I.; Alonso, E. Approaches to the implications of the EU directive on sludge: Analytical methodologies, concentration levels and occurrence of organic pollutants in different types of sewage sludge. In *Sewage Sludge Management: From the Past to Our Century*; Zorpas, A., Ed.; Nova Science Publishers: Hauppauge, NY, USA, 2012; pp. 123–137.
3. Fijalkowski, K. Emerging contaminants in sludge (endocrine disruptors, pesticides, and pharmaceutical residues, including illicit drugs/controlled substances, etc.). *Ind. Munic. Sludge* **2020**, 455–473. [CrossRef]
4. Francou, C.; Poitrenaud, M.; Houot, S. Stabilization of organic matter during composting: Influence of process and feedstock. *Compost Scie Util* **2005**, *13*, 72–83. [CrossRef]
5. Fijalkowski, K.; Rorat, A.; Grobelak, A.; Kacprzak, M.J. The presence of contaminations in sewage sludge: The current situation. *J. Environ. Manag.* **2017**, *203*, 1126–1136. [CrossRef] [PubMed]
6. EU. *Working Document on Sludge, Third Draft*; EU: Brussels, Belgium, 27 April 2000.
7. Aparicio, I.; Santos, J.L.; Alonso, E. Limitation of the concentration of organic pollutants in sewage sludge for agricultural purposes: A case study in South Spain. *Waste Manag.* **2009**, *29*, 1747–1753. [CrossRef] [PubMed]
8. González, M.M.; Martín, J.; Camacho-Muñoz, D.; Santos, J.L.; Aparicio, I.; Alonso, E. Degradation and environmental risk of surfactants after the application of compost sludge to the soil. *Waste Manag.* **2012**, *32*, 1324–1331. [CrossRef] [PubMed]
9. Ozcan, S.; Tor, A.; Aydin, M.E. Investigation on the levels of heavy metals, polycyclic aromatic hydrocarbons, and polychlorinated biphenyls in sewage sludge samples and ecotoxicological testing. *Clean-Soil Air Water* **2013**, *41*, 411–418. [CrossRef]
10. Clarke, R.M.; Cummins, E. Evaluation of 'classic' and emerging contaminants resulting from the application of biosolids to agricultural lands: A Review. *Human Ecol. Risk Assess. Int. J.* **2015**, *7039*, 492–513. [CrossRef]
11. APHA-AWWA-WPCF. *Standard Methods for the Examination of Water and Wastewater*, 21st ed.; American Public Health Association: Washington, DC, USA, 2005.
12. González, M.M.; Santos, J.L.; Aparicio, I.; Alonso, E. Method for the simultaneous determination of the most problematic families of organic pollutants in compost and compost-amended soil. *Anal. Bioanal. Chem.* **2010**, *397*, 222–285. [CrossRef] [PubMed]
13. Martín, J.; Camacho-Muñoz, D.; Santos, J.L.; Aparicio, I.; Alonso, E. Determination of priority pollutants in aqueous samples by dispersive liquid-liquid microextraction. *Anal. Chim. Acta.* **2013**, *773*, 60–67. [CrossRef]
14. European Community (EC). *echnical GUIDANCE Document on Risk Assessment in Support of Commission Directive 93/67/EEC on Risk Assessment for New Notified Substances, Commission Regulation (EC) no. 1488/94 on Risk Assessment for Existing Substances and Directive 98/8/EC of the European Parliament and of the Council Concerning the Placing of Biocidal Products on the Market, Parts I, II and IV*; EUR 20418 EN/1; European Communities: Brussels, Belgium, 2003.
15. Martín, J.; Camacho-Muñoz, D.; Santos, J.L.; Aparicio, I.; Alonso, E. Occurrence of pharmaceutical compounds in wastewater and sludge from wastewater treatment plants: Removal and ecotoxicological impact of wastewater discharges and sludge disposal. *J. Haz. Mat.* **2012**, *239*, 40–47. [CrossRef]
16. Ying, G.G. Fate, behavior and effects of surfactants and their degradation products in the environment. *Environ. Int.* **2006**, *32*, 417–431. [CrossRef] [PubMed]
17. Garrido-Perez, M.C.; Perales-Vargas, J.A.; Nebot-Sanz, E.; Sales-Marquez, D. Effect of the test media and toxicity of LAS on the growth of Isochrysis galbana. *Ecotoxicology* **2008**, *17*, 738–746. [CrossRef] [PubMed]
18. Fenner, K.; Kooijman, C.; Scheringer, M.; Hungerbuhler, K. Including transformation products into the risk assessment for chemicals: The case of nonylphenol ethoxylate usage in Switzerland. *Environ. Sci. Technol.* **2002**, *36*, 1147–1154. [CrossRef] [PubMed]
19. Rhodes, J.E.; Adams, W.J.; Biddinger, G.R.; Robillard, K.A.; Gorsuch, J.W. Chronic toxicity of 14 phthalate esters to Daphnia magna and rainbow trout (Oncorhynchus mykiss). *Environ. Toxicol. Chem.* **1995**, *14*, 1967–1976. [CrossRef]
20. Feijtel, T.C.J.; Struijs, J.; Matthis, E. Exposure modelling of detergent surfactants – Prediction of 90th-percentile concentrations in the Netherlands. *Environ. Toxicol. Chem.* **1999**, *18*, 2645–2652.
21. During, R.A.; Krahe, S.; Gäth, S. Sorption behaviour of nonylphenol in terrestrial soils. *Environ. Sci. Technol.* **2002**, *36*, 4052–4057. [CrossRef] [PubMed]
22. Yu, Y.; Xu, J.; Sun, H.; Dai, S. Sediment-porewater partition of nonylphenol polyethoxylates: Field measurements from Lanzhou Reach of Yellow River, China. *Arch. Environ. Contam. Toxicol.* **2008**, *55*, 173–179. [CrossRef]
23. Bis (2-ethylhexyl) Phthalate (DEHP). Summary Risk Assessment Report. European Comission. Institute for Health and Consumer Protection Toxicology and Chemical Substance (TCS) European Chemicals Bureau I-21027 Ispra (VA) Italy. Available online: http://echa.europa.eu/documents/10162/060d4981-4dfb-4e40-8c69-6320c9debb01 (accessed on 1 December 2021).
24. Alonso Álvarez, E.; Callejón Mochón, M.; .Jiménez Sánchez, J.C.; Ternero Rodríguez, M. Heavy metal extractable forms in sludge from wastewater treatment plants. *Chemosphere* **2002**, *47*, 765–775. [CrossRef]

25. Alonso, E.; Aparicio, I.; Santos, J.L.; Villar, P.; Santos, A. Sequential extraction of metals from mixed and digested sludge from aerobic WWTPs sited in the south of Spain. *Waste Manag.* **2009**, *29*, 418–424. [CrossRef]
26. Camacho-Muñoz, D.; Martín, J.; Santos, J.L.; Alonso, E.; Aparicio, I.; De la Torre, T.; Rodríguez, C.; Malfeito, J.J. Effectiveness of three configurations of membrane bioreactors on the removal of priority and emergent organic compounds from wastewater: Comparison with conventional wastewater treatments. *J. Environ. Monit.* **2012**, *14*, 1428–1436. [CrossRef]
27. Lara-Martín, P.A.; González-Mazo, E.; Brownawell, B.J. Multi-residue method for the analysis of synthetic surfactants and their degradation metabolites in aquatic systems by liquid chromatography-time-of-flight-mass spectrometry. *J. Chromatogr. A* **2011**, *1218*, 4799–4807. [CrossRef] [PubMed]
28. Pakou, C.; Kornaros, M.; Stamatelatou, K.; Lyberatos, G. On the fate of LAS, NPEOs and DEHP in municipal sewage sludge during composting. *Bioresour. Technol.* **2009**, *100*, 1634–1642. [CrossRef] [PubMed]
29. Zheng, G.D.; Wang, T.Y.; Niu, M.J.; Chen, X.J.; Liu, C.L.; Wang, Y.W.; Chen, T.B. Biodegradation of nonylphenol during aerobic composting of sewage sludge under two intermittent aeration treatments in a full-scale plant. *Environ. Pollut.* **2018**, *238*, 783–791. [CrossRef] [PubMed]
30. Moeller, J.; Reeh, U. Degradation of nonylphenol ethoxylates (NPE) in sewage sludge and source separated municipal solid waste under bench-scale composting conditions. *Bull. Environ. Contam. Toxicol.* **2003**, *70*, 248–254. [CrossRef] [PubMed]
31. Gibson, R.W.; Wang, M.-J.; Padgett, E.; Lopez-Real, J.M.; Beck, A.J. Impact of drying and composting procedures on the concentrations of 4-nonylphenols, di-(2-ethylhexyl)phthalate and polychlorinated biphenyls in anaerobically digested sewage sludge. *Chemosphere* **2007**, *68*, 1352–1358. [CrossRef]
32. Poulsen, T.G.; Bester, K. Organic micropollutant degradation in sewage sludge during composting under thermophilic conditions. *Environ. Sci. Technol.* **2010**, *44*, 5086–5091. [CrossRef] [PubMed]
33. Lü, H.; Chen, X.-H.; Mo, C.-H.; Huang, Y.-H.; He, M.-Y.; Li, Y.-W.; Feng, N.-X.; Katsoyiannis, A.; Cai, Q.-Y. Occurrence and dissipation mechanism of organic pollutants during the composting of sewage sludge: A critical review. *Bioresour. Technol.* **2021**, *328*, 124847. [CrossRef] [PubMed]

Article

ASDesign: A User-Centered Method for the Design of Assistive Technology That Helps Children with Autism Spectrum Disorders Be More Independent in Their Daily Routines

Raquel Cañete * and M. Estela Peralta

Escuela Politécnica Superior, Departamento de Ingeniería del Diseño, Universidad de Sevilla, 41011 Sevilla, Spain; mperalta1@us.es
* Correspondence: raqcanyaq@alum.us.es

Abstract: COVID-19 has posed new physical and mental challenges for the population worldwide, establishing social and structural changes in the labor market that could be maintained and implemented permanently. This new reality will require new strategies to improve family and work conciliation, which is especially challenging for families with children suffering from psychological pathologies such as autism spectrum disorder (ASD). These changes have led to more frequent and intense behavioral problems, causing stress, anxiety, and confusion for these children and their families. Thus, the need to have tools that help parents reconcile work with the care of these children, who have low autonomy, is reinforced. This work develops a method for the design of assistive technology and smart products to support children with ASD in following a routine and managing tasks autonomously. In this way, the article analyzes the design problem including the needs and preferences of children with ASD and their parents during confinement in terms of dependence and adaptability; develops a design method for interactive and smart products focused on children with ASD in confinement situations; and validates this method in a case study, in which a robot is developed that makes it easier for children with ASD to follow a routine.

Keywords: assistive technology; smart products; autism spectrum disorder; self-managed tasks; confinement periods; inclusive design

1. Introduction

COVID-19 has posed new physical and mental challenges for the population worldwide. The confinement and disruption of daily routines have led to a series of psychological consequences that society now has to face. This reality is aggravated in those who suffer from psychological pathologies, such as autism spectrum disorder (ASD). Unpredictable changes in routine can lead to serious difficulties for children with ASD. Their behavioral inflexibility causes significant interference with their behavior in different contexts, even for those at the milder end of the spectrum. Additionally, the difficulties they present in planning and organizing skills cause low self-autonomy and the inability to self-manage.

A survey conducted by the University of Verona and King's College London on how children with ASD are experiencing the COVID-19 pandemic concludes that they "are particularly at risk due to their vulnerability to unpredictable and complex changes" [1]. The results of this study show that the experience of confinement led to more frequent and intense behavioral problems, as the unpredictability of the situation caused stress, anxiety, and confusion for these children.

Additionally, this situation has been particularly complex for parents. The need to work from home, together with the difficulties that their children face to self-manage their day-to-day, has generated stressful situations derived from the disruptive behavior of their children. Planning daily activities and reconciling family life has become a chal-

lenge. On top of this, we could add the anxiety caused by the difficulty of these children understanding the situation [2].

Despite this, the confinement experience has raised interesting social changes that could be maintained and definitively implemented in daily life, such as teleworking, online education, and intensive or reduced working hours [3–5]. In addition to this, the new trends in automation and Industry 4.0 are leading to new professional roles which are geographically distributed, requiring more responsibility and flexibility. All this will require families and professionals to adopt new strategies to improve family and work reconciliation, a particularly important challenge for families with children with special needs.

Both in periods of confinement and these future contexts, the need for assistive technology is emphasized [6]. Traditional tools (low-tech), although beneficial, can be inflexible and monotonous, leading to a loss of interest on the part of the children. Moreover, they require constant interaction with an adult. In contrast, there are other tools (mid- to high-tech) that integrate smart behavior and interactively adapt to the child's development, which favors the evolution of the child. Their dynamic content is more attractive, and they autonomously provide constant instructions, facilitating user independence.

It should be noted that an assistive technology product aimed at people with ASD must be accessible from the physical, cognitive, and sensory dimensions. Currently, the variety of frameworks, methodologies, and tools for design specialized in ASD is scarce. There are general frameworks for universal design [7], design processes for people with special needs, and a series of tools focused on participatory and collaborative design. However, there is no single universal methodology for the design of assistive technology specialized in ASD, and not everything available can be applied to smart products. In addition to this, no methodology has been found in the scientific or technical literature aimed at designing interactive and smart products specialized in ASD for confinement or isolation contexts. The identification of the needs of these children and their families during confinement is essential to be able to make an attractive and useful product. With this goal, the following research question was asked:

How can assistive technology help children with autism and their parents to plan and self-manage their day and thus prevent episodes of stress, anxiety, and frustration for both during periods of confinement?

To answer this question, this work develops the ASDesign method. It is a method for the design of assistive technology and interactive and smart products that supports children with ASD in following a routine and managing tasks autonomously in confinement contexts. Within this scope, confinement is understood as a period in which someone is locked up or confined in a certain place or within limits [8]. It can occur for both desirable causes (teleworking, work—family conciliation, vacations, care of relatives, etc.) and undesirable or unexpected causes (rest or sick leave, health emergency, etc.). This method, based on user-centered design, will facilitate the proposal of mid and high-tech solutions that children can use comfortably and safely while alone and with less frequent parental supervision.

To this end, this work is structured as follows; Section 2 presents a context analysis, which justifies the knowledge contribution of this research, including a market study and the new needs arising from the COVID-19 pandemic. Section 3 shows the methods and materials used for the development of the ASDesign method and develops the proposal in detail. Section 4 validates this method in a case study that includes the design of a robot that allows the configuration and monitoring of a daily routine. Finally, Section 5 discusses the results and main conclusions of this work.

2. Context Analysis

The American Psychiatric Association defines ASD as "a complex developmental condition that involves persistent challenges in social interaction, speech, and non-verbal communication, and restricted/repetitive behaviors" [9]. According to Autism Speaks [10], today 1 in 59 children has ASD. There are two fundamental symptoms of autism:

(i) persistent impairments in communication and social interaction and (ii) restrictive and repetitive patterns of behavior, interests, or activities. The areas where they have the most difficulties are social interaction, imagination, and communication. Therefore, people with ASD tend to isolate themselves and show no interest in others. This is accentuated for those who present communication difficulties due to speech delays or, in some cases, the absence of verbal communication. In turn, they avoid eye contact, as well as physical contact, in addition to having difficulty showing their own emotions and understanding those around them. Their social imagination is limited: they tend to avoid symbolic games and often repeat the same game, or even the same movements, continuously. Sudden changes or disruptions in their daily routine can be a great frustration for them. On top of this, they show limited attention, cannot concentrate on a task for too long, and may exhibit sensory processing disorder.

COVID-19 and lockdown have caused serious difficulties for children with ASD and their families. The limitations of physical space and the transformation of their daily routine (including the disruption of diet and sleep, as well as the interruption of teaching, therapies, and sessions in special centers) are generating a very significant mental and emotional impact [2,11–13]. Although several of these services have been offered online, experts argue that they are less effective than those received in person [14]. In addition to this, it should be noted that the online modality has caused interruptions or cancelations of activities due to technical difficulties or digital poverty in some families [11,15,16]. The lack of availability of these resources causes uncooperative and disruptive behavior in children, including periods of anxiety derived from the impossibility of understanding the situation [2]. Fundamentally, the most common consequences have been stress, anxiety, behavioral problems, emotional crises, irritability, obsession, hostility, impulsivity, repetitive behaviors, eating disorders, and insomnia. However, many parents considered satiety one of the biggest problems during the pandemic, stating that there was a "significant reduction in the time children spent with their favorite objects" [12,17]. Several factors have accentuated these consequences: (i) age (the younger the child, the more stress the parents suffer) [18,19]; (ii) degree of disorder (children with a higher degree have suffered more stress, as well as their relatives); (iii) disruption of routine (children at the "mild end" of the spectrum who were used to attending school and other activities suffered stronger behavioral disorders than those who were used to staying at home) [19]; (iv) economic situation (the loss of work and onset of economic difficulties have increased stressful situations, since many families could not afford online treatment or lacked the necessary technology to be able to receive it) [18]; (v) health risk factors (homes with high risk could not accommodate specialized professionals at home) [19]; and (vi) cultural aspects [18].

This situation has also been especially difficult for parents due to the need to combine the workday with the care of their children [17]. Many of the tasks and responsibilities that were previously distributed among several agents (teachers, therapists, etc.) have had to be carried out by parents, assuming the different roles without having the necessary knowledge and resources [11–13,16]. Regarding these difficulties, experts have given a series of guidelines to try to reduce stress and frustration [2,17,20], with some of the most common ones as follows:

- Structure daily activities.
- Divide daily activities to establish new routines.
- Use visual tools to help the child understand and self-manage their day.
- Choose the activities that your child prefers.
- Include serious games that help the child improve certain skills.
- Include parents in activities and avoid isolation from the child.
- Maintain contact with schools, therapists, and other children.
- Leave free time.
- Try to give relaxing resources: relaxing space, physical exercise, or sensory stimuli.
- Explain what COVID-19 is in a simple way.

Most of these guidelines focus on planning the day in a way that helps children follow a clear temporal structure. Despite the efforts made by parents to try to control the situation, they reported the need for more support to preserve their children's mental health, as well as their own [1,12,18]. Therefore, experts state that to help children with ASD during these situations, it is essential to focus the research and intervention efforts not only on the child, but also on the development of appropriate support tools for their parents [12,13,16]. These tools are part of what we know as assistive technology.

The International Classification of Functioning, Disability, and Health (CIF, WHO, Geneva, Switzerland) defines assistive products and technology as "any product, instrument, equipment, or technology adapted or specially designed to improve the functioning of a person with a disability" [21]. This assistive technology differentiates between "low-tech" (traditional tools and methods that use noninteractive products or do not use energy), "mid-tech", and "high-tech".

The low-tech tools used in language therapies are based on augmentative and alternative communication systems (AAC), based on the use of symbols or images as a form of expression. The most common approach within these systems is the Picture Exchange Communication System (PECS), where users communicate their needs and preferences by exchanging images. Usually, these images and drawings are collected in a book that users carry with them. Specifically, there are several tools to assist in task planning, such as visual task sequences, choice boards, first-after sequences, visual schedules, key phrase symbols, and tags [22]. These tools are typically used at home, schools, and special centers to help children with autism know exactly how the day will evolve and avoid episodes of frustration and emotional breakdowns. There is a wide variety of products available for this purpose in the market.

One of the main drawbacks of conventional tools is that they are very rigid and do not adapt to the evolution of the user's needs. Therefore, mid-tech (with simple electronic elements, such as recorders, e-books, headphones, and visual timers [23]) and high-tech products (with electronic and computerized elements that improve efficiency, speed, and accessibility) implement interactive or intelligent actions with multisensory reinforcements, making the user experience self-adaptive, dynamic, and intuitive. The last group also includes mobile applications such as the Happy Kids Timer Chores App [24] and the Choiceworks App [25] or the emerging field of the development of assistive robots. These collaborators can have very positive effects on the development of social skills of children with ASD; in addition, they have an immediate positive effect on communication skills [26].

The repetitive and predictable behavior of these tools provides a greater sense of security to the child [27]. They can also be easier to understand, as they continuously display instructions and comments. Most children with ASD have been shown to exhibit a clear preference for robots [28]; the game is visually more appealing than individual tasks with parents, therapists, or non-robotic toys. In addition to this, the appearance of robots can stimulate social interaction skills, such as eye contact or communication [27], and can motivate the child to talk, learn, or share interests with others, also improving communication [29]. Finally, they can also help detect and understand emotions and social behaviors [30]. To this we must add that many of them do not require the child to own a phone or tablet, avoiding the continuous supervision of the parents. Examples of these robots are Leka [31] and Kaspar [32].

One of the main disadvantages of these products is their affordability; due to their high price, they are mostly used as a professional tool in therapy. On the other hand, most robots focus on improving social skills, and as stated by the Japan Society of Instrument and Control Engineers (SICE, Chiyoda City, Japan) (2019), the development of robots aimed at monitoring the daily routine of children with autism to integrate into real society remains unexplored [33].

Therefore, the line of research in which the ASDesign method is developed is an incipient and interesting area of work that contributes to improving the well-being of children with ASD and their families during confinement periods.

3. Materials and Methods

This section describes the proposal of the ASDesign method and its application context; for its development, the following stages were carried out, which will be explained in the following sub-sections:

- User analysis, context of use, and definition of needs (Section 3.1). Development of the ASD_T1 design module of the method.
- Definition of functional and technical requirements as well as design parameters of assistive technology (Section 3.2). Development of the ASD_T2 and ASD_T3 design modules of the method.
- Definition of the design process for ASDesign and application requirements (Section 3.3)

Due to the fact that the definition of the target user of ASDesign is complex (parents and children with autism in a domestic context in periods of confinement), it was established that the method would include a set of tools or modules, called "ASD_Tnum modules", that would facilitate decision-making throughout the design project. These modules, as will be explained later in the following subsections, allow special needs to be selected and make it easier to establish their relationship with functional requirements and optimal design variables.

3.1. User Analysis and Definition of Needs

Every product design process begins with the analysis of the user's needs. In the design problem solved by the ASDesign method, two groups of needs are identified: (1) the basic needs of ASD and (2) the specific needs derived from confinement.

First, to define basic needs, it is necessary to take into account the severity of children's difficulties, currently classified into three levels of autism [9]. This work focused on the mildest level of the spectrum (level 1). Children in this range may have deficiencies in social communication if they do not have sufficient support, and have little interest (and difficulty) in initiating social interactions, in which they also show atypical or insufficient responses. On the other hand, behavioral inflexibility generates significant disorders in their behavior in different contexts. They have difficulty alternating activities, organizing, and planning, which entails little autonomy and independence. Table 1 shows the basic ASD needs included in the ASDesign method.

Table 1. Definition of basic ASD needs (BN).

Basic Needs of ASD	Objectives
Development, maintenance, and understanding of relationships	Adapt behavior to different contexts
Attention	Obtain and maintain the child's attention
Emotional reciprocity	Express preferences and ideas
	Initiate and maintain social interactions
Restrictive and repetitive patterns of behavior	Disproportionate and repetitive gestures
	Small variety of interests
	Inflexibility in routines
	Stress and frustration to changes
Environment and social awareness	Social and environmental awareness
Motivation	Motivation and goal setting
Disruptive behavior	Behavior management

As a second step, the definition of needs during periods of confinement was carried out through three phases: (1) analysis of research in the socio-health, psychological, and sociological areas, (2) interviews with professionals, and (3) analysis of the behavior of the target user through a questionnaire.

The first phase included an analysis of the state of the art and selection of different statistical studies and research carried out during the confinement period; in general, these studies were based on interviews with parents of children with ASD and professionals in the areas of medicine and psychology [1,2,11–20]. This phase made it possible to identify the greatest difficulties that families experienced during the pandemic, psychosocial risk factors, and main consequences, in addition to a set of expert recommendations to reduce impact. To complete the above information, as a second phase, a semi-structured interview was conducted with a pediatrician from an ASD association. The objective was to collect the specific experiences and intervention activities carried out by specialized centers after the COVID-19 interruption.

As a result of technical analyses 1 and 2, a first classification of the most significant difficulties was obtained, which was later translated into needs for the ASDesign method:

- The attention required by the child.
- Combine work and childcare.
- Calm the child in times of stress.
- Getting the child to focus on an activity or product.
- Explaining the pandemic situation to the child.
- Structuring or planning daily activities.
- Getting the child to do certain activities he or she does not like.

On the other hand, and with the aim of gaining a deeper understanding of the experience of the target user in the periods of confinement, the user's behavior was analyzed through a questionnaire aimed at parents of children with ASD. It was answered by 51 subjects. Most of the children represented in this research are under 10 years old (62%), and most of them received online support from professionals (61%). The questionnaire included three blocks of questions to obtain different information:

- BLOCK 1: Composed of 10 questions about the situation of children with ASD during confinement. It made it possible to identify the greatest difficulties for parents regarding their children and the order of priority of these demands.
- BLOCK 2: Composed of 6 questions on the use of low, medium, and high technology tools. The result was the identification of tools used to plan the routine, the frequency of use, and the evaluation of usability. On top of this, it explored whether the tools were interactive (mid to high-tech), collecting the reasons for preference of use over conventional tools (low-tech).
- BLOCK 3: Composed of 3 questions about purchasing behavior. It led to the aspects that influence the purchasing decision, as well as the priorities of parents when deciding on a product or tool for their children.

In the first block, the results show that 19% of parents did not consider confinement a difficult period for their children, but for themselves. However, on the other hand, this percentage of users argues that one of the biggest difficulties had been to calm their child in emotional crises. That is, parents agree that the confinement was worse for them; however, they verify the existence of emotional crises in their children in this period. On top of this, this block showed that only 29% of the parents combined work and the care of the child (49% had to dedicate themselves exclusively to the care of the child, 16% had to leave home to work onsite, and 6% worked from home while someone else took care of their child). This information is consistent with the conclusions of Phase 1: needs analysis (statistical studies and research published in the scientific literature).

Regarding the identification and prioritization of difficulties, the questionnaire led to the classification of the different demands of the children according to the stress they generated (Table 2). The hierarchy was carried out using a weighting factor to synthesize the results and establish the order of priority of the specific needs of the confinement (CNi). The weighting factor was established following these steps: (I) participants ranked the confinement needs (CNi) from most to least difficult (Positions P1–7); (II) the frequency for how many times each confinement need (CNx) appeared in each position (Pj) was established (Fxj); (III) all frequencies were divided between the number of responses I; (IV) the resulting value for each position (Fxj/R) was multiplied by a scaling factor according to the position (Pj) (P1 = 7, P2 = 6, P3 = 5, P4 = 4, P5 = 3, P6 = 2, P7 = 1); (V) the results were recorded as the weighting factors.

Table 2. Specific needs of the confinement (CNi).

Specific Needs of Confinement—Priority Order		Weighting Factor
N1	Maintain the child's attention and interest in a product or activity.	4.78
N2	Calm my child down in moments of stress.	4.45
N3	The attention required by the child.	4.13
N4	Getting the child to do tasks that he/she does not like.	3.98
N5	Structure and plan daily activities.	3.66
N6	Explain the pandemic situation to the child.	3.57
N7	Combine the workday with the care of the child.	3.36

The answers in block 2 (regarding the tools used and their functionality) showed that those of low technology (or traditional) were the most used and most valued, the most popular ones being: the visual task sequences, the visual daily sequences, the First-Then activities boards, and the choice boards [34]. Parents also added reward panels as a useful tool. Furthermore, only 22% of the families frequently used high-tech in the form of mobile applications to replace conventional tools. Regarding the evaluation of the functionality of the tools, the following design deficiencies were identified: lack of personalization (inserting desired activities) (with 25%), the required supervision (25%), loss of interest of the child (10%), and the need to have a mobile phone for its use (25%). Finally, only 1/51 children used high technology in the form of a robot or a similar product during confinement.

Lastly, block 3 on purchasing behavior identified the design and functional requirements that influence the choice of low, medium, and high technology: have visual support (88%) and hearing support (94%), be easily transportable (98%), allow you to schedule activities in advance (90%), and allow the child to choose the activities at the time of use (94%). On a second level, to the given aspects, some parents added as priorities having different levels of difficulty, being affordable (price), and being intuitive.

Once the three phases of the analysis of the target user were completed, the relationship between the specific needs of confinement (CNi) (Table 2) and the basic needs of ASD (BN) (Table 1) [9] was established. This made it possible to determine which basic needs of the disorder should be worked on for each specific need of confinement. The correct use of cross-relationships ensures that the product design is properly adapted to the complete needs and characteristics of these children. To make it easier for the design team to select cross-needs, the module ASD_T1 was developed (Table 3).

Once the CNi (specific needs of confinement) are known, the design team can easily identify the basic needs of ASD that must be prioritized in the design of the product.

It should be noted that although the specific needs N2 (Calm my child down in moments of stress) and N6 (Explain the pandemic situation to the child) have been included in the ASDesign method, they are too complex to solve with a product. Specifically, as the questionnaire showed, the last stage of the emotional crisis is the most difficult challenge for parents, when the child is already under a lot of stress. Trying to implement a solution at that point is extremely difficult. Instead, the right thing to do would be to identify the causes that led to the crisis in order to avoid the episode in the future. This is grounded on the concept of Positive Behavioral Support of proactive approach [35] and functional analysis, strategies that work on self-regulation to improve independence and autonomy.

Table 3. ASD_T1. Relationship between basic ASD (BN) and confinement-specific (CN) needs.

Basic Needs of ASD		Specific Needs of Confinement						
		N1	N2	N3	N4	N5	N6	N7
Development, maintenance, and understanding of relationships	Adapt behavior to different contexts	X	X	X	X	X	X	X
Attention	Obtain and maintain the child's attention	X	X	X	X	X	X	X
Emotional reciprocity	Express preferences and ideas	X	X	X	X	X	X	X
	Initiate and maintain social interactions		X	X	X	X	X	X
Restrictive and repetitive patterns of behavior	Disproportionate and repetitive gestures	X	X	X	X	X		X
	Small variety of interests	X		X		X		X
	Inflexibility in routines	X	X	X	X	X		X
	Stress and frustration to changes	X	X	X	X	X		X
Environment and social awareness	Social and environmental awareness	X	X	X	X	X	X	X
Motivation	Motivation and goal setting	X		X		X		X
Disruptive behavior	Behavior management		X	X	X	X		X

3.2. Definition of Functional Requirements and Design Parameters

The definition of crossed needs allows designers to translate the areas that we intend to work with or support using technology (signs, symptoms, and difficulties of ASD) into design strategies: (1) functional requirements and (2) design parameters.

First, functional requirements define the functions of the system (technology or components). They specify the behavior between inputs and outputs of resources (information, energy, materials, or space), which are exchanged in the user–product environment. In the design process, it is necessary to have a correct translation of needs into functional requirements. In the ASDesign proposal, for all the crossed relationships of the module ASD_T1 (BN + CN), the functional requirements (FR) to be taken into account in the product design were identified. In addition to this, non-functional or quality requirements (NFR) were defined; these refer to the properties of the product (usability, safety, comfort, adaptability, performance, and transportability). Finally, the smart requirements (SR) for the design of high-tech products were identified. To facilitate the translation of needs (BN + CN) into requirements (FR, NFR, and SR), the module ASD_T2 was developed. Figure 1 shows the structure and its purpose and can be checked in detail in Annex I.

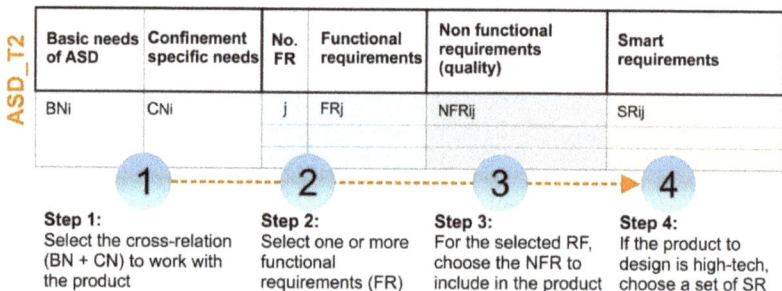

Figure 1. ASD_T2 for the translation of needs (N) into functional requirements (FR).

As a second step, and derived from the relationship between crossed needs (BN + IN)i and functional requirements (FRj, NFRij and SRij), the design parameters (DP) are identified, being the specifications and constraints required in the conceptual and detail design phases to develop the best solution. The definition of design parameters in the ASDesign method was carried out taking into account the special physical, sensory, cognitive, and social characteristics of children with ASD. The definition of design parameters consisted of four phases:

i. An exhaustive review of the scientific and technical literature and the latest research on products aimed at children with ASD [36–40].
ii. Analysis of interactive products [41–47] and robots for children with ASD [48–51].
iii. Market research focused on children with disabilities and developmental disorders [52,53].
iv. Compilation and synthesis of the results of a semi-structured interview conducted with a specialist doctor belonging to an association for children with ASD. Specifically, the interview concluded the routines, tools, methodologies, and other elements that are used in this type of association to maintain the attention and motivation of the children when following a sequence of activities. Most of these proposals referred to conventional methods and low-tech products.

The four previous activities made it possible to identify guidelines for the design (DPjk) and constraints (DCjk) of medium- and high-tech products for these children. To facilitate the selection of DPs, module ASD_T3 was developed. Figure 2 shows the structure of the module and its purpose and can be checked in detail in Annex II of the work.

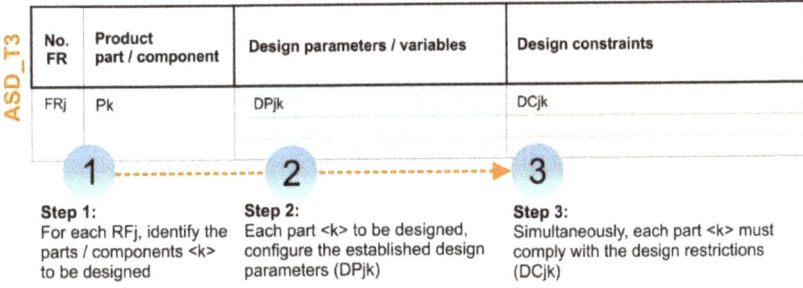

Figure 2. ASD_T3. Design parameters of robots for children with ASD.

Although the module ASD_T3 includes a complete set of DPs, in the conceptual design phase, it is important to take into account the interdisciplinarity and collaboration between different areas. It will be indispensable, in addition to considering appropriate design strategies, to consider the theories, practices, and experiences existing in clinical and educational settings. This is the strategic key to achieving technology-enhanced interventions for users with ASD [54]. Therefore, to make the design project successful, it will be important to use creative techniques (interviews, questionnaires, workshops, research activities, etc.) [55] that allow the integration of different disciplines, such as engineering, psychology, medicine, or education.

3.3. ASDesign Method

This section explains the ASDesign method in detail, specifically the applicability and steps to follow. It should be noted that the modules ASD_T1 (cross-relationships between ASD needs and special confinement needs), ASD_T2 (functional requirements), and ASD_T3 (design parameters) are the three basic tools of the method. These tools will be used throughout the design process of assistive products (medium or high technology) as facilitators in decision-making.

3.3.1. Scope of the ASDesign Method

ASDesign is a user-centered design method; it is applicable in the development of assistive technology, in the form of interactive and smart products to support children with

ASD in following a routine and managing tasks autonomously. The technology designed with this method can be used in periods of confinement; in addition, children can configure it comfortably and safely (and supervised by parents). In the ASDesign method, the term "confinement" is understood as a period in which someone is locked up or confined in a certain place or within limits [8]; this can be due to desirable reasons (teleworking, work-family reconciliation, vacations, etc.) and undesirable or unexpected reasons (rest or leave, illness, health emergency, etc.).

3.3.2. ASDesign Process

Figure 3 sets out the flow chart of the main phases to follow. These phases are not strictly linear. Iteration between stages 5, 6, and 7 will allow the necessary adjustments to be made to design decision-making throughout the entire process.

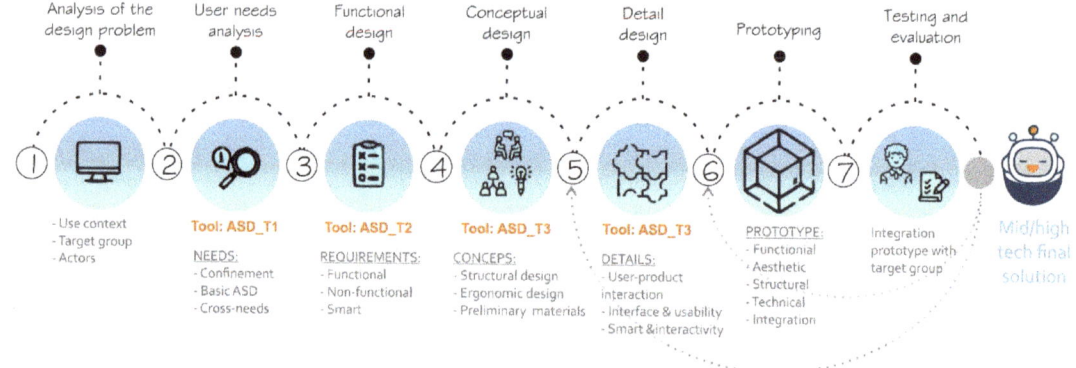

Figure 3. Stages of the proposed method.

Phase 1. Analysis of the design problem: Phase 1 analyzes the design problem, which includes: (1) definition of the characteristics of the context of use (home, therapy, school, indoor or outdoor environments); (2) the target user (ASD picture, degree of ASD to be worked with [9], and the main signs and symptoms to be improved with the product); and (3) the actors involved in the interaction (child with ASD, therapist, parents, other typically developing children, etc.).

Phase 2. User needs analysis: Identification of cross-needs. Once the design problem has been defined, the CNi needs are selected, relating to the difficulties of confinement (Table 2). These are then related to the basic needs of ASd (BNi) (Table 1). Using module ASD_T1, cross-relationships (CNi + BNi) are identified (see Table 3, Section 3.1), which will ensure that the product is correctly adapted to the target user.

Phase 3. Functional design: Identification of functional requirements. Crossed needs (CNi + BNi) are translated into product requirements. In this way, and provided by the tool ASD_T2 (see Figure 1 and Annex I), the Functional Requirements (FRj), Non-Functional Requirements (NFRij), and Smart Requirements (SRij) are obtained according to the needs to be solved. The product's requirements should serve as the basis for the design of the different alternatives of the product.

Phase 4. Conceptual design: Conceptual design includes the development and selection of design alternatives. To facilitate the proposal of appropriate solutions, and grounded on the tool ASD_T3 (see Figure 2, Section 3.2 and Annex II), the PDjk design parameters most in line with the functional, non-functional, and smart requirements specified in Phase 3 are identified. The conceptual design phase must achieve the following goals in product design:

o Structural design: Proposal of the structure of the product and its components.

- Ergonomic design (safety and comfort): An anthropometric, biomechanical, and cognitive study of the child population to ensure the correct adaptation of the product to the user.
- Preliminary selection of materials: Study and selection of materials for the perceptual and sensory adaptability of the product to the user with ASD. In addition, they must ensure the physical and environmental integrity of the product.

Phase 5. Detail design: Once the optimal alternative is selected, the detail design will be finalized, including the following goals:

- User–product interaction design: Development of product states and game modes.
- Interface design and usability analysis: Development of the information exchange system between user and product.
- Design of smart and interactive properties: Development and programming of the intelligent and interactive system.

Phase 6. Prototyping: To validate the design and functions of the technology, different prototypes must be developed:

- Functional prototype: Validates usability in terms of efficiency (goal scope and needs), effectiveness (usage times, usage errors, mental and physical load of the user), and satisfaction (ease of learning, attraction, motivation, attention fixation, ease of recall) in user–product interaction.
- Prototype of aesthetic–formal design (look and feel): Validates the aesthetics and physical and cognitive adaptability (perceptive and sensory) of the design.
- Structural prototype: Validates geometric concordance, shape, and assemblies.
- Technical prototype: Validates the technical feasibility of the solution.
- Integration prototype: A final prototype that integrates the above aspects and aims to be tested with the end-user.

Phase 7. Testing and evaluation: Using the integration prototype of Phase 6, a test is carried out with the target group; this must be composed of children with ASD, children with typical development, and the rest of the users (parents, psychotherapists, and other agents if they exist). It is recommended to prepare a list of assumptions in advance to validate during the test. The results will be useful for the improvement or redesign of the final solution.

4. Case Study: The Robot Pepe

This section applies the ASDesign method with the aim of validating and verifying the proposal. Thus, the robot "Pepe", a high-tech assistive product to support children with ASD to improve the independence of their daily tasks, is designed. Pepe's design made it possible to verify the suitability of the ASDesign method and its applicability to any design project. In addition, the on-site testing with the target user validated the tools ASD_T1, ASD_T2, and ASD_T3 and their suitability to the scope of the method. The following sections summarize the design process and the most representative results.

4.1. Phases 1, 2, and 3: Context Analysis, Definition of Needs, and Functional Design

As discussed in Section 3.3, the design problem is defined in the first phase of the ASDesign method. For this case study of the robot Pepe, the following aspects were considered:

(1) Context of use: Domestic, for which an interactive and smart product in the form of a robot is developed for children with ASD. The robot helps to create and follow a routine, improving self-autonomy. The goal is for children to be more independent of their parents in times of confinement.

(2) Definition of the target user: Children with ASD level 1 (mild end of the spectrum) [9]. They have verbal communication skills and can attend schools with typically developing children.

(3) Actors participating in the interaction: Child or child with ASD and their parents (the therapist is not present).

Once the design problem was defined, in Phase 2 the specific confinement needs (CN) to be solved were selected (see Table 2, Section 3.1). Taking into account that a robot presents high possibilities of customization, as well as self-adaptation to the context and the user, the following CN were selected: N1, N3, N4, N5, and N7, with N2 and N6 being discarded. As discussed in Section 3.1 and according to the principles of the Positive Behavioral Support Methodology [35], although N2 (calming the child in times of stress) is one of the most difficult demands for parents, it is considered important to pay attention to the causes of the problem, that is, to the needs N1, N3, N4, N5, and N7. When these are addressed properly, it is possible to prevent the onset of the crisis. On the other hand, N6 (explain the situation of confinement to the child) is discarded, as it needs to be worked with a completely different approach to the rest of the needs.

Lastly, the specific needs of confinement were related to the basic needs of ASD using module ASD_T1 (Table 3, Section 3.1). In the case study, all the basic needs of ASD were taken into account. These are focused on improving (I) the development, maintenance, and understanding of relationships, (II) attention, (III) emotional reciprocity, (IV) restrictive patterns of behavior, (V), environment and social awareness, (VI) motivation, and (VII) disruptive behavior.

Once the crossed needs to be solved by the Pepe robot were defined, the functional design (Phase 3) was carried out using module ASD_T2 (see Figure 1 or consult Annex I). The tool facilitates the translation of needs (BN + CN) into requirements (FRj, NFRij, and SRij). In this case, the functional requirements RF1–18 were taken into account (see columns 2, 3, 4, and 5 in Annex I).

4.2. Phase 4: Conceptual Design: Ideation, Generation of Alternatives, and Preliminary Solution

The ASDesign method takes an interdisciplinary approach in the ideation phase. To do this, the creative process of the Pepe robot was divided into two phases: (1) semi-structured interview with a specialist in therapy and early care and (2) an interdisciplinary ideation workshop.

The semi-structured interview was conducted with a specialist doctor who belongs to an association for children with ASD. The conclusions obtained from the previous interview (see Section 3.2) were considered and redesigned as interactive and smart in the Pepe robot. In addition, a series of "expert recommendations" were obtained from the interview and later used in the ideation workshop.

As a second phase, in the interdisciplinary ideation workshop, different proposals for the potential product were developed. Idea generation techniques—brief icebreaker ideation activity and divergent thinking (brainwriting diagram) [55]—were applied. Three professionals participated in the workshop, belonging to the areas of product design, psychology, and education. Starting from a set of previously defined [FR j, NFRij, SRij] (see ASD_T2, Annex I), and taking 2 random design parameters (ASD_T3, Annex II) and 1 expert recommendation (see Section 3.1 and Figure 2), participants generated different design ideas. The inclusion of expert recommendations allowed all of these ideas to be aligned with the context, user, and scope of the design. Finally, they were asked to choose a set [FRj, NFRij, SRij, DPjk] and draw or describe a robot design solution. Figure 4 shows the experience of the workshop.

With the results of the semi-structured interview and the interdisciplinary workshop, seven alternatives to robot design were generated. They were then analyzed considering the design parameters of module ASD_T3 (see Figure 2 and Annex II). All alternatives (Figure 5) were designed not only as an AT product but also as a toy, meaning that technology, in addition to improving, developing, and working on certain skills of the child, is also fun, safe, age-appropriate, and attractive for the user. Therefore, alternative 4 presented better results, thus being the final solution.

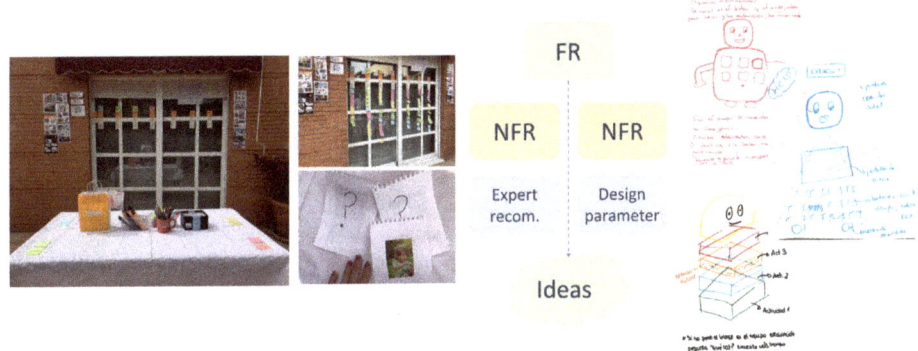

Figure 4. Interdisciplinary ideation workshop.

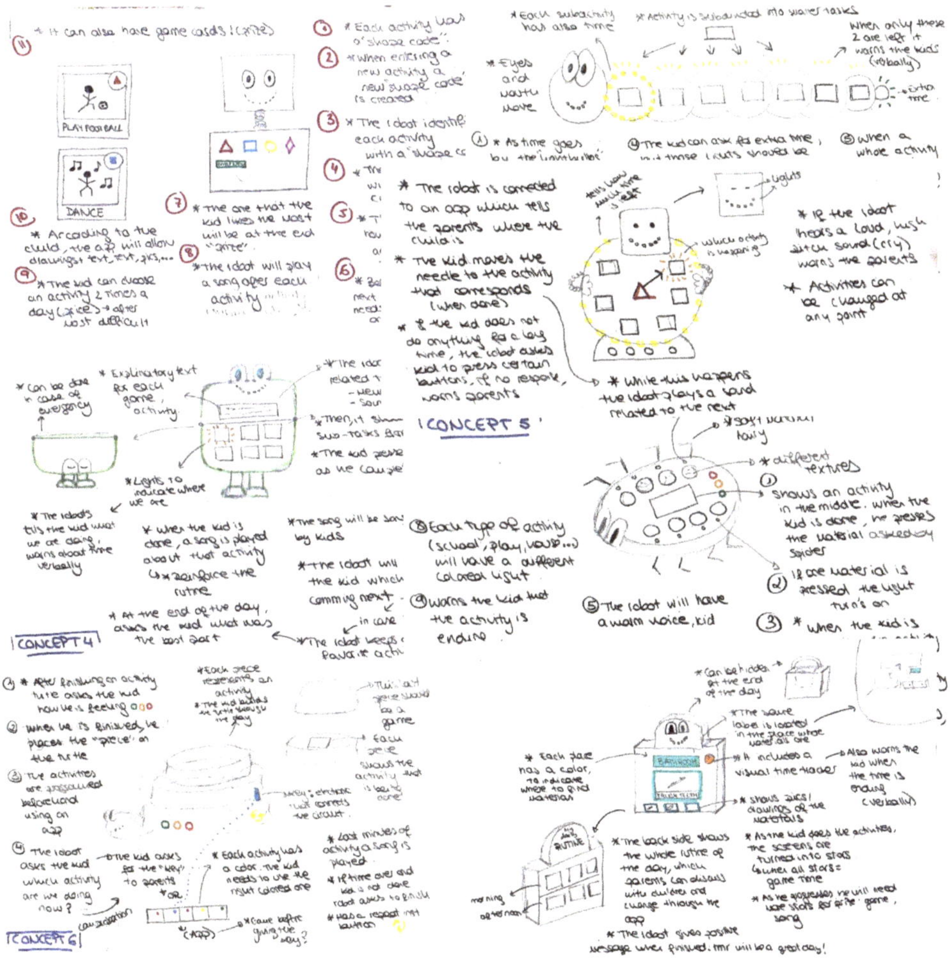

Figure 5. Generation of design alternatives.

As indicated in Phase 4 of the method, for the selected alternative (see design), the ergonomic design and the preliminary choice of materials were developed using the tool ASD_T3 (see Figure 2 and Annex II). The Pepe robot consists of a structure 40 cm high and 25 cm wide. It has four wheels at the bottom that allow it to be easily turned around; in addition, the upper semi-cylinder can be hidden. Some physical elements are elastic, allowing tangibility and mobility of elements. It was designed with a small size to avoid intimidation and also to facilitate manipulation by the child. It is made up of simple geometric shapes to avoid frustration. Pepe has an identity that is easy to understand for a child with ASD: includes an anthropomorphic design (eyes, mouth, and arms) with interactive expressions of emotions. The idea behind giving it a "humanoid" appearance is grounded on the thought that interacting with robots that have "humanoid" elements and interaction features might help children with ASD with their social and communication skills, as well as to maintain attention [26]. In addition, identity makes the child see the product as a playmate (and not as a supervisor); this will create a strong bond between the two and encourage the child to use it with a playful attitude.

As the ASDesign methodology indicates, one of the most important features in smart products is perceptual and sensory stimuli. To achieve a correct adaptation, Pepe includes traditional (low-tech) and computational (high-tech) elements; the objective is to create a product flexible to the user's development but adapted to their sensory characteristics. The low-high tech combination is supported by Mikael Wiberg's approach to "material movement" [56]. Therefore, those elements of traditional tools (low-technology) that are suitable for the user and the satisfaction of their needs were integrated with the ICT and electronic elements, thus configuring the design of the robot. Regarding visual reinforcements, Pepe includes 7 screens that show drawings, images, or text, depending on the child's skills and the game. Compared to conventional tools, including messages through ICT (information and communications technology), elements make the product more adaptable and flexible (both in terms of content and in the way it is presented). In addition to this, Pepe includes a set of physical interaction elements (pushbuttons and sliders). Touch screens were discarded because of the need for tangibility and materiality in the product. These types of physical elements can be manufactured with different textures and materials that create a multisensory experience (greater comfort, acceptability, and agreeability in the task according to the basic needs of the ASD); in addition, they facilitate interaction by involving varied movements of fine (sliding, squeezing, pressing, holding, grasping, gripping, clamping, turning) and gross (dragging and turning) motor skills. The design of low-tech elements is based on the conventional tools used in therapies, such as: (1) having arrows that the child can move and point to different scenes that help him stay focused, and visually understand the next task or predict the next information; or (2) the fact that Pepe's head (semi-cylinder top) can hide, embedding the "Turtle Tool" [22] commonly used to teach children with ASD that, in times of crisis, it is important to disconnect and think before reacting impulsively.

Lastly, the correct selection of materials allows the robot to adapt to the sensory needs of the child, where the materials were prioritized according to the characteristics of their surface (texture, color, and brightness) avoiding the use of reflective materials; pastel colors were prioritized (see Figure 6). Tactile stimuli are very beneficial in capturing the child's attention and making him enjoy a certain product. In addition, according to the design parameters of the ASDesign method (see Annex II), it is also advisable to include soft, non-toxic materials, which are suitable for outdoor environments. Therefore, the selected materials were as follows: PET of different textures (smooth, bubbles, metallic), felt, curl, nylon, and foam.

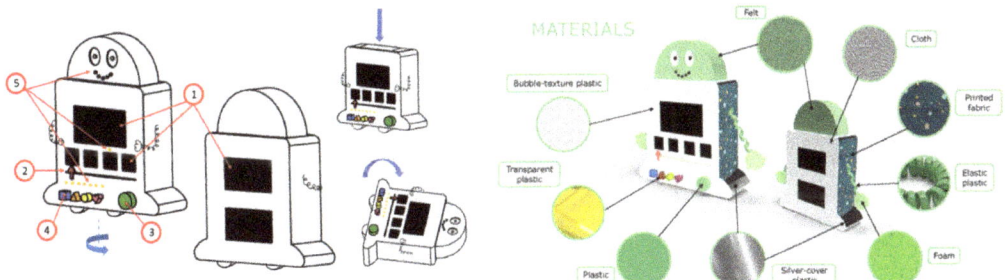

Figure 6. Preliminary design of the Pepe robot.

4.3. Phase 5: Detail Design

Following Phase 5 of ASDesign, (1) the design of the user–product interaction (in this case robot states and game modes), (2) the design of the interface and usability analysis (information-action exchange system between user and robot), and (3) design of the intelligent and interactive properties of the product were carried out.

Pepe is a robot designed to help follow a routine previously planned by the parents and the robot itself. It is designed to guide the child throughout the day and work on needs (N1, N3, N4, N5, and N7), making use of different activities based on gamification. In this way, Pepe suggests a routine to parents for the next day, including main activities and subtasks based on the child's performance with the product itself. It is able to record which activities the child prefers, which are more difficult, and which ones he/she has already learned (depending on how long they take the child). On top of this, it can change the level of support given to the child, adding and removing elements progressively. Scheduled activities are presented throughout the day using the different interface screens; the child will interact with the robot through different interface elements (see Figure 6, elements 2, 3, and 4). Pepe includes voice, sounds, music, lights, and textures as supporting elements to make the experience more engaging and interactive. The main goal is to keep the child involved in the different activities by creating a reward system and making a game out of the routine (gamification). The robot and its functionality will evolve along with the child as he/she progresses in the development of self-autonomy. These strategies are based on the Positive Behavioral Support methodology [35]. Figure 7 shows the functions and stages of the robot.

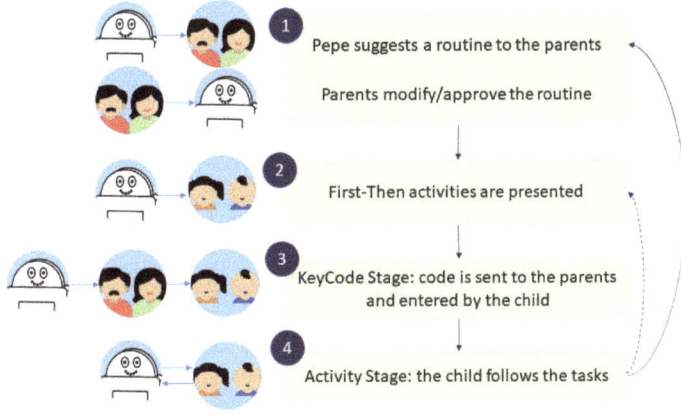

Figure 7. Functionality of the Pepe robot. Main stages.

As shown in Figure 7, the design has four main stages that allow for creating a dialogue between parents, child, and robot. These stages are described below:

1. **Stage 1. Establishment and approval of the routine**: The robot suggests a possible routine to parents for the next day. Since the robot can record and interpret the child's data (time spent on each activity, how the activity made him feel, and which tasks proved more difficult), Pepe suggests an adapted routine. This means that depending on the child's performance, he/she will add or remove tasks, if necessary, combine the activities that the child likes or dislikes in the correct order, add games to help the child memorize the tasks, and change the images according to the level of difficulty in interpretation (going from <images> to <pictograms> to <text>). Parents, through a mobile application, will be able to modify and approve the routine. At first, Pepe will have a more rigid approach (the control of the activity depends on the parents); however, as the child progresses, Pepe will be more flexible and allow the child to choose more activities, gradually giving him more self-autonomy (see Figure 8a).

2. **Stage 2. Presentation of First-Then activities**: The robot gives the child a "welcome message" verbally and says it is time to start with the activities. It asks the child to turn it around. The back will always show which activity is going to be done first and which one is going to be done next. The robot will ask the child what activity is going to be performed next and ask for the "KeyCode" of the next activity (see Figure 8b).

3. **State 3. Identification and introduction of the activity code**: The "KeyCode" consists of a sequence of "shapes" that is connected to each activity. This code will be displayed in the app (to parents). Once the child has identified which activity is going to be performed "now" (first screen on the back), he/she must find out the code and enter it into the robot. The child enters the sequence using the SHAPE buttons on the robot to "unlock" the next activity. This stage will serve as a checkpoint (see Figure 8c).

4. **Stage 4. Activity and subtasks**: When Pepe detects the correct code, the screens display the main activity (large) and subtasks (small), along with lights on the timeline. The robot warns the child verbally about the beginning of the activity and indicates the first subtask. The first subtask will light up. As time passes, the lights on the timeline will begin to turn off. When the timeline approaches the next subtask, the robot will tell the child that it is time to move on to the next one. As the child finishes with one subtask, he or she moves the arrow to the next one (which will light up) and will press the YES button. When the YES button is pressed, Pepe verbally congratulates the child, and a star will appear in the completed subtask. The stars will be recorded in the app for parents and children to see. A certain number of stars will imply a reward. The app also informs parents of the child's progress on tasks and subtasks (see Figure 8d).

Figure 8. Human-interface interaction: main stages of the robot. (**a**) Creating a routine; (**b**) First-Then activities; (**c**) activity code; (**d**) activity and subtask.

Figure 8 shows a simplified drawing of the user-interface interaction for these stages.

Pepe has an interactive interface (see interactive elements in Table 4) to control the different stages and interact through the different activities.

Table 4. Interactive elements of the interface.

Interaction Design Elements		Function
Screens		Shows First-Then activities
		Shows images of activities; show the subdivision of tasks
		Shows the reward system to the child
Red arrow		Allows the child to point to a screen to mark the activity that is being carried out
YES button		Allows the child to confirm a response or action; affirmation
SHAPE buttons		Allows the child to enter the code (Key Code)
LED lights		Time count: indicates the time remaining for each activity
		Enhances what is being done at the moment
		Positive reinforcement
		Facial expressions, identity
Speaker	Child's voice	Guides through the time of use
		Positive reinforcement
		Time count
		Identity
	Music	Indicates if the answer is correct
		Positive reinforcement

Lastly, Table 5 details the user–product interaction, relating the functions and subfunctions (FR, NFR, and SR) and the corresponding actions of the user.

Table 5. List of product functions and user actions.

Product Functions			User Actions P: Parent/Supervisor
Functions		Subfunctions/Functional Requirements	C: Child with ASD
Stage 1. Establishment and approval of the routine	Routine proposal to parents	The application is activated FR 1	Activate the application (P)
			Find desired command (P)
		Presents routine proposal (application) FR 8, 9, 11, 12	Request routine for X time (P)
			Check routine (P)
			Decide whether to modify or approve (P)
	Modification/ routine approval	The routine is modified (app) FR 5, 9	If the parent modifies the routine
		The routine is approved (app) FR 5, 9	If the parent approves of the routine
		The routine is recorded and sent to the robot FR 6, 8, 12	
Turn robot ON		Robot ON FR 4	Press the ON/OFF button (C)
		Verbal welcome message FR 2, 3, 6	Check that the toy is on (C)
		The expression "SMILE" lights up FR 2, 3, 4, 6	

Table 5. Cont.

Product Functions		User Actions
Functions	Subfunctions/Functional Requirements	P: Parent/Supervisor C: Child with ASD
Stage 2. Presentation of First-Then activities	Verbal message: start routine FR 2, 4, 6, 10	Receive info: start the routine (C)
	Ask the child verbally to turn it around FR 2, 3, 6, 10	Turn robot around (C) Check position (C)
	Shows activities First-Then FR 2, 4, 10, 11, 13	Receive info: tasks to perform (C)
	Verbal message: identify the first activity and enter the code FR 3, 6, 10, 14	Identify the first activity (C) Ask the supervisor for KeyCode (C)
Stage 3. Identification and introduction of the activity code	The code is displayed in the app FR 3, 10, 15	Identify the activity (P) Communicate the code to the child (P)
	Allows code input FR 2, 3, 7, 15	Receive supervisor info: code (C) Enter the code (C) Check code (C)
	Acquire emotion "HAPPY" FR 2, 4, 6, 16	If the code is correct
	Verbal message: "VERY GOOD" FR 2, 4, 6, 16	
	Plays "SUCCESS" melody FR 2, 4, 6, 16	
	Acquire "SAD" emotion FR 4, 6, 17	If the code is incorrect
	Plays "FAILURE" melody FR 4, 6, 17	
Stage 4. Activity and subtasks	Warns that the activity will begin FR 2, 4, 6, 10	Verify that the activity is going to start (C)
	Displays tasks and subtasks FR 2, 4, 10, 11, 13	Receive info: task and subtasks to perform (C)
	The light for the first subtask comes on FR 2, 4, 10, 11	Identify the first subtask to perform (C)
	Timeline lights up FR 2, 3, 4, 10, 13, 14	Receive info/check: start first subtask (C) Start with the first subtask (C)
	Time lights turn off progressively FR 2, 3, 4, 10, 13, 14	Receive/check info: time is passing (C)
	Verbal message: time to switch subtasks FR 2, 3, 4, 10, 13, 14	Receive info: switch subtasks (C) Move the arrow to the next subtask (C) Check arrow position (C)
	The next activity light turns on FR 2, 3, 4, 7, 10	Press the YES button (C)
	Verbal message: "VERY GOOD" FR 2, 3, 4, 6, 16	
	Plays "SUCCESS" melody FR 2, 4, 6, 16	
	Remember how long the subtask took FR 6, 8, 12	
	A star appears in the completed subtask FR 2, 4, 6, 16, 17	Verify: the activity is complete (C)
	The star is saved in the app FR 16, 17	Receive info: number of stars achieved (P, C)

Table 5. *Cont.*

Product Functions		User Actions P: Parent/Supervisor
Functions	Subfunctions/Functional Requirements	C: Child with ASD
	Update the parent (app) FR 1, 14, 18	Receive info: what the child is doing (P) 🟨
	Warn the parent FR 7, 8	If the child does not state the progression in the subtasks
	Plays FINISHED melody FR 2, 4, 6, 10, 16	Check: activity is over (C) 🟩
LEGEND. Categorization of activities		
Action: includes attention, perception, decision-making, and action		🟦
Verification: includes the attention, perception, and evaluation of the task itself and the state of the product		🟩
Retrieval: includes attention, perception, decoding information, and interpretation		🟨
Communication: includes attention, perception, decision-making, and data communication		🟧
Selection: includes attention, perception, and decision-making between two or more options		🟪

4.4. Phases 6 and 7: Prototyping and Testing

Once the detail design was established, the prototyping phase allowed the final proposal to be validated. As indicated by the ASDesign methodology, the prototype aims to validate the following aspects: (1) aesthetic (look and feel), (2) structural, (3) functional, (4) technical, and (5) integrated. For this work, technical, functional, and integration prototypes (where the look and feel aspects were also included) were developed.

The technical prototype was manufactured using an Arduino and the Arduino software. It included buttons, photoresistors, and a clock module as sensors, and LED lights, a speaker, and a buzzer as actuators. This made it possible to test the following elements: (3) KeyCode State and (4) Activity State. All these functions, together with the corresponding lights and sounds, were tested. Thus, the electronic circuit was established to create the electronic skeleton of the robot (Figure 9, left). Subsequently, this skeleton was integrated into a working prototype to test the different user–product interactions. For this prototype, the "Wizard of Oz" prototyping method [57] was used for the screens, in order to get a complete perspective of the entire functionality (see Figure 9, right).

Finally, an integration prototype (see Figure 10) was manufactured, including elements of materiality and appearance (look and feel) combined with the electronic skeleton. This was developed with the following materials: foam boards, cardboard, cork, felt, foam, bubble wrap, PET spiral bracelets, aluminum foil, and cellophane paper. This last prototype was tested with a child with ASD, a typically developing child, and an adult.

For the test with the integration prototype (Figure 11), a list of assumptions to be validated was created: (1) structure and resistance, (2) visual elements, (3) sound elements, (4) motor action elements and (5) learning and understanding. From this test, it was concluded that the user was attracted to the robot, its shape, colors, and textures. The lights and sounds did not disturb him and the textures were attractive, especially the arms, as they quickly attracted attention. Facial expressions were also a key element in capturing the user's attention. The way to interact with the different elements of the interface (arrow and buttons) was understood. The images (in this case pictograms) were understandable. Choosing answers in the interface, as well as entering the "Key Code", was intuitive. Additionally, the sequence of moving the arrow and pressing the green button was clear to the user. The user was willing to play as long as they understood the flow of the interaction. This last aspect suggests that, for the early stages, an adult should be present and help the child.

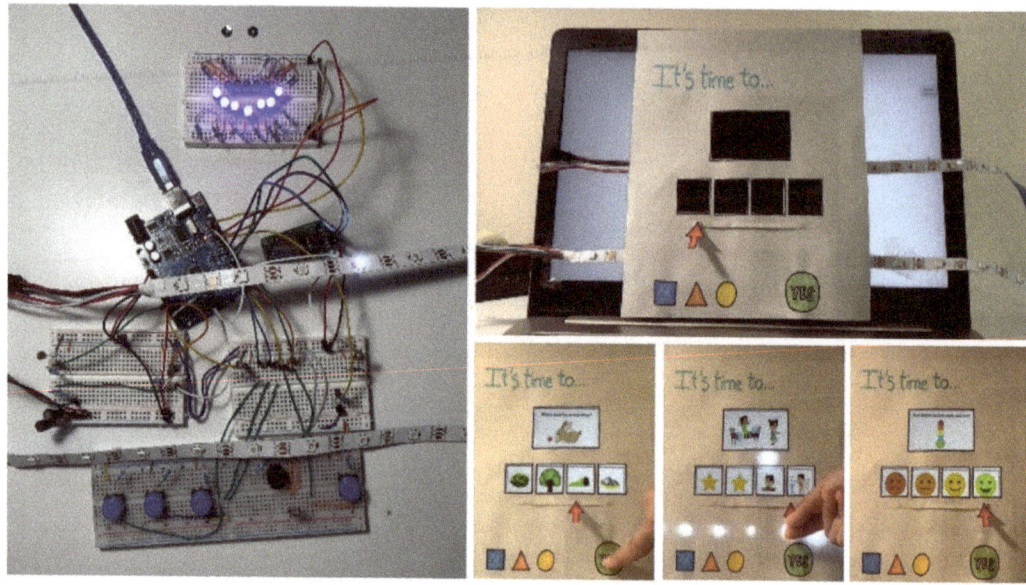

Figure 9. Technical (**left**) and functional (**right**) prototype.

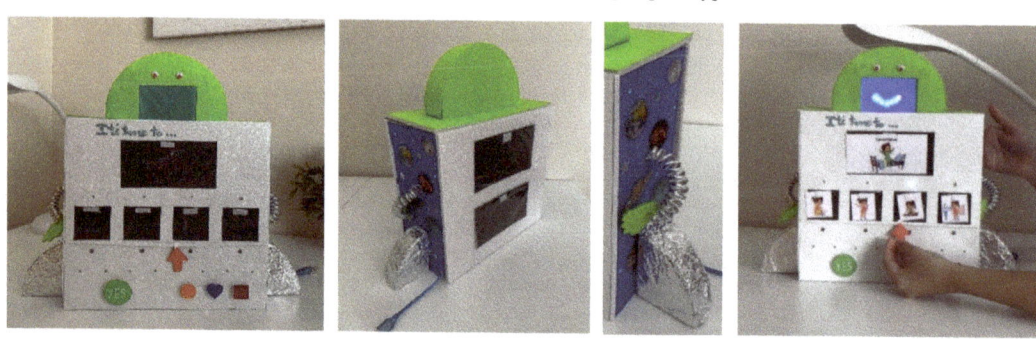

Figure 10. Integration prototype.

Figure 11. Testing the prototype with the user.

4.5. Design Validation

The characteristics of the Pepe robot make it an interactive and smart product. This section evaluates the design, making use of different principles and data from the scientific and technical literature. This evaluation, together with the verification of the design through prototyping and testing with the target user, establishes the suitability of the ASDesign method.

Through its different functions and elements, Pepe integrates the following perspectives of interactivity: (1) Dialogue interaction (interaction is perceived by children as a dialogue, where the different elements are used to provide information and wait for a response; in this case, the dialogue includes three actors: parents, children, and Pepe [58]); (2) System interaction (the process of learning and interpretation of information by the robot, to subsequently shape and establish how the dialogue will evolve) [58]; (3) Transmission interaction (present information to the child for him to learn and interpret) [59]; and (4) Agent-based interaction (based on the instructions given by the user to perform a specific task that determines how the user–product interaction will develop) [60]. The combination of these perspectives makes the interaction adaptive to the child's development, resulting in an evolution of dialogue, thus being interconnected perspectives.

The fact that it is also smart was demonstrated by Cagiltay et al. (2014), who argued that "smart toys are new forms of toys that incorporate tangible objects and electronic chips to provide two-way interactions that lead to purposeful tasks with behavioral or cognitive merit" [41].

According to Rijsdijk and Hultink, the seven dimensions of the smartness of a product are autonomy, adaptability, reactivity, multifunctionality, the ability to cooperate, human interaction, and personality [61]. Thus, the degree of smartness of a product is measured to the extent that it has one or more of the following characteristics. Table 6 explores these dimensions. This check validates the ASD_T2 and ASD_T3 modules.

Table 6. Smart dimensions of a product.

Dimension	Definition	Implementation with ASD_T2 and ASD_T3
Autonomy	The extent to which a product is able to operate independently and meet objectives without user interference.	Although the collaboration of parents is required to insert preferences, Pepe guides the child throughout the day independently and without the need for supervision.
Adaptability	The ability of a product to improve the match between its functioning and its environment [62].	Pepe is able to save and interpret information about the child's performance (time required by a task, how it makes him or her feel, which visual and auditory supports work best, etc.), continuously adapting its interactions to the needs of the child.
Reactivity	The ability of a product to react to changes in its environment [63].	Pepe is able to react to the child's actions through the different sensors—buttons, light sensors, sound sensors, or time modules—and adapt its behavior.
Multifunctionality	A single product fulfills multiple functions [64].	Although it fulfills the main function of guiding the child through a routine, Pepe includes different modalities and subfunctions to meet this objective: establishment of a routine, the introduction of codes, monitoring of activities, monitoring of the child's achievements, and help system.
Ability to cooperate	Ability to cooperate with other devices to achieve a common goal.	In the case of this robot, it can be connected to the mobile application so that parents can insert their preferences, modify the routine, and be aware of what their children are doing.
Human interaction	The degree to which the product communicates and interacts with the user in a natural and humane way.	The Pepe robot communicates with the user visually and verbally, in an intuitive way, using familiar expressions, facial expressions, and positive reinforcements that motivate the child.
Personality	The ability of the product to show the properties of a believable character [63].	The Pepe robot has a believable personality thanks to its material elements that give it a humanoid appearance (eyes, mouth, arms), as well as a child voice that gives it the personality of a playmate.

Table 7 compiles the set of results obtained from the testing according to the user experience analysis parameters as established by Donald Norman [65–67].

Table 7. Testing results: user experience analysis.

Basis	Usability Analysis
Visibility	All interface elements are visible. Having two different sides with screens creates two separate interfaces, avoiding distractions in the different stages. The design is safe.
Feedback	It is achieved with visual, perceptual, and proprioceptive stimuli. Every action of the product has immediate feedback, which keeps the child engaged. The lights and sounds do not disturb the child.
Affordance	All the affordances have an intuitive meaning that is easy for the user to understand. The green button and red arrow are very familiar.
Mapping	Having the screens in a horizontal disposition is intuitive for the child. The action–reaction consistency of musical sounds and lights is adequate. The "light count down" was not completely understood but could be due to the need of having a learning period.
Constraints	Correct. They reduce the probability of human error. The different simultaneous stimuli of the interface help to focus attention on the tasks and the product; the mechanical restrictions and design of buttons and arrow avoid unwanted actions and help to reduce errors by restricting the type of interactions that the user can carry out.
Consistency	Aesthetic, functional, internal, and external consistency is correct. The grip, movement, and stereotype patterns are correct.
Reinforcements	Correct. Through visual (light and color), auditory (musical sounds), and proprioceptive (textures, movements, grip, pulsation, pressure, and force generation) stimulation.

Table 8 compiles the set of results obtained from the testing regarding effectivity, efficiency, and satisfaction [68]. The results have been divided into two phases: (1) the learning period, which involved following simple instructions, and (2) following the activity, which involved understanding the complete flow of the interaction.

Table 8. Testing results: effectivity, efficiency, and satisfaction.

Effectivity	Phase 1	**Completed of activities:** The child was able to complete simple instructions: move the arrow, press a sequence of buttons, turn the robot around. **Failures:** The child did not make many mistakes regarding simple instructions. **Number of times that the child stopped the activity:** When the child understood the instructions, there were no interruptions. **Number of times that the child lost control over an activity:** The child lost control because of a loss of attention, but not due to misunderstanding.
	Phase 2	**Completed of activities:** The child struggled to understand the flow of the interaction in such a short period. Only simple activities could be performed. **Failures:** The child was not able to follow the complete flow of an activity, reinforcing the need of a learning period. **Number of times that the child stopped the activity:** The activities had to be stopped several times because of a loss of attention. This could be because the prototype did not integrate real screens. **Number of times that the child lost control over an activity:** The child lost control because of a loss of attention and misunderstanding.
Efficiency	Phase 1	**Time to complete the task:** The simple instructions were completed immediately, which proves that the child understood the functionality of the different elements. **Time loss due to errors:** When the child understood the tasks, the tasks were completed quickly. **The number of elements used from the interface:** All elements on the interface were used. **Time spent to learn:** The simple instructions were easily understood and performed. The child was already familiar with the elements and sequence of use.
	Phase 2	**Time to complete the task:** The child was not able to complete the flow of an activity. **Time loss due to errors:** The loss of attention led to time loss. **The number of elements used from the interface:** All elements on the interface were used. **Time spent to learn:** Half an hour was not enough for the child to learn the flow of an activity.
Satisfaction	Phase 1	**Positive reactions:** The user enjoyed the sensory stimulation from the product: lights, music, tactile stimuli. He or she was attracted to the product. **Times that the user encountered problems:** The user did not encounter problems with simple instructions.
	Phase 2	**Positive reactions:** The user enjoyed the sensory stimulation from the product: lights, music, tactile stimuli. He or she was attracted to the product. **Times that the user encountered problems:** The user encountered problems understanding the whole activity, which led to a loss of attention.

5. Results and Discussion

ASDesign and the presented case study (the Pepe robot) fit the development of a user-centered methodology for assistive technology design in a domestic context that helps children with ASD plan and self-manage a daily routine. This methodology is based on the principles of Positive Behavioral Support; that is, it is committed to products that allow a long-term development and negotiation process in which the child with ASD learns to self-regulate progressively, with the ultimate goal of independence.

In addition to this, the scope of the ASDesign method allows one to create products that can be used in periods of confinement. Although the line of work arises from the situation derived from the pandemic, the method extends the context of use to the new normal in which the reconciliation of the working day with the care of the child with ASD can be a challenge for many families. Thus, ASDesign is a methodology aimed at the design of assistive technology for children with ASD, which gradually helps them to become more independent in the follow-up of a routine, with the intention of improving their self-autonomy in situations in which they need support that cannot be granted given the social context in which they find themselves. Including desirable periods within the term "confinement" is of great interest. Several studies show that the experience of confinement caused by the health crisis has led to social and structural changes in the labor market that could be definitively implemented in daily life, such as telework, online training, or intensive or reduced working days, to which professionals and families must adapt [3–5]. In addition, the new trends in automation and industrialization 4.0 will generate positions with more responsibility, geographical distribution, and flexibility, which also means an alteration of the working schedule.

The ASDesign method allows one to integrate into mid- and high-tech products the properties of multifunctionality, smartness, and interactivity, complying with the following principles:

(1) *Adaptability*: Design of products that adapt to the needs, desires, and abilities of the child by collecting information from his performance with the product and interacting with parents. This adaptability allows assistive products to interact with the child and adapt to different needs; that is, technology changes its behavior as the child progresses.

(2) *Agent Design*: It is important that products have the interaction design based on agencies [56] as an essential element. This will allow for a negotiation of responsibilities, giving control to the parents during the early stages and allowing the child to make decisions progressively. Products must be adapted to the needs and preferences of children, gradually giving them autonomy and therefore the independence they need for their social and personal development.

(3) *Sensoriality*: Assistive technology should be designed based on the accentuated needs and special sensory characteristics that children with ASD have. ASDesign relies on extensive user analysis to design products that present and collect information in an intuitive and engaging way for these children. In this way, it proposes to move away from the limitations of 2D technologies and combine traditional (low) and computational (mid-high tech) elements and materials to ensure the safety, comfort, and correct physical, cognitive (perceptual, sensory, and emotional) adaptation of children with ASD.

(4) *Learning through play:* ASDesign focuses on the planning and structuring of routines in a domestic context. Therefore, making use of the advantages offered by medium/high assistive technology, this methodology proposes a playful approach, in which routine gamification is put into practice by having a reward system to keep the child motivated.

With these four principles, ASDesign can be applied to high-tech products that aim to help children with ASD be more independent in their daily routine. It has all the design stages, the task planning is correct, and the modules ASDT_T1, 2 and 3 have been validated.

Lastly, as future lines of work, it would be interesting to expand the scope of the ASDesign methodology to the spectrum levels (2 and 3), enabling the design of interactive and smart products for other difficulties and needs of autism. Another future development would involve the creation of a module for the evaluation of design solutions, which would provide the design team with quantitative indicators of the adequacy of the potential design. Finally, the development of a computer application that facilitates the use of the ASDesign methodology is proposed, linked to a database that makes the update of the ASD_T2 and T3 tools more flexible.

6. Conclusions

Although some parents have seen "positive" effects of confinement, for example, seeing their children be more relaxed by not feeling the pressure that exists in education centers for having to socialize with others or perform certain tasks, most studies show that negative effects have prevailed [11,15]. Children with ASD have experienced a loss of independence and communication skills [12]. Social isolation, lack of services due to the economic downturn, and job loss further accentuate the inequality of opportunity gap for children with ASD and their families [11,19]. On the other hand, many of the new daily and work practices derived from confinement (such as telework or the globalization of the labor market) could be implemented indefinitely, which will require new tools to address family reconciliation.

This work develops the ASDesign method, a user-centered method for the design of assistive technology and smart products that helps children with ASD follow a routine and manage tasks autonomously. The method has the scope of user-centered technology design; that is, it can be used and configured comfortably and safely by children (and supervised by parents), thus helping them to plan a routine and be more independent in their day-to-day. This methodology has been applied to the Pepe Robot case study, which has obtained very good results when tested with the target user, thus validating the proposed method.

This work aims to improve the living conditions, health, and social welfare of groups that have been negatively affected by the pandemic. This research contributes to engineering and design fields, and more specifically to the development of science that serves society, with the end goal of improving well-being and social sustainability.

Author Contributions: Conceptualization, R.C. and M.E.P.; formal analysis, R.C.; investigation, R.C. and M.E.P.; methodology, R.C. and M.E.P.; software, R.C.; design and prototyping: R.C.; supervision, M.E.P.; validation, R.C. and M.E.P.; writing—original draft, R.C. and M.E.P.; writing—review and editing, R.C. and M.E.P. All authors have read and agreed to the published version of the manuscript.

Funding: This research received no external funding.

Institutional Review Board Statement: Not applicable.

Informed Consent Statement: Informed consent was obtained from all subjects involved in the study. Written informed consent has been obtained from the patient(s) to publish this paper.

Data Availability Statement: The data presented in this study (Annex I and II) are available upon request from the corresponding author.

Acknowledgments: We thank TuYTEA, Asociación TAJIBO, Asociación Autismo Sevilla, Centro Psicopedagógico Impulso, el Gabinete LegoDUO, Asociación Seta, Neurointegra, and AOSA-TEA for their interest, commitment, and collaboration in the project.

Conflicts of Interest: The authors declare no conflict of interest.

References

1. Colizzi, M.; Sironi, E.; Antonini, F.; Ciceri, M.L.; Bovo, C.; Zoccante, L. Psychosocial and Behavioral Impact of COVID-19 in Autism Spectrum Disorder: An Online Parent Survey. *Brain Sci.* **2020**, *10*, 341. [CrossRef]
2. Smile, S.C. Supporting children with autism spectrum disorder in the face of the COVID-19 pandemic. *CMAJ* **2020**, *192*, E587. [CrossRef]

3. Baert, S.; Lippens, L.; Moens, E.; Sterkens, P.; Weytjens, J. *How Do We Think the COVID-19 Crisis Will Affect Our Careers (If Any Remain)*; Global Labor Organization: Essen, Germany, 2020.
4. Baert, S.; Lippens, L.; Moens, E.; Sterkens, P.; Weytjens, J. *The COVID-19 Crisis and Telework: A Research Survey on Experiences, Expectations and Hopes*; Global Labor Organization: Essen, Germany, 2020.
5. Brynjolfsson, E.; Horton, J.J.; Ozimek, A.; Rock, D.; Sharma, G.; Tuye, H.-Y.; Upwork, A.O. *COVID-19 and Remote Work: An Early Look at US Data*; National Bureau of Economic Research: Cambridge, MA, USA, 2020.
6. Chia, G.L.C.; Anderson, A.; McLean, L.A. Use of Technology to Support Self-Management in Individuals with Autism: Systematic Review. *Rev. J. Autism Dev. Disord.* **2018**, *5*, 142–155. [CrossRef]
7. Iwarsson, S.; Ståhl, A. Accessibility, Usability and Universal Design—Positioning and Definition of Concepts Describing Person-Environment Relationships. *Disabil. Rehabil.* **2003**, *25*, 57–66. [CrossRef]
8. Merriam-Webster Confined | Definition of Confined. Available online: https://www.merriam-webster.com/dictionary/confined (accessed on 21 October 2021).
9. American Psychiatric Association. *Diagnostic and Statistical Manual of Mental Disorders: DSM-5*; American Psychiatric Association: Washington, DC, USA, 2013; ISBN 0890425558.
10. Resource Guide | Autism Speaks. Available online: https://www.autismspeaks.org/resource-guide (accessed on 5 March 2020).
11. Spain, D.; Mason, D.; Capp, J.S.; Stoppelbein, L.; White, W.S.; Happé, F. "This May Be a Really Good Opportunity to Make the World a More Autism Friendly Place": Professionals' Perspectives on the Effects of COVID-19 on Autistic Individuals. *Res. Autism Spectr. Disord.* **2021**, *83*, 101747. [CrossRef]
12. Tokatly Latzer, I.; Leitner, Y.; Karnieli-Miller, O. Core Experiences of Parents of Children with Autism during the COVID-19 Pandemic Lockdown. *Autism* **2021**, *25*, 1047–1059. [CrossRef]
13. Yılmaz, B.; Azak, M.; Şahin, N. Mental Health of Parents of Children with Autism Spectrum Disorder during COVID-19 Pandemic: A Systematic Review. *World J. Psychiatry* **2021**, *11*, 388–402. [CrossRef] [PubMed]
14. Baweja, R.; Brown, S.L.; Edwards, E.M.; Murray, M.J. COVID-19 Pandemic and Impact on Patients with Autism Spectrum Disorder. *J. Autism Dev. Disord.* **2021**, *10*, 1–10. [CrossRef] [PubMed]
15. Mumbardó-Adam, C.; Barnet-López, S.; Balboni, G. How Have Youth with Autism Spectrum Disorder Managed Quarantine Derived from COVID-19 Pandemic? An Approach to Families Perspectives. *Res. Dev. Disabil.* **2021**, *110*, 103860. [CrossRef]
16. Fergusson, E.F.; Jimenez-Muñoz, M.; Feerst, H.; Vernon, T.W. Predictors of Satisfaction with Autism Treatment Services during COVID-19. *J. Autism. Dev. Disord.* **2021**, 1–12. [CrossRef]
17. Degli Espinosa, F.; Metko, A.; Raimondi, M.; Impenna, M.; Scognamiglio, E. A Model of Support for Families of Children with Autism Living in the COVID-19 Lockdown: Lessons from Italy. *Behav. Anal. Pract.* **2020**, *13*, 550–558. [CrossRef]
18. Alhuzimi, T. Stress and Emotional Wellbeing of Parents Due to Change in Routine for Children with Autism Spectrum Disorder (ASD) at Home during COVID-19 Pandemic in Saudi Arabia. *Res. Dev. Disabil.* **2021**, *108*, 103822. [CrossRef]
19. Manning, J.; Billian, J.; Matson, J.; Allen, C.; Soares, N. Perceptions of Families of Individuals with Autism Spectrum Disorder during the COVID-19 Crisis. *J. Autism Dev. Disord.* **2020**, *51*, 2920–2928. [CrossRef]
20. Narzisi, A. Autism Spectrum Condition and COVID-19: Issues and Chances. *Humanist. Psychol.* **2020**, *48*, 378–381. [CrossRef]
21. WHO. *International Classification of Functioning, Disability and Health*; World Health Organization: Geneva, Switzerland, 2014.
22. Lentini, R.; Vaughn, B.J.; Fox, L. University of South Florida Teaching Tools for Young Children with Challenging Behavior. In *Early Intervention Positive Behavior Support*; University of South Florida: Tampa, FL, USA, 2005.
23. Hopkins, J. Assistive Technology: 10 things to know. *Libr. Media Connect.* **2006**, *25*, 12–14.
24. Establish Morning or Evening Routine | Happy Kids Timer Chores during COVID-19. Available online: https://happykidstimer.com/ (accessed on 2 April 2021).
25. Choiceworks App. Available online: http://www.beevisual.com/ (accessed on 2 April 2021).
26. Syriopoulou-Delli, C.K.; Gkiolnta, E. Review of Assistive Technology in the Training of Children with Autism Spectrum Disorders. *Int. J. Dev. Disabil.* **2020**. [CrossRef]
27. Castillo, J.; Goulart, C.; Valadão, C.; Caldeira, E.; Bastos, T. Robótica Móvil: Una Herramienta Para Interacción de Niños Con Autismo. In Proceedings of the VII Congreso Iberoamericano de Tecnologías de Apoyo a la Discapacidad-IBERDISCAP, Santiago de Los Caballeros, Dominican Republic, 28–29 November 2013.
28. Bekele, E.; Crittendon, J.A.; Swanson, A.; Sarkar, N.; Warren, Z.E. Pilot Clinical Application of an Adaptive Robotic System for Young Children with Autism. *Autism* **2014**, *18*, 598–608. [CrossRef] [PubMed]
29. Robins, B.; Dautenhahn, K.; Te Boekhorst, R.; Billard, A. Robotic Assistants in Therapy and Education of Children with Autism: Can a Small Humanoid Robot Help Encourage Social Interaction Skills? *Univers. Access Inf. Soc.* **2005**, *4*, 105–120. [CrossRef]
30. Shamsuddin, S.; Yussof, H.; Mohamed, S.; Hanapiah, F.A. Design and Ethical Concerns in Robotic Adjunct Therapy Protocols for Children with Autism. *Procedia Comput. Sci.* **2014**, *42*, 9–16. [CrossRef]
31. Leka. Available online: https://leka.io/en/product.html (accessed on 30 March 2021).
32. University of Hertfordshire. Kaspar the Social Robot. Available online: https://www.herts.ac.uk/kaspar/the-social-robot (accessed on 30 March 2021).

33. Attawibulkul, S.; Asawalertsak, N.; Suwawong, P.; Wattanapongsakul, P.; Jutharee, W.; Kaewkamnerdpong, B. Using a Daily Routine Game on the BLISS Robot for Supporting Personal-Social Development in Children with Autism and Other Special Needs. In Proceedings of the 58th Annual Conference of the Society of Instrument and Control Engineers of Japan, SICE 2019, Hiroshima, Japan, 10–13 September 2019; pp. 695–700. [CrossRef]
34. Meadan, H.; Ostrosky, M.M.; Triplett, B.; Michna, A.; Fettig, A. Using Visual Supports with Young Children with Autism Spectrum Disorder. *Teach. Except. Child.* **2011**, *43*, 28–35. [CrossRef]
35. Lucyshyn, J.M.; Albin, R.W.; Horner, R.H.; Mann, J.C.; Mann, J.A.; Wadsworth, G. Family Implementation of Positive Behavior Support for a Child with Autism: Longitudinal, Single-Case, Experimental, and Descriptive Replication and Extension. *J. Posit. Behav. Interv.* **2007**, *9*, 131–150. [CrossRef]
36. Cibrian, F.L.; Mercado, J.; Escobedo, L.; Tentori, M. A Step towards Identifying the Sound Preferences of Children with Autism. In Proceedings of the 12th EAI International Conference on Pervasive Computing Technologies for Healthcare, New York, NY, USA, 21–24 May 2018; ACM International Conference Proceeding Series. pp. 158–167. [CrossRef]
37. Dickie, V.A.; Baranek, G.T.; Schultz, B.; Watson, L.R.; McComish, C.S. Parent Reports of Sensory Experiences of Preschool Children With and Without Autism: A Qualitative Study. *Am. J. Occup. Ther.* **2009**, *63*, 172–181. [CrossRef] [PubMed]
38. Garrido, F.J.; Benjamín García, B. La Contaminación Acústica En Nuestras Ciudades. *Colección Estud. Soc.* **2003**, *12*, 1–248.
39. Seeman, L.; Cooper, M. Techniques for the The Cognitive and Learning Disabilities Accessibility Task Force (COGA). Available online: https://w3c.github.io/coga/techniques/ (accessed on 14 April 2021).
40. Seeman, L.; Cooper, M. Cognitive Accessibility User Research. Available online: https://www.w3.org/TR/coga-user-research/#autism (accessed on 14 April 2021).
41. Cagiltay, K.; Kara, N.; Cigdem, C. *Smart Toy Based Learning: Handbook of Research on Educational Communications and Technology*, 4th ed.; Springer: Berlin/Heidelberg, Germany, 2014; pp. 1–1005. [CrossRef]
42. Caro, K.; Tentori, M.; Martinez-Garcia, A.I.; Alvelais, M. Using the FroggyBobby Exergame to Support Eye-Body Coordination Development of Children with Severe Autism. *Int. J. Hum. Comput. Stud.* **2017**, *105*, 12–27. [CrossRef]
43. Chen, J.; Wang, G.; Zhang, K.; Wang, G.; Liu, L. A Pilot Study on Evaluating Children with Autism Spectrum Disorder Using Computer Games. *Comput. Hum. Behav.* **2019**, *90*, 204–214. [CrossRef]
44. Cibrian, F.L.; Peña, O.; Ortega, D.; Tentori, M. BendableSound: An Elastic Multisensory Surface Using Touch-Based Interactions to Assist Children with Severe Autism during Music Therapy. *Int. J. Hum. Comput. Stud.* **2017**, *107*, 22–37. [CrossRef]
45. Contreras, V.; Fernández, D.; Pons, C.; Contreras, V.; Fernández, A.; Fabiana Pons, C. Interfaces Gestuales Aplicadas Como Complemento Cognitivo y Social Para Niños Con TEA. *Rev. Iberoam. Tecnol. Educ. Educ. Tecnol.* **2016**, *17*, 58–66.
46. Kara, N.; Cagiltay, K. Smart Toys for Preschool Children: A Design and Development Research. *Electron. Commer. Res. Appl.* **2020**, *39*, 100909. [CrossRef]
47. Malinverni, L.; Mora-Guiard, J.; Padillo, V.; Valero, L.; Hervás, A.; Pares, N. An Inclusive Design Approach for Developing Video Games for Children with Autism Spectrum Disorder. *Comput. Hum. Behav.* **2017**, *71*, 535–549. [CrossRef]
48. Dattolo, A.; Luccio, F.L. A Review of Websites and Mobile Applications for People with Autism Spectrum Disorders: Towards Shared Guidelines. In *Lecture Notes of the Institute for Computer Sciences, Proceedings of the Social-Informatics and Telecommunications Engineering*; Springer: Berlin/Heidelberg, Germany, 2017; Volume 195, pp. 264–273.
49. Dautenhahn, K.; Bond, A.; Cañamero, L.; Edmonds, B. *Socially Intelligent Agents. Creating Relationships with Computers and Robots*; Kluwer Academic Publishers: Dordrecht, The Netherlands, 2002; ISBN 978-1-4020-7057-0.
50. Javed, H.; Burns, R.; Myoughoon, J.; Howards, A.M.; Hyuk Park, C. A Robotic Framework to Facilitate Sensory Experiences for Children with Autism Spectrum Disorder: A Preliminary Study. *ACM Trans. Hum.-Robot Interact.* **2019**, *9*, 26. [CrossRef]
51. Pavlov, N. User Interface for People with Autism Spectrum Disorders. *J. Softw. Eng. Appl.* **2014**, *7*, 128–134. [CrossRef]
52. Diaz, R.; Peralta, M. Commercial Products for Children with Special Needs. In *Future Challenges for an Inclusive Sector*; University of Seville: Seville, Spain, 2021.
53. Hernández, A.; Peralta, M.E. *Market. Research for Special Needs Products*; University of Seville: Seville, Spain, 2020.
54. Porayska-Pomsta, K.; Frauenberger, C.; Pain, H.; Rajendran, G.; Smith, T.; Menzies, R.; Foster, M.E.; Alcorn, A.; Wass, S.; Bernadini, S.; et al. Developing Technology for Autism: An Interdisciplinary Approach. *Pers. Ubiquitous Comput.* **2012**, *16*, 117–127. [CrossRef]
55. Sanders, E.B.N.; Stappers, P.J. *Convivial Toolbox: Generative Research for the Front End of Design*; Laurence King Publishing: London, UK, 2012.
56. Wiberg, M. Interaction, New Materials & Computing—Beyond the Disappearing Computer, towards Material Interactions. *Mater. Des.* **2014**, *90*, 1200–1206. [CrossRef]
57. Buxton, B. *Sketching User Experiences: Getting the Design Right and the Right Design*; Elsevier: San Francisco, CA, USA, 2007; ISBN 978-0-12-374037-3.
58. Iversen, O.S.; Krogh, P.G.; Petersen, M.G. *Proceedings of the Fourth Danish Human-Computer Interaction Research Symposium*; Aalborg University: Aalborg, Denmark, 2003.
59. Hornbæk, K.; Oulasvirta, A. What Is Interaction? In Proceedings of the 2017 CHI Conference on Human Factors in Computing Systems, Association for Computing Machinery, Denver Colorado, CO, USA, 6–11 May 2017; Association for Computing Machinery: New York, NY, USA, 2017; pp. 5040–5052. [CrossRef]
60. Wiberg, M. *The Materiality of Interaction*; MIT Press: Cambridge, MA, USA, 2018.

61. Hultink, E.J.; Rijsdijk, S.A. How Today's Consumers Perceive Tomorrow's Smart Products. *J. Prod. Innov. Manag.* **2009**, *26*, 24–42.
62. Nicoll, D. *Taxonomy of Information Intensive Products*; The University of Edinburgh Management School: Edinburgh, Scotland, UK, 1999.
63. Bradshaw, J.M. An Introduction to Software Agents. In *Software Agents*; AAAI Press: Palo Alto, CA, USA, 1997.
64. Poole, S.; Simon, M. Technological Trends, Product Design and the Environment. *Des. Stud.* **1997**, *18*, 237–248. [CrossRef]
65. Norman, D.A.; Nielsen, J. *The Definition of User Experience (UX)*; Nielsen Norman Group: Fremont, CA, USA, 2016; Volume 2.
66. Norman, D.A.; Nielsen, J. Gestural Interfaces: A Step Backward in Usability. *Interactions* **2010**, *17*, 46–49. [CrossRef]
67. Norman, D.A. *Design of Everyday Things*; Basic Books: New York, NY, USA, 2013; ISBN 9780465050659.
68. *ISO 9241-11:2018*; Ergonomics of Human-System Interaction—Part 11: Usability: Definitions and Concepts. Asociación Española de Normalización: Madrid, Spain, 2018.

Article

Research on Collaborative Innovation Mode of Enterprise Group from the Perspective of Comprehensive Innovation Management

Wei Feng [1,*], Ling Zhao [1] and Yue Chen [2]

1 School of Economics and Management, Beijing University of Posts and Telecommunications, Beijing 100876, China; zling@bupt.edu.cn
2 Marketing Department, Tus-Holdings Co., Ltd., Beijing 100085, China; alice_china2021@163.com
* Correspondence: fengwei@bupt.edu.cn

Abstract: At present, collaborative innovation has become an integral part of corporate group strategy. However, there are few collaborative innovation research pieces focusing on corporate groups. This article takes Tus-Holdings, a model enterprise in the field of science and technology services, as the research object, uses case study methods, and systematically analyzes the corporate group's strategy, customers, R&D, management, finance, talent, and other factors from the strategic, business, and support levels under the framework of total innovation management research on the collaborative innovation model of management and the form of cooperative surplus. The research found that the collaborative innovation model is an important support for enterprise groups to build a comprehensive, collaborative innovation system; the internal collaborative innovation model of enterprise groups shows nonlinearity and diversity; collaborative surplus performance is closely related to the collaborative innovation mode, and different collaborative innovation modes will produce a different collaborative surplus. These research results have important theoretical value and practical significance for modern enterprise groups to correctly implement collaborative innovation strategy and improve the efficiency of collaborative innovation.

Keywords: total innovation; group enterprises; collaborative innovation model; cooperative surplus

1. Introduction

Under the background of open innovation, the complexity and integration of technological innovation have become increasingly intensified, and collaborative innovation has become an inevitable choice for the survival and development of enterprises [1,2]. Since the emergence of the enterprise organizational form represented by the Adam Smith nail factory, after more than 200 years of development, the enterprise organizational form has undergone many changes. The enterprise group formed with property rights as a link has become one of the most common and most dynamic organizational forms under the framework of the modern enterprise system and has become one of the important collaborative innovation forces. Enterprise groups can achieve performance improvement through collaborative innovation of the industry–university research model with scientific research institutions, governments, and other subjects. For example, research in the field of industry–university research has made rich achievements in the collaborative innovation mechanism, influencing factors and performance evaluation of enterprises, scientific research institutions, and governments [3–5], and can also rely on external innovation networks to obtain relevant innovation resources [6,7]. In addition to the above two ways, the enterprise group also realizes that different member enterprises within the group share specific 'resources', and can also rely on collaborative innovation to achieve the overall value greater than the sum of the value of each part, thereby enhancing the competitiveness of the enterprise group [8]. From the perspective of the network, the internal network of an enterprise group is a strong relationship network, which can bring resource effects to member enterprises [9]. Further research finds that the internal network of enterprise groups has a significant role in promoting inward open innovation. In the practice of enterprise groups, a collaborative innovation strategy can enhance the competitiveness of the enterprise group once it has formed a consensus; thus, the internal collaborative innovation as important planning of an enterprise group's strategy is actively explored. Most of the current studies incorporate innovation synergy into the research framework of technological innovation and place technological innovation at the core of collaborative innovation relationships. However, innovation research in the non-technical field is also an important part of the innovation field [10]. Only when the two kinds of innovation cooperate with each other can the innovation performance be the best [11]. In addition, non-R&D enterprises and non-R&D-intensive enterprises also play an important role in innovation in economic and innovation strategic management [12]. Compared with other organizational structures, enterprise groups have entered a stable period of development. The importance of innovation in strategy, business, management, and other elements cannot be ignored. The collaborative innovation of enterprise groups should be balanced and comprehensive, and the research results from this perspective need to be further enriched. Collaborative innovation is conducive to the formation of 'new resources', resulting in knowledge spillover, bringing 'collaborative surplus' to enterprises, and breaking through resource constraints [13]. The cooperative innovation model is an important method for various elements of enterprise groups to obtain collaborative surplus through nonlinear organic combinations [14].

The existing research results in the field of collaborative innovation models within enterprise groups are rarely involved. Based on the existing research results, the main purpose of this paper is to answer the following questions: 'First, what types of collaborative innovation models mainly exist within enterprise groups. Second., what are the characteristics of these collaborative innovation models. Third, these collaborative innovation models will bring what kind of collaborative surplus'.

2. Theoretical Basis and Research Framework

2.1. Connotation, Motivation, and Research Level of Collaborative Innovation

The founder of Synergetics, Harken, proposed that "Synergetics is to study a complex system composed of a large number of subsystems inter acting in a complex way. Under certain conditions, the synergy phenomenon and synergy effect are produced among the subsystems through nonlinear interaction, so that the system forms a self-organizing structure of space, time or space-time with certain functions" [15]. "The value of the enterprise as a whole may be greater than the sum of the values of each part, and the effectiveness of the collaborative model partly stems from the benefits brought by economies of scale", which reflects the economic significance of Synergetics [16]. Collaborative innovation describes an organization's capability to create, integrate, and transform diverse knowledge, brainstorms, perspectives, and ideas into innovations in the context of value co-creation, which brings benefits to all participants [17–19]. The collaborative innovation mode has the advantages of risk sharing, mutual benefit, win–win and sustainable innovation, which can significantly reduce the risk and threshold [20].

2.1.1. Connotation and Motivation of Collaborative Innovation

Innovation is gradually transforming into a systematic and networked paradigm. Individual innovation activities, even small-scale and single-level cooperative innovation, are difficult to meet the needs of technological innovation. The accumulation of external knowledge requires enterprises to break through the innovation of the original single enterprise and carry out collaborative innovation [2,21]. The main driving forces for enterprises' collaborative innovation can be summarized as two aspects: breaking through resource constraints and improving enterprise performance: (1) Collaborative effect helps enterprises break through their own resource constraints and industry boundaries and bring them many benefits, such as higher innovation efficiency, lower innovation risk, high-quality tacit knowledge spillovers, etc. [22,23]. The interaction between heterogeneous innovation subjects breaks the barriers in the original innovation process and forms a more open collaborative innovation network [24]. From the perspective of the resource base, enterprises carry out collaborative innovation to meet their respective strategic resource needs; that is, the complementarity of resources is the core driving force for enterprises to carry out cross-border collaborative innovation [25]. Enterprises can obtain resources such as technology, equipment, capital, business networks, and intellectual property rights by relying on the external relationships of various partners [26]. In short, collaborative innovation is conducive to the formation of "new resources", resulting in knowledge spillovers, bringing "Cooperative Surplus" to enterprises, breaking through resource constraints [13]; (2) Different from general cooperative innovation, collaborative innovation is the embodiment of win–win cooperation and overall optimization. The synergy highlights the phenomenon that heterogeneous innovation subjects use their own advantage to speed up the technological innovation process through complementary advantages and resource integration so as to achieve performance improvement in the overall network. A good collaborative innovation relationship in an open communication environment will increase the innovation performance of enterprises [27,28]. At the same time, collaborative innovation can also affect the innovation performance of innovation subjects through the knowledge integration mechanism [29]. Collaborative innovation is a complex strategy used for organizational innovation. According to "network theory", on the one hand, collaborative innovation is the organic collection of various elements, such as R&D, human resources and capital, processes, and systems [30]. On the other hand, it is the dynamic integration of complementary resources to achieve mutual complementarities between all the partners [28].

2.1.2. Research Level of Collaborative Innovation

The research level of collaborative innovation is divided into the macro level, which is the national level, the meso level, which is the industrial level, the regional level, and the micro level, which is the enterprise group level. Constrained by the difficulty of information collection, the research on the micro level is similar to a black hole; there are few research results. The current research level of collaborative innovation is mainly concentrated on the macro level and the meso level. At the macro level, national collaborative innovation research mainly focuses on the characteristics and evolution trend of national innovation networks [31]; the research on the industry–university research collaborative innovation at the national level focuses on the division of labor, positioning, cooperation mode and distribution mechanism of enterprises, scientific research institutions, and government [32,33]. At the meso level, industrial research focuses on innovation evolution, ability evaluation, risk control, enterprise competition and cooperation relationship, influencing factors, policy formulation, and so on. Regional research focuses on the evolution of industrial collaboration paths, strategy formulation, capability evaluation, operating mechanisms, and risk assessment agglomerations [34]. At the micro level, some scholars have conducted research on how to carry out collaborative innovation within enterprise groups from the aspects of corporate customers, markets, and R&D [35]. However, there are few research results at the micro level, and research needs to be strengthened.

2.2. Theoretical Paradigm and Research Progress of Comprehensive Innovation Management

Under the background of the increasingly complex and systematic trend of innovation management, in order to solve the lack of systematic and in-depth research and empirical analysis of comprehensive innovation thought, with the deepening research of multidisciplinary fields such as complexity theory, system science theory, and ecological view, in order to solve the problem of paying too much attention to technical factors in innovation, Academician Xu Qingrui systematically proposed a new paradigm of innovation management in 2006: Total Innovation Management Theory (TIM). This theory breaks the previous pattern of management innovation theory focusing on technological innovation, establishes a new concept of ecological innovation, and expands the innovation elements and the space-time range. Comprehensive innovation management takes all-factor innovation, all-staff innovation, all-time and space innovation, and comprehensive combination as the core characteristics and takes the organic combination and collaborative innovation of various innovation factors such as technology, organization, market, strategy, management, culture, and systems as the means, and combines innovation management mechanisms, innovation management methods, and innovation management tools to achieve the state of innovation for everyone, innovation for everything, innovation at all times, and innovation everywhere, so as to achieve the improvement of competitiveness [36].

On this basis, Zheng Gang scholars put forward a five-stage comprehensive, collaborative process model (C^3IS); the model first proposed the concept of "comprehensive collaboration" between technology and non-technical elements in the innovation process. The model includes five stages: contact/communication-competition/conflict-cooperation-integration-cooperation; its meaning is that under the framework of full participation and full-time airspace, strategy, organization, culture, institution, technology, market, and other elements are coordinated and matched in an all-round way to achieve "2 + 2 > 5" through collaborative innovation, and then improve innovation performance [37]. Some scholars have studied the processing development model of the palm industry under the framework of comprehensive innovation management theory and proposed meaningful improvement measures [38].

Existing collaborative innovation generally places technological innovation at the core position. For enterprise groups, it generally enters a stable period of development. Technological innovation is important, but at the same time, the importance of strategic planning, business expansion, management improvement, and other factors also cannot be ignored. Non-technical innovation and technological innovation should be in the same important position. The collaborative innovation of enterprise groups should be balanced and comprehensive, and the research results in this field need to be further enriched.

2.3. Collaborative Innovation Mode and Collaborative Surplus

The factors of innovation resources freely flow among different subjects and dynamically select the best combination to achieve greater innovation efficiency. The difference between the total synergy effect of innovation of each innovation subject and the total effect generated by the independent innovation of each innovation subject is the collaborative surplus. Its manifestations are the invention of new products, new processes, and new equipment, the improvement of innovative human capital, the enhancement of innovation ability, the realization of economic profits, and the formation of scientific research achievements [39]. Collaborative innovation is conducive to the formation of 'new resources', generating knowledge spillover, bringing 'collaborative surplus' to enterprises, and breaking through resource constraints [14]. In addition, the collaborative innovation of enterprise groups can obtain collaborative surplus [40] so as to improve the innovation performance of enterprise groups. Therefore, obtaining collaborative surplus is the core motivation for enterprise groups to carry out collaborative innovation, and it is also the key for enterprise groups to use collaborative innovation to form competitive advantages at the group level. Chen Jin, Yao Yanhong, and other scholars have conducted in-depth research on the motivation, mechanism, and manifestation of collaborative surplus. In view of the mechanism of collaborative surplus, Chen Jin believed that enterprise groups have the effects of technical collaboration, production collaboration, and market collaboration under the influence of organizational structure, incentive mechanism, collaborative atmosphere, horizontal interaction, information bridge, and other factors. Collaborative surplus is manifested as the reduction of new product development cost and the shortening of the development cycle [35]. Although the above research has expounded and discussed the generation mechanism and measurement method of collaborative surplus, they do not elaborate in-depth on the realization path of collaborative innovation; that is, the existing theory does not explain clearly what path enterprise groups achieve better performance than a single innovation.

The collaborative innovation model is an important method for enterprise groups to obtain collaborative surplus through a nonlinear organic combination of various factors [15]. The collaborative innovation model is closely related to enterprise innovation performance and enterprise synergy effect and has played a positive impact [41]. By mining the benefits of economies of scale, enterprise groups make nonlinear combinations of technology, strategy, organization, culture, institution, market, and other factors [37,42,43], thus forming a variety of collaborative innovation models and affecting the performance of collaborative innovation. In addition, as the command center, the enterprise group headquarters promotes internal collaborative innovation through business expansion, shared collaboration, business management, and support services [44]. It can be seen that the collaborative innovation model is the bridge of resource matching mechanism of enterprise group members, the link of various implementation mechanisms of enterprise group integration, and the key to obtaining collaborative surplus. Different collaborative innovation models will produce a different collaborative surplus so that collaborative surplus has diverse manifestations. The aggregate collection of collaborative innovation mode constitutes the formation path of collaborative surplus, so the study of collaborative innovation mode is the core link to analyzing the path of collaborative innovation.

Research on collaborative innovation of internal resources for enterprise groups has important theoretical and practical significance. Existing research fails to conduct in-depth research on the main collaborative innovation modes, mode characteristics, and obtaining collaborative surplus of enterprise groups. In view of this, this paper, based on the existing research results, takes Tus-Holdings as the research object, adopts the single case study method, and systematically studies the types and characteristics of collaborative innovation modes of enterprise groups from the perspective of comprehensive innovation management, and analyzes the manifestations of the collaborative surplus of enterprise groups.

2.4. Analysis Framework

In recent years, the research results have laid a certain foundation for the study of this paper and provided new ideas. However, the existing research has not systematically expounded on the types and main characteristics of the enterprise group collaborative innovation model and the forms of collaborative surplus through the collaborative innovation model. From the perspective of comprehensive innovation management theory, based on collaborative innovation theory, this article uses case study analysis methods to construct a comprehensive, collaborative innovation model of enterprise groups with Tus-Holdings as the research object. The integrated analysis framework of this research is shown in Figure 1.

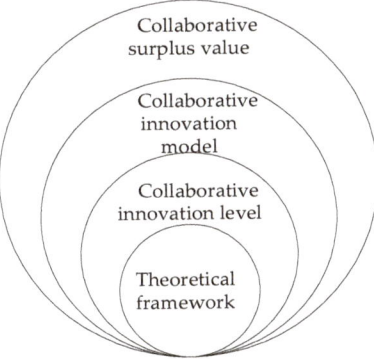

Figure 1. Integrated analysis framework.

3. Research Design

This research follows the process of case study: review theory → design case study plan → collect case data → analyze case data → research conclusion [45].

3.1. Research Methods

Case study method is one of the important methods in management research. The use of a single case or multiple cases in-depth analysis of "how" and "why" and other mechanisms to reveal the nature of the problem is of great significance in the discovery of new theories, enriching the existing theory, which is confirmed by many scholars' research [46,47]. In view of the existing theoretical gaps, this paper aims to study the main types, characteristics, and surplus of collaborative innovation model within enterprise groups. The existing literature does not have in-depth answers, nor does it involve, so an exploratory case study is needed [48,49]. Related literature also used the exploratory single-case study to study the collaborative innovation and reached some valuable conclusions [50,51]. This paper aims at the collaborative innovation model within the enterprise group and takes the ownership of the enterprise group as the link of management as the research boundary, so it is more suitable to adopt the single case study method.

3.2. Case Selection

The selection of typical cases is a common practice based on case study methodology [52]. This study believes that it is appropriate to take Tus-Holdings Co., Ltd. (Beijing, China) (hereinafter referred to as Tus-Holdings) as a case study, which is mainly based on the following considerations [53]: First, Tus-Holdings is an enterprise group that attaches great importance to innovation, and puts forward innovation theories such as "Four Gatherings", "eight major factors" and "three-dimensional triple helix". Relying on the accumulation in the field of innovation, it has become the leader of the domestic science technology service industry and is the first science and technology service enterprise group in China with assets exceeding CNY 100 billion. At present, the management assets exceed CNY 200 billion, which is representative of the industry. Second, since its establishment, Tus-Holdings has lasted for 20 years. It has developed from a single science and technology park Development and Construction Enterprise into a large enterprise group involving five strategic emerging industries such as "environmental protection, new energy, great health, digital economy, and new materials", and supporting industries such as education, culture, and style. It has gone through many different stages. The characteristics, problems, and solutions of each stage are rich and interesting, and the historical data are relatively complete. Third, Tus-Holdings develops under the guidance of the strategic policy of "pattern, strategy and synergy", which regards collaborative innovation as the core competitiveness of enterprises. At present, more than 800 enterprises have been listed and practiced in the field of collaborative innovation mode for many years, which provides a good sample for this study.

3.3. Case Enterprise Situation

Tus-Holdings Co., Ltd. (hereinafter referred to as Tus-Holdings) was established in July 2000. It is a science and technology service enterprise group focusing on management innovation and emphasizing on science and technology services field and relying on Tsinghua University. After more than 20 years of development, Tus-Holdings has built a "Science and Technology Park, Science and Technology Industry, Science and Technology Finance" trinity, collaborative cluster innovation business pattern, and more than 800 consolidated enterprises, the total assets under management exceed RMB 200 billion. Tus-Holdings has formed a complete industrial chain layout and built a complete scientific and technological innovation ecosystem in five strategic emerging industries of environmental protection, new energy, big health, digital economy, and new materials, as well as supporting industries of education, culture, and sports. Under the guidance of the strategic thinking of "pattern, strategy, and coordination", Tus-Holdings has formed some typical collaborative innovation models according to its own business characteristics: collaborative innovation models at the strategic level, collaborative innovation models at the business level, and collaborative innovation models at the support level. These collaborative innovation models show the strong competitiveness of Tus-Holdings in the fierce market competition.

3.4. Data Collection and Analysis

This study collected various forms of data through different methods, including interviews, internal company information, and public information. (1) Public information, such as corporate public number and debt market announcement, etc.; (2) Corporate archives information, such as enterprise summary, company archives, internal meeting records, etc.; (3) In-depth interviews and semi-structured interviews; (4) Participation in Company meetings and other activities. This paper uses a multi-level, multi-data source data collection method to form a triangular verification to enhance the accuracy of the research results [52].

4. Case Analysis and Discovery

In the process of its own development, Tus-Holdings has gradually formed a relatively mature collaborative innovation model through strategic guidance. Collaborative innovation is implemented at the whole group level and involves a wide range of aspects. Member enterprises have adopted a collaborative innovation model suitable for their own development according to their own conditions. Therefore, a variety of collaborative innovation models with different characteristics have been formed within the group, showing the characteristics of all-around and all-level. The collaborative innovation model is mainly manifested in strategic collaborative innovation, strategic customer collaborative innovation, collaborative business innovation, R&D collaborative innovation, collaborative management innovation, collaborative talent innovation, and collaborative financial innovation.

4.1. Strategic Collaborative Innovation Model

TUS-EST and Tus-Clean Energy are both independent parts of Tus-Holdings; each has an independent and complete business system and conducts business independently. TUS-EST focuses on the comprehensive management of the environment, and Tus-Energy focuses on the transformation of scientific and technological achievements in the field of clean energy. In 2018, Tus-Holdings proposed the strategic concept of "energy + environmental protection". Through the organic combination of the two blocks of business of TUS-EST and Tus-Clean Energy, it achieved the strategic goal of "permanent cure by energy treatment and temporary solution by environmental protection" and built a leading energy and environmental integration group in China. In September 2018, Mr. Wen Hui, Chairman of Tus-Clean Energy, began to serve as the director of the ninth board of directors and secretary of the General Party Branch of TUS-EST. In April 2019, Wen Hui, Chairman of Tus-Clean Energy, was elected as Chairman of the ninth Board of Directors of TUS-EST. At the same time, he announced the overall upgrading of the main industry strategic layout direction and industrial planning orientation and promoted the integration and coordination of the overall operation and strategy of the company. In January 2019, Tus-Clean Energy Research Institute and TUS-EST Research Institute merged to form Tus-Energy and Environment Joint Research Institute so as to promote the integration of energy and environmental protection technologies and realize the collaborative innovation of energy and environmental protection technologies. In July 2019, TUS-EST changed from "Tus-Sound Environmental Resources Co., Ltd. (Beijing, China)" to "TUS-EST", reflecting the strategic layout of "energy + environmental protection". At the same time, the company is positioned as an integrated environmental management service provider of integrated energy and environmental protection. Under the strategic framework of energy and environmental protection integration, TUS-EST and Tus-Clean Energy achieve strategic synergy by distinguishing strategic positioning: TUS-EST as a leading enterprise in the environmental protection industry orients "zero carbon and waste-free city builder", and in the business model, it mainly invests in environmental protection projects and other forms of heavy asset investment; as a platform for the transformation of scientific and technological achievements in the field of energy, Tus-Clean Energy is positioned as the "global clean energy messenger". In the business model, technological innovation services and other forms of light asset operations are given priority. In September 2019, TUS-EST and Tus-Clean Energy jointly established Beijing Tus-Energy Zero Carbon Technology Co., Ltd. (Beijing, China), focusing on clean, comprehensive energy services and zero-carbon energy business areas, giving full play to the synergistic innovation role of the two. By August 2020, Beijing Tus-Energy Zero Carbon Technology Co., Ltd. invested a total of four holding subsidiaries, with an investment amount of 70 million yuan, and the integration strategy of energy and environmental protection has been rapidly implemented.

Tus-Holdings has recombined the environmental sector and the energy sector through collaborative innovation to create advanced domestic energy and environmental protection integrated group. Through differentiated strategic positioning, concurrent chairmanship, integrated technology research, and joint business investment, the strategic combination was quickly implemented. The characteristics of strategic collaborative innovation mode and collaborative surplus can be summarized as follows: Under the guidance of major strategic objectives, the enterprise group takes the executive subjects represented by the chairman and vice president as the core and combines the optimization and combination of internal strategic resources of the enterprise group. It uses strategic positioning adjustment, high-level personnel mobilization, technical research integration, joint business investment, and other methods to obtain the collaborative surplus: rapid realization of strategic objectives and rapid formation of strategic advantages in the core business field.

4.2. Strategic Customer Collaborative Innovation Model

In May 2019, Qingdao Municipal People's government and Tus-holdings held the signing ceremony of the strategic cooperation agreement in Qingdao Municipal Organization Conference Center (Qingdao, China). The two sides conduct in-depth cooperation in the fields of civil–military integration, artificial intelligence, marine science and technology, energy and environmental protection, education and training, as well as building an industrial incubation system, building a science and technology park, and promoting international scientific and technological cooperation. Under the framework of strategic cooperation between the two sides, Tus-Holdings relies on its own advantages of a diversified business system to concentrate resources and provide a package of cross-market services for strategic customers through collaborative innovation, so as to help the Qingdao municipal government to upgrade in an all-round way. As of July 2020, there are more than 70 companies and projects invested and introduced by Tus-Holdings in Qingdao. The landing business involves the field of digital security, talent education, fund, rural revitalization, marine fishery, etc., which provides the best practice cases for Qingdao to carry out clustering and industrial chain investment. The key projects are shown in Table 1.

As a new science and technology "department store", Tus-Holdings uses a diversified business system to provide self-selection services for strategic customers. When facing strategic customers, it provides cross-market, comprehensive, and integrated services for strategic customers through diversified business collaborative innovation so as to provide services for strategic customers at the strategic level, improve customer experience and obtain a high evaluation from strategic customers. The characteristics of the market collaborative innovation model and collaborative surplus can be summarized as follows: The enterprise group relies on its advantages of multiple business systems, with the general manager, deputy general manager, and market director as the main executives and middle managers' special groups, through matching resources with the strategic objectives of customers, providing a cross-market package of services for strategic customers from the strategic level, accurately docking the internal business of the enterprise group with the specific needs of strategic customers, helping the realization of strategic customer objectives, providing customers with super expected service experience to win customer trust, and improving brand awareness.

Table 1. List of cooperation projects between Tus-holdings and Qingdao.

Time	Important Issues
May 2019	Qingdao Municipal Government and Tus-Holdings signed a Strategic Cooperation Agreement
June 2019	Jimo District and Qingdao Blue Valley signed 14 key projects with the subsidiaries of Tus-Holdings, involving clean energy, science and technology education, biomedicine, marine industry, big data, artificial intelligence and other fields.
July 2019	Chengyang District of Qingdao City and Tus Triple Helix Co., Ltd. (Beijing, China) held a signing ceremony. Tus Triple Helix will move its headquarters to Chengyang after the signing of the contract.
August 2019	Jimo District and Tus-Holdings signed 5 project cooperation agreements, covering industrial parks, energy, sports technology, new materials, science and technological innovation services and other fields. The Qingdao branch of Tus Business School was unveiled at the same time and settled in Tus-Holdings Building.
September 2019	Huatong Group and Tus-Holdings subsidiary companies signed a strategic cooperation agreement. The two sides will carry out in-depth cooperation in the fields of civil–military integration, artificial intelligence, marine science and technology, big data operation, education and training, etc, as well as building industrial incubation system, building science and technology park, and industrial funds.
September 2019	Qingdao Laixi Municipal Government's Economic Development Zone, Education and Sports Bureau, Culture and Tourism Bureau, Health and Health Bureau, and City Investment Office respectively signed 5 enterprise landing agreements and four cooperation framework agreements with Tus-Holdings subsidiaries.
September 2019	Qingdao North District and Tus-Holdings hold a comprehensive project doking meeting and signing ceremony of the strategic cooperation framework agreement. The two sides will cooperate around in artificial intelligence, robots, big data, cloud computing, block chain, intelligent manufacturing, civil–military integration, biomedicine, new energy and new materials and other strategic emerging industries.
October 2019	Qingdao North District Government, the Fourth Highway Engineering Bureau of CCCC and Tus-subsidiaries signed a framework cooperation agreement on the construction of an artificial intelligence industrial park.
February 2020	Tus-Holdings and Qingdao City Co-Resistance Epidemic: Tus Cloud established an intelligent monitoring and management system for isolated personnel in Chengyang District of Qingdao. Tus Guoxin provided overall solutions for remote mobile office for medical personnel and government personnel. TUS-EST responded to the epidemic by carrying out medical waste and kitchen projects in Qingdao. Tus Hailiang assisted the unattended urban management in north District through the dual-pronged approach of intelligent temperature measurement and video AI.
May 2020	Tus Middle School was approved to establish three campuses in Laoshan, Shibei and Chengyang, and the spring teacher recruitment examination was completed.
July 2020	Tus-Tech City Group Co., Ltd. (Beijing, China) participated in the mixed-ownership reform of Double Star Group, and obtained 35% equity of Double Star Group by means of capital increase and equity expansion.

4.3. R&D Collaborative Innovation Model

In order to seize the opportunity of the coal-to-clean energy policy and open the heating market of household biomass heating furnaces, Tsinghua Solar Co. Ltd. (Beijing, China) uses its own solar thermal utilization and resource accumulation in the coal-to-electricity market to achieve new business expansion by developing new types of household biomass furnaces. Tsinghua Solar Co., Ltd. has accumulated rich experience in government relations, product promotion, engineering installation, and other fields through its own development. However, due to its late involvement in the research and development of biomass stoves, its accumulated experience is not rich enough; the independent research and development time is long and difficult. Beijing Nowva Energy Technology Co., Ltd. (Beijing, China) is a technology company focusing on the development and efficient solid fuel energy utilization technology and the implementation of the project. It has a full understanding of the combustion characteristics of biomass fuels and has rich experience in the field of furnace core structure design. In the early stage, Beijing Nowva Energy Technology Co.,

Ltd. has invested a lot of research and development manpower and material resources in the field of biomass heating furnaces and carried out relevant technical reserves, but it has not been successfully promoted to the market for various reasons. In May 2019, Tsinghua Solar Co., Ltd. decided to expand the business of biomass combustion stoves. In order to speed up the input of new products, it released the demand for collaborative research and development of new products within Tus-Holdings. In July 2019, Tsinghua Solar Co., Ltd. and Beijing Nowva Energy Technology Co., Ltd. reached a collaborative R&D intention through contact and communication and signed a technology development agreement. After the collaborative R&D matters were determined, the two sides established a collaborative R&D team to carry out R&D work for household biomass heating furnaces that meet the latest standards. Tsinghua Solar Co., Ltd. Company is responsible for the research and development of biomass fuel, smoke exhaust system, and thermal cycle system, while Beijing Nowva Energy Technology Co., Ltd. is responsible for the research and development of combustion system and fire sealing system. The coordination group held regular discussion meetings to inform each other about the R&D situation, communicate with each other on key points and difficulties in a timely manner, discuss the work arrangements in the next stage, determine detailed work plans, and ensure that both parties complete relevant R&D work according to the time node. In case of major emergencies, the collaborative R&D team will hold a special meeting on emergency issues for special treatment to ensure R&D efficiency. The collaborative R&D division of the two sides is shown in Table 2.

Table 2. Division of collaborative R&D.

Collaborator	R&D Achievements	Collaborator
Tsinghua Solar Co., Ltd.	Biomass Fuel	
Tsinghua Solar Co., Ltd.	Smoke Exhaust System	
Tsinghua Solar Co., Ltd.	Thermal Cycle System	
	Combustion system	Beijing Nowva Energy Technology Co., Ltd.
	Fire sealing system	Beijing Nowva Energy Technology Co., Ltd.

After one year of collaborative research and development, in July 2020, Tsinghua Solar Co. Ltd. successfully launched a biomass combustion stove with intellectual property rights. After testing by a qualified third-party testing organization, the product meets the requirements of the national standard GB-13271-2014 in terms of particulate matter, sulfur dioxide, nitrogen oxide, and other air pollutants emissions. At the same time, it also meets the current standards of the National Energy Administration in terms of thermal efficiency and cooking power. Compared with the existing products, it has the characteristics of sufficient combustion and more environmentally friendly emissions, showing strong competitiveness.

Tsinghua Solar Co., Ltd. and Beijing Nowva Energy Technology Co., Ltd. have successfully developed biomass combustion stoves for only one year through the research and development of a collaborative innovation model. At the same time, all core indicators are in line with the current standards of the National Energy Administration and are superior to existing market products. They have participated in the formulation of NB/T 34006-2020 "Technical Conditions for Clean Heating Stoves" issued by the National Energy Administration and established the leading position in the industry. Tsinghua Solar Co., Ltd. and Beijing Nowva Energy Technology Co., Ltd. successfully realized the research and development of new products through complementary R&D technology. Collaborative R&D not only greatly shortens the R&D cycle of Tsinghua solar co. for its new products but also revitalizes the reserve technology of Beijing Nowva Energy Technology Co., Ltd., and realizes the win–win cooperation. The characteristics of the R&D collaborative innovation model and collaborative surplus can be summarized as follows: with the help of abundant

R&D resources in the group, when the member enterprises of the enterprise group face market opportunities, the staff joint working groups, the middle-level managers with R&D director, and R&D personnel as the main force, give full play to their respective advantages and uses complementary advantages to obtain collaborative surplus: revitalize reserve technology, shorten the R&D cycle of new products, and quickly establish the leading industrial status of technology leadership.

4.4. Management Collaborative Innovation Model

Hot Spring Project Department and the Xian'an Project Department are two independent project departments with urban sanitation as their core business. The two project departments are independently managed and independently accounted for. Since the business of the two project departments is highly similar, the requirements for personnel, mechanical equipment, and operation methods are the same. In order to meet the operating requirements, the two projects are equipped with a variety of operating equipment and multi-post operators. Due to the limited operating area, the personnel are bloated, the work is not full, the operating machinery costs are high, the idle time is long, and the project performance is poor. At the end of 2018, the two project departments began to explore management collaborative innovation in order to achieve performance improvement. Management collaborative innovation is mainly carried out from two aspects: personnel streamlining and job optimization: (1) The two project departments are optimized and adjusted into a management team, readjust the organizational structure, optimize management responsibilities, sort out and reengineer the process, reduce duplication of work, and improve work efficiency. As of 30 June 2019, the cumulative optimization of 22 managers, including 10 captains and 12 managers, reduced management costs by 980,000 per year. By increasing the salary of front-line workers and encouraging front-line workers to take on more jobs, the "three-person, four-salaries and five-posts" model was explored, and a total of 30 front-line workers were optimized, including 9 drivers, 4 car attendants, and 17 cleaners. The annual labor cost was reduced by CNY 1.05 million, and the total cost was reduced by CNY 2.03 million through the optimization of managers and front-line workers. (2) After the two project departments carried out collaborative management innovation, the scope of mechanical operation is adjusted from small regional operations to large regional operations, and the vertical division of labor is adjusted to comprehensive, collaborative operations. The sprinkler and sweeper routes are optimized in the overall region to improve the operation efficiency of mechanical equipment, reduce the number of equipment used, and reduce the idle time of mechanical equipment, thereby reducing the cost of equipment procurement and maintenance. By June 2019, the number of mechanical equipment decreased by 5, and the idle time of mechanical equipment decreased by 50%. The first-line operation mode is adjusted from the original independent operation of sprinkler, cleaning, and dust suppression to collaborative operation. The three types of operators form a collaboration group to achieve seamless connection through the three-dimensional operation method. Road sprinkler, cleaning, and dust suppression are completed at one time to ensure the quality of road operation, reduce the frequency of rework operation and improve the operation efficiency. Through the optimization of operation mode, the quality of operations has increased significantly: in the first half of 2019, the cumulative deduction of environmental sanitation operation service assessment was 99,000 yuan, which was 73% lower than the deductions of sanitation operation service assessment of 368,000 yuan in 2018. The overall management collaborative innovation effect of the two project departments is shown in Table 3.

Through collaborative innovation, the two project departments enhance the overall management level and carry out work from two aspects of personnel streamlining and job optimization, achieving a significant decline in management costs, labor costs, and assessment deductions, thereby significantly improving performance. The characteristics of management collaborative innovation mode and collaborative surplus can be summarized as follows: With the help of advanced management experience sharing, the management personnel of the enterprises affiliated to the enterprise group can improve the management level and change the business thinking. The front-line employees can achieve one specialty and multiple abilities by improving the comprehensive quality so as to obtain the collaborative surplus by optimizing and simplifying redundant personnel and changing the operation mode: the management cost, labor cost, and assessment deduction are reduced so as to improve the business performance.

Table 3. Achievements of management collaborative innovation.

Type	Management Cost	Labor Cost	Assessment Deduction
Improved amount	decreased by ¥980,000	decreased by ¥1,050,000	decreased by ¥269,000

4.5. Talent Collaborative Innovation Model

The appointment and appointment of the board secretary of the vertical system of the board secretary of the Tus-Holdings are jointly and jointly managed by the Department of Strategic Planning and Investment Development (hereinafter referred to as "the Department of Strategic Investment"), and it is an important function to cultivate and reserve excellent expatriate board secretaries for the group member enterprises. The expatriate board secretaries need to have the comprehensive professional ability in strategic planning implementation, investment evaluation implementation, risk control management, and so on and need to have good communication and coordination ability, so they have higher requirements for employees. From 2017 to 2019, the strategic investment department of Tus-Holdings appointed the secretary of the board of directors to the members of Tus Blockchain Group, Tus-Clean Energy Group, Tus Triple Helix Group, Tus New Materials Group, and Tus Digital Group and other enterprises. All these expatriate board secretaries are graduated from 985/211 colleges and universities, most of them have a master's degree or above and have an excellent comprehensive ability. At the same time, because their tenure in the strategic investment department is more than one year, they have a global vision and are familiar with the overall situation of the group so as to better help subordinate enterprises. By sending the secretaries of the board of directors to export exports high-end talents for member enterprises, Tus-Holdings improves the management level, thereby promoting the overall management level of the group. In 2019, the TUS-EST began to layout the park area for energy business development under the integrating energy and environmental protection strategy. None of the senior managers are involved in the park area, so they need to introduce relevant personnel. Tus Science and Technology City Group is a professional platform for Tus-Holdings in the planning, construction, and operation of science and technology parks. Its senior managers have rich industry experience and industry resources in the park area. In June 2020, the executive vice-president of Tus Science and Technology City Group was transferred to the executive vice-president of TUS-EST, responsible for business expansion in the park area. The business of Guangxi Fangchenggang Company (Fangchenggang, China) is power distribution services. In order to meet the needs of business development, a certain number of professionals are needed. Because it has not been able to recruit suitable personnel for a long time, which affects the business development, it seeks collaborative support from Tus member enterprises. In order to support Fangchenggang Company in carrying out business, Tus member enterprises have successively delivered one chief engineer, one financial director, one senior financial supervisor, one engineering supervisor, and three electricians to it. Fangchenggang Company has completed the formation of a professional team in a short period of time through personnel coordination among member enterprises, saving a lot of time and personnel recruitment and training costs for the rapid development of the business. The EBA project of Tus-Holdings Business School has been held for six consecutive years and has trained more than 300 middle and high-level managers for Tus-Holdings. Each period, EBA students will undergo 9 months of training. Tus-up member enterprises distributed around the world provide field visits, research, and other collaborative support for the training students. For example, in the sixth period, students, through the collaborative support of member enterprises, have successively entered the Tus business landing areas such as Hangzhou, Xi'an, Heze, Zhengzhou, Beijing, Liuzhou, Tengchong, and Moscow for exchange and learning.

Through collaborative talent innovation, the vertical and horizontal flow of talents is realized within the Tus-Holdings, which promotes the improvement of the management level of member enterprises and reduces the cost of talent acquisition. Selecting talents at all levels within the group for a centralized training mechanism promotes the exchange and integration of talents. The characteristics of talent collaborative innovation mode and collaborative surplus can be summarized as follows: With the help of abundant talent resources within the group, the member enterprises of the enterprise group solve the problem of high-end talent recruitment of the subordinate enterprises through the outpost of the group, realize the optimal matching of personnel and posts and the sharing of human resources in the vertical and horizontal flow of talents within the group, and promote the exchange and integration of talents by means of training. The listing measures cover all the subjects, such as executives, middle managers, and employees, and obtain the collaborative surplus: maximize the value of talents, improve the management level of member enterprises, reduce the cost of talent acquisition, and provide effective support for performance improvement.

4.6. Financial Collaborative Innovation Model

Tus-Holdings as the group headquarters to play the role of fund collection and collaborative distribution. In 2019, Tus-Holdings collected funds from five enterprises, such as Hefei Tus Technology Industry Center Management Center (Limited Partnership) (Hefei, China), paid interest of 57,668,700 yuan, and loaned funds to five enterprises, such as Nanjing Tus Smart Technology City Investment and Construction Co., Ltd. (Nanjing, China) with interest income of 118,358,100 yuan [54], as shown in Table 4.

Table 4. Capital occupation of related parties of Tus-holdings.

Affiliated Parties	Trading Content	Amount
Nanjing Tus Intelligent Technology City Investment and Construction Co., Ltd. (Nanjing, China)	Interest income	¥63,006,422.41
Beijing Tus Legend Film Media Co., Ltd.	Interest income	¥8,575,663.52
Yadu Technology Group Co., Ltd. (Beijing, China)	Interest income	¥11,522,510.48
Shaanxi Tus Science Park Development Co., Ltd. (Xi'an, China)	Interest income	¥2,420,152.78
Tusinceie Technology City Investment Group Co., Ltd. (Beijing, China)	Interest income	¥32,833,333.33
Hefei Tus Technology Industry Management Center (Limited Partnership)	Interest exchange	¥28,254,947.56
Enlightenment Huizhi (Beijing) Investment Management Co., Ltd. (Hefei, China)	Interest exchange	¥115,058.49
Suzhou Saide Investment Management Co., Ltd. (Suzhou, China)	Interest exchange	¥46.650,555.55
Qihong United Enterprise Management (Tianjin) Partnership (Limited Partnership) (Tianjin, China)	Interest exchange	¥4,838,028.79
Qitong Jiarong (Zhuhai) Equity Investment Fund (Limited Partnership) (Zhuhai, China)	Interest exchange	¥24,460,711.10

Superior companies provide guarantees for subordinate companies so as to help them more easily obtain loans from financial institutions. For example, Tus-Holdings provides guarantees for Tuspark Forward Co., Ltd. (Cayman Islands) and Tus-Technology Services Co., Ltd. (Beijing, China), and Tus-Tech City Group Co., Ltd. (Beijing, China) provides guarantees for Jurong Tus-Tech Development Co., Ltd. (Jurong, China) TUS-EST (000826) signed a loan agreement with the controlling shareholder Tus Technology Services Co., Ltd. on 17 November 2017, 8 December 2018, 28 March 2019, and 23 January 2020, respectively. The single loan amount is up to 1.5 billion yuan, and the loan interest rate is not more than 6.5%. Borrowing does not require any form of guarantee such as guarantee, mortgages, and pledges from the TUS-EST. Borrowing provided by the controlling shareholder timely supplements the liquidity required for the daily operation of the TUS-EST. The TUS-EST moved to Tsinghua Science and Technology Park and its surrounding areas as a whole in 2018. By cooperating with Tus-Holdings, Tus-Clean Energy, Huaqing Property, and other Tus-Holdings member enterprises to own or lease property, the relocation was realized in a short time, and the related intermediary costs were saved. In order to solve the problem of insufficient vehicles in the company, Tus Qingyun leased vehicles from brother companies with the help of Tus-Holdings controlling members. Joint investment is also an important part of financial coordination. In November 2019, Tus-Techology City Group Co., Ltd. and TUS-EST jointly invested in Hebei Xiong'an Tus-Zero Carbon Technology Co., Ltd. (Xiong'an, China) in Xiong'an New District, with a registered capital of 10 million yuan. The main business is new energy technology promotion services, environmental protection engineering construction, environmental protection consulting, environmental engineering special design services, environmental protection testing, etc.; Zhengzhou Tus-Donglong Science and Technology Development Co., Ltd. (Zhengzhou, China) and TUS-EST jointly invested in Henan TUS-EST Energy Development Co., Ltd. (Zhengzhou, China) in Zhengzhou in November 2019. The registered capital is 1 billion yuan, and the main business is new energy technology, urban domestic waste removal and transportation, power engineering design and construction, etc.

Through the financial collaborative innovation model, Tus-Holdings improves the efficiency of capital utilization by collecting and allocating funds at the group level. Subsidiary companies are more likely to obtain funds through shareholder guarantees and shareholder funding, and other ways, and asset sharing and joint investment are realized among member enterprises. It can be seen that the financial collaborative innovation model realizes the efficient use of funds and assets within the group. The characteristics of the financial collaborative innovation model and collaborative surplus can be summarized as follows: The members of the enterprise group make use of the advantages of financing at the group level and various financing channels to collect and distribute funds within the enterprise group, shareholder assistance, mutual guarantee, asset sharing, joint investment, etc., to obtain collaborative surplus: help members of the enterprise to obtain low-cost financial support, timely supplement operating funds, reduce investment risk, and realize the efficient use of funds and assets within the group.

5. Research Conclusions and Enlightenment

5.1. Research Conclusion

Through a longitudinal case study, this paper systematically studies and analyzes the collaborative innovation mode of Tus-Holdings, and constructs the path framework for obtaining the collaborative surplus of enterprise groups (as shown in Figure 2). Through the research on the collaborative innovation mode of strategy, customer, business, R&D, management, personnel, finance, and other factors, this paper analyzes the main types, characteristics, and manifestations of collaborative innovation modes and draws the following conclusions:

Figure 2. Collaborative Innovation model of enterprise group.

(1) The main types of enterprise group internal collaborative innovation model are strategic collaborative innovation model, strategic customer collaborative innovation model, business collaborative innovation model, R&D collaborative innovation model, management collaborative innovation model, personnel collaborative innovation model, financial collaborative innovation model. As a concentrated reflection of systematic collaborative innovation activities, the collaborative innovation mode has both technical innovation and non-technical innovation [8], which runs through the whole collaborative innovation system of enterprise groups.

(2) The internal collaborative innovation model of enterprise groups shows nonlinearity and diversity. In order to achieve the goal of collaborative innovation, the enterprise group invests in collaborative innovation activities by executives such as the chairman, supervisors and other middle managers, employees, and other subjects. Using the internal resources of the enterprise group, such as strategy, market, technology, management, talent, finance, and so on, combined with strategic combination, unified market, joint research and development, management improvement, personnel mobility, financial sharing, through nonlinear combination, a variety of collaborative innovation models are formed.

(3) Collaborative surplus performance is closely related to collaborative innovation mode; different collaborative innovation modes will produce a different collaborative surplus. Various collaborative innovation models have created multiple collaborative surpluses, such as reducing the time cost of strategic transformation, improving strategic customer satisfaction, improving business operation efficiency, reducing R&D costs, reducing talent acquisition and training costs, and improving capital and asset efficiency. These collaborative surpluses cannot be realized by a single enterprise, so collaborative innovation enhances the internal competitiveness of enterprise groups, improves the performance of enterprise groups, and wins the competitive advantage at the group level. This conclusion supports that the collaborative innovation model is an important path for enterprise groups to obtain collaborative surplus [14].

5.2. Theoretical Contribution and Practical Significance

The theoretical contribution of this paper mainly lies in the following three aspects: First, the research on collaborative innovation within enterprise groups in this paper enriches the collaborative innovation theory. The existing collaborative innovation research mainly focuses on the collaborative innovation mode and mechanism among different actual controllers [3–7]. The research in this paper makes up for the lack of research attention in the field of collaborative innovation of enterprise groups in recent years. It is a good supplement to the strategic alliance model [55] and the industry–university research model [56] formed between different controllers and contributes to the research results of collaborative innovation theory. Second, this paper contributes new cases and knowledge to comprehensive innovation management theory. According to the comprehensive innovation management theory proposed by Xu Qingrui and other scholars based on the long-term research on Haier's innovation activities [36], this paper takes Tus-Holdings as the research object, summarizes the collaborative innovation model of enterprise groups, and puts forward the collaborative residual acquisition framework. This study is not only the development of the "five-stage comprehensive collaborative process model" proposed by scholar Zheng Gang [37] but also the supplement of 'technical elements as the core, and the non-technical elements collaborative model' [8]. It contributes new cases and knowledge to the comprehensive innovation management theory and verifies the universality of the theory. Third, this paper provides a new idea for the subsequent research on the performance evaluation of collaborative innovation. The selection of existing achievements for collaborative innovation performance indicators mainly focuses on the speed of technological development, the success rate, the degree of upgrading, and the number of new products as the core indicators of collaborative innovation performance, while ignoring the non-technical factors, such as the collaborative innovation achievements in the fields of strategy, customer and management, which leads to the incomplete research perspective. However, some scholars have realized the existence of this problem, selected new or improved products, new or improved processes, new or improved management practices, new or improved marketing methods, and the other indicators as collaborative innovation performance [2]. The results of this study support this evaluation standard and further refine the evaluation index of collaborative innovation, which provides a valuable reference for subsequent collaborative innovation performance research.

Although our empirical study is situated in the context of China, the research findings may also apply to a larger scale of emerging economies because China is representative of a wide range of emerging economies [57,58]. The practical significance of this paper is as follows: through in-depth analysis of the collaborative innovation model formed by Tus-Holdings in the long-term enterprise practice, this paper reveals the path to obtain collaborative residuals, and vividly demonstrates the manifestation of collaborative residuals of enterprise groups, providing case reference for other enterprise groups: (1) Enterprise groups can learn from the collaborative innovation model of Tus-Holdings, and form special collaborative innovation committees through organizational structure optimization [59], such as Strategic Collaborative Innovation Committees, Market Collaborative Innovation Committees, and Technical Collaborative Innovation Committees, through the top-level design and guidance, explore the collaborative needs within enterprise groups, and organize and implement collaborative innovation activities; (2) By combining their own resource endowments, business groups uses the flow and combination of factors such as strategy, market, technology, management, talent, finance to form collaborative innovation mode, optimize the collaborative innovation process and improve the collaborative innovation efficiency through process reengineering [60]; (3) Enterprise groups should unswervingly promote collaborative innovation strategy, carry out comprehensive and systematic design through the adjustment and distribution of internal resources, encourage full collaborative innovation through system construction and incentive mechanism design [61], stimulate full collaborative innovation enthusiasm, explore potential collaborative innovation opportunities, and obtain more collaborative surplus, so as to enhance the overall competitive advantage of enterprise groups.

5.3. Limitations and Future Research Directions

Although this study tries its best to select representative enterprise groups in the field of collaborative innovation practice for typical analysis, due to the limitations of the case study itself, the universality of the conclusions of this study has become one of the limitations that cannot be ignored. Future research can be expanded from the following aspects: on the one hand, future research can expand the research background to other enterprise groups with collaborative innovation as the core competition through multi-case studies to verify the universality of the research conclusions; on the other hand, the efficient identification of collaborative innovation opportunities within enterprise groups is an important part of collaborative innovation capability, and it is also the basis for the formation of collaborative innovation model. Subsequently, the identification model of collaborative innovation opportunities is constructed through relevant research to further enrich the research results in the field of collaborative innovation.

Author Contributions: Conceptualization, W.F. and L.Z.; methodology, W.F.; validation, W.F., L.Z. and Y.C.; formal analysis, W.F.; investigation, W.F. and Y.C.; resources, Y.C.; data curation, W.F.; writing—original draft preparation, W.F.; writing—review and editing, W.F. and Y.C.; supervision, L.Z.; project administration, W.F. All authors have read and agreed to the published version of the manuscript.

Funding: This research received no external funding.

Institutional Review Board Statement: Not applicable.

Informed Consent Statement: Not applicable.

Data Availability Statement: Not applicable.

Conflicts of Interest: The authors declare no conflict of interest.

References

1. Grossman, G.M.; Helpmann, E. *Innovation and Growth in the Global Economy*; The MIT Press: Cambridge, MA, USA, 1993; pp. 1–26.
2. Xie, X.; Wu, Y.; Devece, C. Is collaborative innovation a double-edged sword for firms? The contingent role of ambidextrous learning and TMT shared vision. *Technol. Forecast. Soc. Chang.* **2022**, *175*, 121340. [CrossRef]
3. Li, M.; Zhang, M.; Agyeman, F.O.; Ud Din Khan, H.S. Research on the Influence of Industry-University-Research Cooperation Innovation Network Characteristics on Subject Innovation Performance. *Math. Probl. Eng.* **2021**, *2021*, 4771113. [CrossRef]
4. Meissner, D.; Zhou, Y.; Fischer, B.; Vonortas, N. A multilayered perspective on entrepreneurial universities: Looking into the dynamics of joint university-industry labs. *Technol. Forecast. Soc. Chang.* **2022**, *178*, 121573. [CrossRef]
5. Shvetsova, O.A.; Lee, S.K. Living Labs in University-Industry Cooperation as a Part of Innovation Ecosystem: Case Study of South Korea. *Sustainability* **2021**, *13*, 5793. [CrossRef]
6. Wang, H.; Shu, C. Constructing a Sustainable Collaborative Innovation Network for Global Manufacturing Firms: A Product Modularity View and a Case Study From China. *IEEE Access* **2020**, *8*, 173123–173135. [CrossRef]
7. Liu, W.; Yang, J. The Evolutionary Game Theoretic Analysis for Sustainable Cooperation Relationship of Collaborative Innovation Network in Strategic Emerging Industries. *Sustainability* **2018**, *10*, 4585. [CrossRef]
8. Shi, H.-B.; Cui, Y.-C.; Tsai, S.-B.; Wang, D.-M. The Impact of Technical–Nontechnical Factors Synergy on Innovation Performance: The Moderating Effect of Talent Flow. *Sustainability* **2018**, *10*, 693. [CrossRef]
9. Xu, P.; Dong, M.; Bai, G. A research on the open innovation under the frame of business group. *Sci. Res. Manag.* **2019**, *4*, 92–102.
10. Nouman, M.; Yunis, M.S.; Atiq, M.; Mufti, O.; Qadus, A. The Forgotten Sector': An Integrative Framework for Future Research on Low- and Medium-Technology Innovation. *Sustainability* **2022**, *14*, 3572. [CrossRef]
11. Daft, R.L. A dual-core model of organizational innovation. *Acad. Manag. J.* **1978**, *21*, 193–210.
12. Som, O.; Kimer, E. Low-tech Innovation Competitiveness of the German Manufacturing. In *How Do Traditional Enterprises Shape Sustainable Competitiveness with Non-R&D Innovation*; Posts & Telecom Press: Beijing, China, 2016.
13. Fu, L.-P.; Zhou, X.-M.; Luo, Y.-F. The Research on Knowledge Spillover of Industry-University-Research Institute Collaboration Innovation Network. In Proceedings of the 2012 IEEE 19th International Conference on Industrial Engineering and Engineering Management (IE&EM 2012), Changsha, China, 27–29 October 2012.
14. Shyu, J.Z.; Chiu, Y.-C.; Yuo, C.-C. A cross-national comparative analysis of innovation policy in the integrated circuit industry. *Technol. Soc.* **2001**, *23*, 227–240. [CrossRef]
15. Synergetics, H.H. *Cooperative Phenomena in Multi-Component Systems*; B. G. Teubner: Stuttgart, Germany, 1973; pp. 9–21.
16. Ansoff, H.I. *Corporate Strategy: An Analytic Approach to Business Policy for Growth and Expansion*; McGraw-Hill: New York, NY, USA, 1965.

17. Skippari, M.; Laukkanen, M.; Salo, J. Cognitive barriers to collaborative innovation generation in supply chain relationships. *Ind. Mark. Manag.* **2016**, *62*, 108–117. [CrossRef]
18. Heil, S.; Bornemann, T. Creating shareholder value via collaborative innovation: The role of industry and resource alignment in knowledge exploration. *R&D Manag.* **2018**, *48*, 394–409. [CrossRef]
19. Shen, B.; Xu, X.; Chan, H.L.; Choi, T.M. Collaborative Innovation in Supply Chain Systems: Value Creation and Leadership Structure. *Int. J. Prod. Econ.* **2021**, *235*, 108068. [CrossRef]
20. Kang, W.; Zhao, S.; Song, W.; Zhuang, T. Triple helix in the science and technology innovation centers of China from the perspective of mutual in-formation: A comparative study between Beijing and Shanghai. *Scientometrics* **2019**, *118*, 921–940. [CrossRef]
21. Mueller, E.; Syme, L.; Haeussler, C. Absorbing partner knowledge in R&D collaborations—The influence of founders on potential and realized absorptive capacity. *R&D Manag.* **2020**, *50*, 255–276. [CrossRef]
22. Okamuro, H.; Kato, M.; Honjo, Y. Determinants of R&D cooperation in Japanese start-ups. *Res. Policy* **2011**, *40*, 728–738. [CrossRef]
23. Wang, C.; Hu, Q. Knowledge sharing in supply chain networks: Effects of col-laborative innovation activities and capability on innovation performance. *Technovation* **2020**, *94*, 102010. [CrossRef]
24. Zhou, P.; Zhang, D.; Yao, S. Research on Collaborative Innovation and Multiple Interactions of Enterprises. *J. Manag. World* **2013**, *8*, 181–182.
25. Vuola, O.; Hameri, A.-P. Mutually benefiting joint innovation process between industry and big-science. *Technovation* **2006**, *26*, 3–12. [CrossRef]
26. Schwartz, M.; Peglow, F.; Fritsch, M.; Günther, J. What drives innovation output from subsidized R&D cooperation?—Project-level evidence from Germany. *Technovation* **2012**, *32*, 358–369. [CrossRef]
27. D'Angelo, A.; Baroncelli, A. An investigation over inbound open innovation in SMEs: Insights from an Italian manufacturing sample. *Technol. Anal. Strateg. Manag.* **2020**, *32*, 542–560. [CrossRef]
28. Dyer, J.H.; Singh, H.; Hesterly, W.S. The relational view revisited: A dynamic perspective on value creation and value capture. *Strateg. Manag. J.* **2018**, *39*, 3140–3162. [CrossRef]
29. Bercovitz, J.; Feldman, M.P. Entpreprenerial Universities and Technology Transfer: A Conceptual Framework for Understanding Knowledge-Based Economic Development. *J. Technol. Transf.* **2006**, *31*, 175–188. [CrossRef]
30. Chen, J.; Yin, X.; Mei, L. Holistic Innovation: An Emerging Innovation Paradigm. *Int. J. Innov. Stud.* **2018**, *2*, 1–13. [CrossRef]
31. Liu, H.; Liu, Z.; Lai, Y.; Li, L. Factors Influencing Collaborative Innovation Project Performance: The Case of China. *Sustainability* **2021**, *13*, 7380. [CrossRef]
32. Li, T.; Zhou, X. Research on the Mechanism of Government–Industry–University–Institute Collaborative Innovation in Green Technology Based on Game–Based Cellular Automata. *Int. J. Environ. Res. Public Health* **2022**, *19*, 3046. [CrossRef]
33. Ji, H.; Miao, Z. Corporate social responsibility and collaborative innovation: The role of government support. *J. Clean. Prod.* **2020**, *260*, 121028. [CrossRef]
34. Liu, N.; Wang, J.; Song, Y. Organization Mechanisms and Spatial Characteristics of Urban Collaborative Innovation Networks: A Case Study in Hangzhou, China. *Sustainability* **2019**, *11*, 5988. [CrossRef]
35. Chen, J.; Xie, F.; Jia, L. Mechanism of Synergetic Innovation inside Enterprise groups. *Chin. J. Manag.* **2006**, *3*, 733–740.
36. Xu, Q.; Zheng, G.; Chen, J. Theoretical Trace and Framework of Overall Innovation on Management—Theoretical Origin and Framework. *Chin. J. Manag.* **2006**, *3*, 135–142.
37. Zheng, G.; Zhu, L.; Jin, J. Comprehensive collaborative innovation: A five-stage comprehensive collaborative process model—A case study of Haier Group. *J. Ind. Eng. Eng. Manag.* **2008**, *22*, 24–30.
38. Assadi, N.B.; Samari, D.; Hosseini, S.J.F.; Najafabadi, M.O. Development model for palm processing industries with emphasis on total innovation management (TIM) in Kerman province. *Heliyon* **2021**, *7*, e07587. [CrossRef]
39. Yao, Y.; Xia, D. Synergetic Innovation Motivation—Synergetic surplus: Formation Mechanism and Promotion Strategy. *Sci. Technol. Prog. Policy* **2016**, *30*, 1–5.
40. Song, R. Research on Collaborative Innovation Model and Mechanism of Medium and Low Technology Enterprises. *Sci. Manag. Res.* **2019**, *37*, 104–108.
41. Xie, X.; Liu, S. Impact Mechanism of Collaborative Innovation Modes on Collaborative Effect and Innovation Per-formance. *J. Manag. Sci.* **2015**, *28*, 27–39.
42. Qu, H. Research Review and Prospect of Collaborative Innovation Model. *J. Ind. Technol. Econ.* **2013**, *32*, 132–142.
43. Zhang, F.; Tao, J. A research of the relationship between synergy of internal elements and innovation performance. *Sci. Res. Manag.* **2016**, *37*, 20–28.
44. Wang, C.; Sun, Q.; Xu, J.; Zhou, W. Research on Headquarters Value Creation Mechanism under the Network Environment: A Case Study Based on the Dual Embeddedness Perspective. *Manag. Rev.* **2019**, *31*, 279–294.
45. Zhou, W.; Li, B.; Zhou, Y.; Chen, L. The Impact of Entrepreneurial Platform's Empowerment on Entrepre-neurial Performance: A Case Study Based on 'Haier + Thunderbot'. *Manag. Rev.* **2018**, *30*, 276–284.
46. Cao, X.; Ouyang, T.; Balozian, P.; Zhang, S. The Role of Managerial Cognitive Capability in Developing a Sustainable Innovation Ecosystem: A Case Study of Xiaomi. *Sustainability* **2020**, *12*, 7176. [CrossRef]
47. Dawson, C.; Dargusch, P.; Hill, G. Assessing How Big Insurance Firms Report and Manage Carbon Emissions: A Case Study of Allianz. *Sustainability* **2022**, *14*, 2476. [CrossRef]

48. Yin, R.K. *Case Study Research: Design and Methods*; Sage Publications: London, UK, 2013.
49. Myers, M.D. *Qualitative Research in Business & Management*; Sage: Los Angeles, CA, USA, 2009.
50. Zhou, F.; Liu, Y.; Chen, R. Research on Collaborative Innovation of Intelligent Connected Vehicles Industry Based on Test Field: Embedded Case Study from the Perspective of Open Innovation. *Sustainability* **2021**, *13*, 5880. [CrossRef]
51. Aalbers, R.; Whelan, E. Implementing digitally enabled collaborative innovation: A case study of online and offline interaction in the German automotive industry. *Creat. Innov. Manag.* **2021**, *30*, 368–383. [CrossRef]
52. Yin, R.K. *Case Study Research*; Sage Publications: London, UK, 2008.
53. Glaser, B.G.; Strauss, A.L. *The Discovery of Grounded Theory: Strategies for Qualitative Research*; Transaction Publishers: Piscataway, NJ, USA, 2009.
54. Tus-Holdings Group Co., Ltd. *2019 Annual Report*; Tus-Holdings Group Co., Ltd.: Beijing, China, 2020; Available online: https://www.chinamoney.com.cn/chinese/qwjsn/?searchValue= (accessed on 15 December 2021).
55. Yu, B.; Xu, H.; Dong, F. Vertical vs. Horizontal: How Strategic Alliance Type Influence Firm Performance? *Sustainability* **2019**, *11*, 6594. [CrossRef]
56. Li, Z.; Zhu, G. Knowledge Transfer Performance of Industry-University-Research Institute Collaboration in China: The Moderating Effect of Partner Difference. *Sustainability* **2021**, *13*, 13202. [CrossRef]
57. Bruton, G.D.; Su, Z.; Filatotchev, I. New Venture Performance in Transition Economies from Different Institutional Perspectives. *J. Small Bus. Manag.* **2018**, *56*, 374–391. [CrossRef]
58. Cai, L.; Chen, B.; Chen, J.; Bruton, G.D. Dysfunctional competition & innovation strategy of new ventures as they mature. *J. Bus. Res.* **2017**, *78*, 111–118. [CrossRef]
59. Chaurasia, S.S.; Kaul, N.; Yadav, B.; Shukla, D. Open innovation for sustainability through creating shared value-role of knowledge management system, openness and organizational structure. *J. Knowl. Manag.* **2020**, *24*, 2491–2511. [CrossRef]
60. Mandych, O.; Mykytas, A.; Ustik, T.; Zaika, S.; Zaika, O. The development of theoretical, methodological and practical recommendations of the innovative development vectors of business process reengineering and strategic management of enterprises. *Technol. Audit. Prod. Reserves* **2021**, *6*, 62. [CrossRef]
61. Ai, F.; Zhang, L. Incentive mechanism system of the management of IC design enterprises. *J. Ambient Intell. Humaniz. Comput.* **2021**. [CrossRef]

Article

Design and Evaluation of a Heterogeneous Lightweight Blockchain-Based Marketplace

Javier Antonio Guerra [1,*], Juan Ignacio Guerrero [1,2], Sebastián García [1], Samuel Domínguez-Cid [1], Diego Francisco Larios [1,2] and Carlos León [1]

[1] Department of Electronic Technology, EPS, Universidad de Sevilla, 41011 Sevilla, Spain; juaguealo@us.es (J.I.G.); sgarcia15@us.es (S.G.); sdcid@us.es (S.D.-C.); dlarios@us.es (D.F.L.); cleon@us.es (C.L.)
[2] Department of Electronic Technology, ETSII, Universidad de Sevilla, 41012 Sevilla, Spain
* Correspondence: jgcoronado@us.es

Abstract: The proposal of this paper is to introduce a low-level blockchain marketplace, which is a blockchain where participants could share its power generation and demand. To achieve this implementation in a secure way for each actor in the network, we proposed to deploy it over efficient and generic low-performance devices. Thus, they are installed as IoT devices, registering measurements each fifteen minutes, and also acting as blockchain nodes for the marketplace. Nevertheless, it is necessary that blockchain is lightweight, so it is implemented as a specific consensus protocol that allows each node to have enough time and computer requirements to act both as an IoT device and a blockchain node. This marketplace will be ruled by Smart Contracts deployed inside the blockchain. With them, it is possible to make registers for power generation and demand. This low-level marketplace could be connected to other services to execute matching algorithms from the data stored in the blockchain. Finally, a real test-bed implementation of the marketplace was tested, to confirm that it is technically feasible.

Keywords: blockchain; marketplace; Internet of Things

Citation: Guerra, J.A.; Guerrero, J.I.; García, S.; Domínguez-Cid, S.; Larios, D.F.; Leon, C. Design and Evaluation of a Heterogeneous Lightweight Blockchain-Based Marketplace. *Sensors* **2022**, *22*, 1131. https://doi.org/10.3390/s22031131

Academic Editor: Naveen Chilamkurti

Received: 13 December 2021
Accepted: 30 January 2022
Published: 2 February 2022

Publisher's Note: MDPI stays neutral with regard to jurisdictional claims in published maps and institutional affiliations.

Copyright: © 2022 by the authors. Licensee MDPI, Basel, Switzerland. This article is an open access article distributed under the terms and conditions of the Creative Commons Attribution (CC BY) license (https://creativecommons.org/licenses/by/4.0/).

1. Introduction

The smart grid has become the solution used to integrate issues such as the appearance of renewable energy resources, changes in consumption patterns, the inclusion of electric vehicles, etc. In this sense, new technologies under the auspices of the Smart Grid concept have appeared to help manage power grids [1,2], with the management of the Distributed Energy Resources (DER) being a complex problem that presents important challenges [3].

In this scenario, it is important to consider that the installation of home photovoltaic (PV) systems has increased drastically in the last few years [4], mainly to reduce electricity bills, but also to reduce carbon footprints and to reduce dependency of the grid. In any case, these consumers could require electricity at some moments of the grid, but in other instances they could have a power excess, so they could feed this power surplus into the power grid, acting as producers. Therefore, these clients can be considered as DERs, following the prosumer paradigm [5]. All these actors make the smart grid difficult to manage, including precision aspects, such as grid synchronization [6] or prosumer energy matching [7].

Solutions to these problems present some scalability issues for managing power surplus management, especially with many unpredictable PV systems spread over the network [8]. Ideally, the systems need to keep the network stable. This implies trying to balance the surplus of power in the network, with its power demand using a matching algorithm. This could make other actors benefit and allow them to take advantage of this power surplus if they need it. Thus, both benefit from this power exchange.

The use of blockchain technology can be seen as a natural solution to overcome these problems [9–13]. Blockchain, by its definition, is a scalable, distributed, and decentralized ledger without the need for a central authority. Blockchain also allows the use of Smart Contracts (SCs) to execute algorithms within the network [14].

In this sense, this paper proposes a real implementation of a low-level power marketplace. This marketplace has been deployed using a blockchain over low-performance devices in the smart grid, which would even be installed inside the smart meters. This marketplace registers power production and demand on the blockchain through SCs. In this way, registered consumption and generation are stored in a public repository. Thus, the proposed marketplace could be applied to high-level or existing blockchains in established markets, and these main markets could be based on matching algorithm-based platforms or novelty markets based on blockchains, by means of the usage of sidechains. As described above, this blockchain is deployed over lightweight devices, which would be existing Internet of Things (IoT) devices used to process and generate data on the smart grid, allowing integration between these technologies.

However, to perform this, it is necessary to consider that the number of actors in the proposed blockchain could vary: prosumers, facilities, or external regulators, which could lead to the appearance of some problems in the blockchain network, such as synchronization faults or network overload [15]. In this paper, a proposal will be addressed that will be scalable, to avoid these risks.

A connection is proposed as close as possible between the smart meters of prosumers and the blockchain, even using the existing hardware inside the smart meter if possible. This approach involves taking the data closest to the data generation [16], ensure its veracity, and setting the data into the blockchain as quickly as possible, without the need for expensive computers or servers, which may have a high power consumption. However, it forces the use of low-performance economic low-performance devices on prosumer, aggregator, or retailer installations to maintain the feasible installation.

In this sense, this paper analyzes the application of a lightweight blockchain algorithm on low-performance devices for the design of a decentralized electrical marketplace, where prosumers and energy retailers can exchange directly and automatically among themselves. For that, it takes advantage of low-performance IoT nodes to act as blockchain nodes, by implementing a consensus protocol that does not require a high processing time or a graphic processing unit (GPU) to calculate complex and expensive calculus but is able to provide security and guarantee the quality and traceability of the data inside the blockchain. Furthermore, some SCs are implemented to automatize the power generated in the marketplace. All this architecture was deployed over real devices, tested, and compared over a similar blockchain network deployed on a personal computer.

The rest of this paper is organized as follows. First, Section 2 provides a brief overview of the integration of blockchain into IoT systems and some existing applications in the power sector. Later, Section 3 has an approach of different ways to implement the proposed solution. Then, the proposed architecture for the creation of the marketplace is shown in Section 4. After that, Section 5 shows the devices that are used in the designed marketplace. Then, how the marketplace is deployed in a real environment is presented in Section 6. Section 7 shows the obtained metrics and the comparison of these with a typical blockchain network. Finally, in Section 8 some conclusions are made, and possible future work derived from this paper are discussed.

2. Review of Blockchain Applied to Energy Markets

Energy retailers have been using blockchains more and more frequently in the last few years [17]. This technology allows them to save and trace electrical information about consumers [18]. Recently, the usual consumer profile has changed [19]. Many users install new renewable energy resources, such as PV panels, in their homes. Consumers who have installed resources such as these have seen their energy bills reduced, as part of this, demand for power is compensated by the generated one [20]. In addition, some of them now have no dependency on the power supplied, which means they are no longer actors on the power grid. Furthermore, in some cases, renewable energy resources produce so much power that is partially returned to the grid. In this scenario, the consumer has evolved into a generator and consumer, called a prosumer [21]. Therefore, the original purpose of the blockchain, which is to register only consumers in the smart grid, has become obsolete in many cases. The proposed marketplace overcomes this problem, allowing for generation and demand records.

Another challenge in DER management is the implementation of the electric vehicle, which could work in the local smart grid as a load or as a battery, known as a vehicle-to-grid paradigm [22,23]. This, as a new actor in the system or a new source for an existing user, requires new solutions, in which the blockchain could act as a guarantor of security or traceability, as [24] shows. The authors used blockchain technology to create a secure identity mechanism to implement a secure charging system. In the proposed marketplace, the electric vehicle could be part of any actor on the blockchain, and there would be no difference in the behavior of the network.

The integration of blockchains into power grids is not a new challenge. It is widespread in the literature, where many challenges have been exposed. Ref. [25] exposes some of the challenges blockchain needs to overcome in the power sector and briefly shows the advantages of using it on IoT platforms. That work also mentioned the Brookling Microgrid [26], one of the first studies of a real microgrid, which demonstrates that blockchain is a suitable technology to provide microgrid marketplaces. They proposed a market divided into seven components, using blockchain as an information system in the most specific component of the market. In [15], the authors expose some use cases for the implementation of blockchain networks in smart grids, such as peer-to-peer, power trading, and power distribution. Throughout this paper, it is shown how most of the challenges have been overcome, such as the increase of disk size among the blockchains, or the use of heavy consensus algorithms, which will be discussed and partly solved in this paper.

There have been some attempts to merge IoT and blockchain with different approaches. Some research has been conducted on the hybridization of both technologies [27]. In this sense, the blockchain–Internet of Things paradigm has been analyzed, such as in [28], to reach trends in industry and academia, using innovative review methods. The authors in [29] go a step further and introduce 5G networks to the scope of the IoT to study their convergence with Blockchain. Regarding the practical applications for both technologies together, in [30] the authors provide the execution of SCs for access control on IoT devices, using blockchain technology to guarantee secure accesses. However, the blockchain network is implemented over servers, using a heavy consensus protocol. The authors of [31] use Raspberry Pi as an example of an IoT device for a scalable Blockchain-into-IoT scenario for secure information sharing. Despite this, it continues to use non-IoT devices as nodes of the blockchain, leaving Raspberry Pi as an IoT-device server. More similar to the approach of this paper is the use of Raspberry Pi as blockchain nodes proposed in [32], with devices connected to a blockchain network. Unfortunately, the use of IoT devices such as these has no more use than as nodes in the network, wasting its potential to work in other applications, as proposed in Section 1.

A close implementation of IoT nodes on blockchain is [33], where the authors proposed a framework for distributed applications in an IoT network with blockchain. That solution introduces another layer to its architecture, with the use of a specific server to deploy the framework. Multi-layer architecture was also needed for the investigation in [34], although it is deployed over a Field Programmable Gate Array (FPGA) device and does not provide a lightweight protocol to the blockchain network. In this paper, it is explained that there is no need for servers or complex devices to deploy a blockchain for the use of IoT at the same time. The same conclusion is reached in [35]. In that, a power trade system using SCs, using Raspberry Pi as nodes for their test network is proposed. Although they mention the use of whisper as a messaging service that the blockchain provides, it is not very clear which algorithm they used to reach consensus in their network, as they talk of Proof-of-Work and mining blocks. This made the nodes unusable for other purposes, more than supporting the blockchain.

3. Advantages of the Use of a Blockchain-Based IoT Marketplace

As mentioned above, the main technology on which this whole proposal is based is blockchain, which is defined as a distributed ledger technology. In blockchain, as in traditional ledgers, all information recorded can be considered true and immutable. The blockchain is composed of blocks, which are a set of transactions with some metadata information (i.e., the information of each transaction codified through hash functions) to prevent public access to all data and guaranteed anonymity. The main properties of a blockchain are as follows:

- Decentralization

  Blockchain was born as a decentralized technology to avoid any unique authority in the network. Instead, it is distributed into different nodes, with multiple connections among them, as shown in Figure 1. This creates a solid and reliable network with many benefits, such as efficiency and data propagation. This implies that if a node is down for some reason, the rest of the marketplace will continue working without fault.

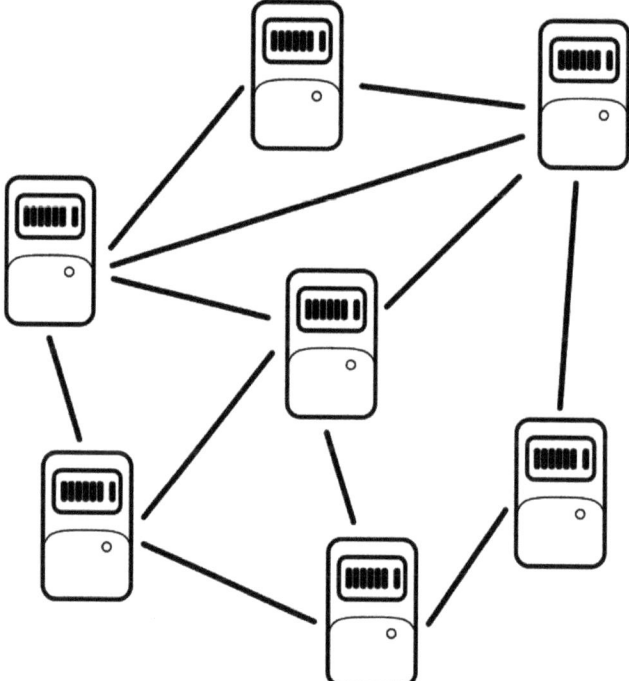

Figure 1. Nodes in a decentralized network.

- Immutability

  The moment that a block is validated by the network and linked to the chain, this information is very difficult to modify; the difficulty of doing so increases with each new block added to the network. Therefore, it is very difficult to modify a validated block in a chain. With application to the marketplace, it is guaranteed that the power data in the blockchain will not be modified by a third party unless that party manages to gain control of almost all nodes.

- Distributed consensus

  It is only possible to add a new block to the blockchain if most of the nodes agree on adding it. This is called validation of a block. Thus, a consensus is needed between the nodes of the network. This consensus means that the responsibility is decentralized, as there is no main individual responsible for a block's creation. Decisions in the blockchain network are only made by a consensus in the network. This reduces possible bottlenecks due to the dependence of a unique validator. The consensus also helps to increase network reliability because, as mentioned earlier, it would be necessary to attack all validators to modify the data in the blockchain. This implies that the consensus of the network is distributed, and the truth of the information is the same for the network. Thus, if a malicious node in the marketplace changes its own information and tries to share it with the rest of the network, it will be discharged, and the misinformation will have not be distributed to the network.

- Traceability

  The information in all of the nodes of the blockchain is public. Therefore, it is possible to access any transaction recorded in a block and obtain its hashed information, such as the user who performed the transaction, the user to whom it was sent, or the timestamp of it. Thus, if a user knows the hash of some data, it is possible to obtain all transactions related to these data and the block where they are stored. This ensures the traceability of any data in the network, so it is possible to trace the power data of a user in the marketplace.

- Speed of transactions

  Integration of the IoT infrastructure on the same nodes as the blockchain network allows the possibility of communication between them. This communication could be performed using less information than in a conventional blockchain, which increases the speed of transactions and, at the same time, the time needed to create a block.

- Reduced cost

  Although it is not relevant from a technological point of view, it is true that integrating IoT and blockchain in the same device would result in the use of fewer devices than using both technologies separately in the same scenario. Therefore, fewer devices implies fewer failure options and less maintenance. This is directly related to a decrease in the cost of implementing the technology.

The number of different blockchain networks and applications referring to the access to these networks is counted by dozens [36,37]. Therefore, it is quite a difficult decision about the optimal blockchain network that meets the objective of the marketplace proposed by this paper, as is a blockchain that runs for short periods of time, allowing the device to be used also as an IoT node.

In this sense, several consensus algorithms have been evaluated, with its main characteristics in the following subsection.

Consensus Algorithm

A consensus algorithm is named as the algorithm required to validate the data contained in a block by all nodes before adding to the tail of the blockchain. By this, it is guaranteed that the data inside the block have been defined as true by the majority of nodes in the network, which means that they have no false or manipulated data before introducing them into the blockchain. Due to that, consensus algorithms have to be complex, involving a large percentage of nodes in the network, whether creating the proposed block or validating it. These consensus algorithms have already been studied, obtaining properties such as the number of rogue nodes to fail [38].

As the blockchain network has nodes, such as low-performance devices, it is necessary that the consensus algorithm is secure, fast, and lightweight. Some of the most extended consensus algorithms and their suitability for the proposed architecture are presented below. Although normally blockchain networks have their own consensus protocols, most of them are based on these consensus algorithms.

- Proof of Work
  The first consensus algorithm proposed for the blockchain was Proof-of-Work (PoW), defined for the Bitcoin blockchain. PoW based its effectiveness on using all available computational resources to solve a cryptography problem to create a valid block in the network. The solution to this problem consists of finding a nonce, a number that makes the hash function of the whole block match any condition, which differs according to the difficulty of the network at that moment. Once a block is created, the rest of the network is responsible for validating it. Although it is very costly to create a block, it is quite simple to check if it is valid or not, due to the use of hash functions as part of the cryptography problem. With N denoting the nonce, P_{hash} the hash of the previous block, T_x all the transactions inside the block to be created, and $Target$ the difficulty of the network, the PoW requirement in Bitcoin is defined as the following equation:

$$H(N || P_{hash} || T_x || T_x || \ldots T_x) < Target \tag{1}$$

  The above equation was obtained from [39].
  Due to the fact that PoW needs to use the most resources of any algorithm, any node has to find the nonce to create a block [40], this consensus algorithm is discharged in the marketplace.

- Proof of Stake
  The Proof-of-Stake (PoS) consensus algorithm is based on the following premise: nodes that have the most coins in a blockchain network have the least interest in attacking the network [41]. On the basis of this, PoS does not need to solve any complex problem to create a block. Instead, PoS bases block creation on a stake of coins for each node in the network. Thus, the node that stakes the largest amount of coins is declared the creator of the block and is in charge of adding the data to the new block and linking it to the main blockchain.
  Besides all its advantages, PoS has a main problem. As the consensus in this algorithm is reached by staking coins, it would be possible that some nodes join themselves and transfer funds between them to have enough coins to win any stake in the network. Due to that and some other related problems [42], PoS is not a suitable consensus algorithm for the proposed marketplace.

- Proof of Authority
  Another alternative to reach consensus between nodes is the Proof-of-Authority (PoA) consensus algorithm. In PoA, the main difference from other consensus algorithms is that the trust of the blockchain network is based on identity and reputation [43] and does not requires any currency exchange. Nodes of the network need to be firstly authorized as voting nodes to avoid new nodes tampering the next steps of the algorithm. Then, the voting nodes decided, based on the reputation of the nodes, which nodes are voting to be selected as validator nodes. The selected validator nodes finally need to show themselves to the network, in return for being able to add new blocks to the blockchain. In this consensus algorithm, as for the PoS consensus algorithm, a block is said to be signed instead of mined, and it does not require a high computational capability per GPU. In case the remaining nodes detect a rogue block, incorporated by a malicious node, the node will be identified and removed from the network. The main disadvantage of the PoA algorithm is that if an authority could have control of half plus one node of the network, the consensus algorithm will be tampered with, similar to PoW. Therefore, increasing the number of nodes in the network (i.e., with more prosumers in the proposed market), the security of the system increases.

- Practical Byzantine Fault Tolerance
  The practical Byzantine Fault Tolerance (pBFT) is an algorithm that ensures the unalterability of the network messages. Its consensus is based on state machines' replication, i.e., the consensus is reached when two thirds plus one of the nodes answers the same

state for the request sent by a client, which is another node. So, it is possible to support rogue nodes if they are fewer than a third of the total nodes in the blockchain network. As any client needs to communicate to all nodes in the network in each request, this consensus algorithm generates excessive network traffic, which is against the purpose of creating a lightweight blockchain proposed in this paper.

- Proof of Elapsed Time

  The Proof of Elapsed Time (PoET) is based on random time windows, in which a participant of the consensus competes to generate a block, develop a cryptographic proof, and send it to a controller. The controller verifies the proof and accepts the block to proceed. The controller then creates a new time window, and the participants compete again. Although the consensus algorithm is free to use, it is developed using technology introduced by Intel hardware. Due to this, it is not suitable for use in the proposed lightweight devices.

- Proof of Importance

  The Proof of Important (PoI) consensus algorithm seems like the PoS algorithm, with a main difference: the selection of the node is made by the score any node has. This score is related to the quality of the node and refers to the number of recent transactions, the vest of currency for the creation of new blocks, the network traffic, and more. Due to the fact that the more network traffic a node has, the better score it is able to reach, this consensus algorithm favors this feature, which is against the proposed lightweight blockchain.

As the blockchain network has nodes, such as low-performance devices, it is necessary that the consensus algorithm be secure, fast, and lightweight. There are other consensus algorithms based on disk usage, specific hardware chips, or other properties. Due to the hardware and software restrictions of the devices used for implementation, all of these alternatives have been discharged.

Finally, the consensus algorithm for the marketplace is PoA. It does not require many processing resources and could act during short periods of time, leaving the rest of the time for other purposes of the device. As will be seen later, PoA has solid, reliable, and supported alternatives in the proposed devices, which reinforces the decision about which consensus algorithm to use.

4. Proposed Architecture of the Distributed Marketplace

The general architecture of the proposed marketplace is depicted in Figure 2. In this figure, prosumers and energy retailers have access to the blockchain marketplace, in which resources are available to the participating actors, either in terms of generation or power offers. The registers of this generation or power offers are stored at the source, establishing a high grade of quality and traceability of energy in a permissioned ledger (only the clients with the device could access to the platform). Thus, this is very useful for a marginal power market, providing an additional and decentralized veracity.

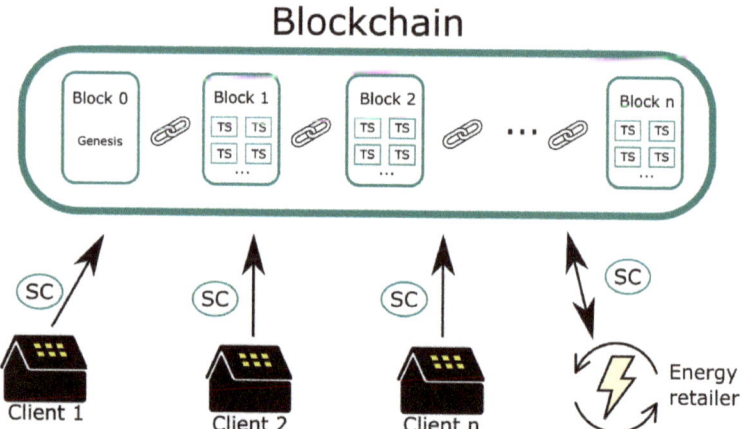

Figure 2. Marketplace scheme.

The definitions of the participants in the marketplace are as follows:

- Client or prosumer: as described above, they should act as a consumer or a generator. In both cases, the proposed node will be deployed on the smart meter (or be the smart meter by itself) to obtain the energy data each fifteen minutes, in the best case. This node will send the data to the energy retailer database and trigger the SC to publish both powers: surplus or demand.
- Energy retailer: It is responsible for the installation of the nodes and the connection between them. It may have a register of all information registers in the blockchain network within its database.

The data flow of the information in the marketplace results as follows. Inside the client supply, there is one of the proposed devices, working as an IoT and a blockchain node.

1. The IoT node behavior registers any power data, sent each fifteen minutes periodically. This information includes the power surplus of the prosumer. This will be a positive number, in the case of power generation, or a negative for power demand. In the remaining time, all nodes work as blockchain nodes, reaching a consensus of the blocks in the network using the PoA consensus algorithm [44]. These blocks contain transactions that consist of hashed data and SC information.
2. At the moment a surplus of power data is received, the node checks through a SC if the user is registered into the blockchain to send data. If not, an SC is called to register the user on the blockchain. If the user exists in the blockchain network, the blockchain node will send the power surplus data inside a generation SC or a demand SC, depending on the sign of the power surplus data. The data are also sent to the energy retailer to store them in its database, before the new data is registered. It is important to note that the marketplace does not perform any data processing, only stores the data sent by the SCs, which consist of hashed data and not the data itself.
3. The high-level or primary blockchain periodically executes (each 15 min or each time required by the retailer) a matching algorithm. It matches the generation registers with demand registers in the blockchain, checking the source of generation and offers.

As can be seen, the blockchain is a crucial part of the marketplace, as it is the basis of the traceability of the entire marketplace and needed in the matching process that can act with the independence of the retailer.

In this sense, it is important to find a valid architecture for the proposed marketplace, searching for a trade-off between computational needs, consumption, and cost. To reach the correct implementation of the blockchain marketplace, five main architectures have been analyzed:

1. Scalable sidechain
   In this case, it would be proposed to have only one sidechain in the whole marketplace. A sidechain is defined as a parallel blockchain, which works while the original blockchain properties are locked so that the common data between those two blockchains have no potential for conflict. This sidechain has its main advantage in its scalability property, which allows it to reach all the actors in the energy retailer. Unfortunately, it would require developing a specific sidechain core, compliant with ARM architectures, this is required, as ARM is currently state-of-the-art among computational capabilities and consumption. Although it has been studied, this restriction provides a non-valid solution for this marketplace.
2. Using Hyperledger-based distributed ledger technology.
   This option requires distributed ledger technology (DLT) that uses some of the solutions provided by the Hyperledger blockchain [45]. Some Hyperledger applications do not have a non-native ARM implementation, such as Hyperledger Besu [46], so it is not known if it would be possible to adapt it to the requirements of this paper.
3. Using alternatives based on consensus algorithms by capacity, space, or time.
   This kind of consensus algorithm seems to be appropriate for this project, as it reduces computing and consumption of the nodes. Due to the limited storage that nodes have, it would be difficult to implement this alternative with effectiveness. Therefore, it is discharged.
4. Using IoT-oriented alternatives
   There are alternatives that are not blockchains, but would be considered blockchains or DLTs, such as IOTA [47]. It has been tested in a private network (Tangle) over a virtual machine in a PC, which allows the installation of a Hornet service in lightweight nodes. However, after that, some shortcomings merged in terms of compatibility and interoperability. Although it appears that this problem will be solved in the next version of the client, right now it is not a real solution for the purposes of this paper.
5. Using Ethereum-based solutions
   There are some Ethereum-based blockchains which are able to be used in ARM architectures and also different options related to Ethereum that could be used. Moreover, the main structure for this type of network eases the interoperability and the connection with other sidechains if necessary. In fact, commercial solutions exist for this purpose. Therefore, it is the most viable alternative for this paper, and the one we finally used.

Once it is decided to deploy a blockchain based on Ethereum, there are two main options available to be implemented on the proposed devices. Both options involve using a lightweight client to deploy the blockchain written in the Go programming language. Therefore, a decision is needed between an Ethereum-based blockchain, such as Quorum, or the Ethereum client, with its implementations called GoQuorum and Go Ethereum. Both alternatives have the option of using a consensus protocol based on Proof-of-Authority, called raft and clique, respectively. The decision to use one of them is based on its maturity state, where Go Ethereum is older and more developed than Go Quorum. Therefore, Go Ethereum is the chosen implementation for the marketplace. In all nodes, a standalone Go Ethereum client, called Geth, has been installed. Geth provides enough tools to develop a full Ethereum node with the desired consensus protocol and the deployment of SCs on different nodes.

Geth only allows the execution of the blockchain on the CPU, as it is not guaranteed that all devices could have a dedicated GPU for this purpose. Moreover, it is not required for a PoA. The chosen consensus mechanism would also have a low impact in the case of GPU processing, as clique has been designed to be lightweight, in contrast to protocols based on PoW.

5. Energy Market Testbed Description

As described before, to deploy the blockchain network, we decided to use low performance devices, which act as an IoT node and a blockchain node at the same time, without loss of performance. Although many devices exist which have a special purpose for IoT, we decided to use some generic and popular devices that have a reasonable price. This is a frequent requirement, especially for large deployments, as the proposed marketplace can affect production. These devices have been implemented on their own smart meters, or very close to them, in several projects [48–50].

The three devices used as nodes for the marketplace are as follows:

- Raspberry Pi 3 B+ (RPi3), a low-cost single-board computer. Its central processing unit (CPU) has been developed over an ARM architecture and has 1GB of random access memory (RAM). Its disk size depends on the size of the memory card inserted into it. Its consumption is low, and it is commonly used in several applications, such as edge computing in an IoT context [51] or sensor network management [52].
- Raspberry Pi 4 (RPi4) is an upgraded and more recent version of RPi3. The main differences are that RPi4 has better CPU performance and different versions with different sizes (RAM), in exchange for slightly higher consumption.
- Jetson Nano (JNano) is an embedded computing board with a target-embedded IoT application and robotics [53]. Its CPU is also ARM-based, the same as the other devices, but JNano has some extra hardware features, such as a GPU. It has more power consumption than the other devices but consumes power without excess.

Table 1 shows an overview of the relevant specifications for the three devices used for this paper.

Table 1. Specifications of the selected devices.

	RPi3	RPi4 [54]	JNano [54]
CPU	Quad-core ARM Cortex-A53 @ 1.4 GHz	Quad-core ARM Cortex-A72 @ 1.5 GHz	Quad-core ARM Cortex-A57 @ 1.42 GHz
Memory	1 GB LPDDR2	8 GB LPDDR4	4 GB LPDDR4
Power mean	2.1 W	2.56 W	5 W
Power peak	5 W	7.30 W	10 W

All three devices used in this paper have an ARM architecture. There are many blockchain applications that have been deployed only for x86 or x86_64 architectures, so it would be very difficult to run this network inside any of the devices mentioned above. Fortunately, there are also some kinds of blockchain that have a native implementation for ARM architectures [55]. That implementation is highly different from the one used in servers or specific devices for mining or signing blocks, such as application-specific integrated circuit (ASIC) devices [56].

Apart from that, the existence of an ARM-architecture client is not enough to ensure the blockchain will work properly. Both RPi3 and RPi4 are built on an ARMv7 architecture, whereas JNano has an ARMv8 architecture. Although this may be seen as a small difference, it would make any blockchain client suitable for RPi3 or Rpi4 and not fully compatible with JNano. Sometimes, it is possible to avoid this restriction from building from the source. However, other times, even building from its source code does not guarantee that the blockchain runs properly. Therefore, to meet the heterogeneous blockchain network, some blockchain clients are discharged due to this restriction.

These devices are used for validation purposes. Their main goal is to test the suitability of developing a blockchain for a marketplace with low-cost, low-power consumption and limited computational capability hardware. Commercially, the marketplace deployment can be performed on ad hoc devices with processors based on the tested ones, or similar, as a function of their production costs.

In this sense, it is important to note that it would be possible to deploy the blockchain marketplace using the same device for all nodes. However, this paper tries to show that the proposed solution is heterogeneous, without dependence on a single commercial solution. If the proposed marketplace can be fully executed on a device with lower computational requirements, it can be executed on the devices of other manufacturers if they have equal or superior computational capabilities.

6. Marketplace Testbed and Results

The deployment of the experimental marketplace testbed is divided into three parts. First, the optimal deployment of the architecture to perform the blockchain marketplace and its nodes is shown. Then, it is defined as the genesis block of the blockchain that allows the network to be lightweight, fast, and secure. Finally, the Smart Contract (SC) used in the marketplace is commented on.

The proposed network is shown in Figure 3. As can be seen, clients (i.e., the prosumers) have their nodes connected to the smart meter. To guarantee security and avoid other risks derived from the use of public networks for the communication between nodes, it is a necessary, efficient, private, and alternative connection. Due to that, all these nodes exist in a virtual private network (VPN), enabling the required exchange of information in the blockchain and being the retailer responsible to maintain this network. Each client must have an internet connection, with different providers and different internet restrictions, such as firewalls or Carrier Grade Network Address Translation (CG-NAT). However, for the execution of the selected consensus algorithms, the clients must have a direct connection among them, to exchange information. In this sense, more complex communication protocols can be implemented, but VPN solves these problems in a natural way, allowing direct communication.

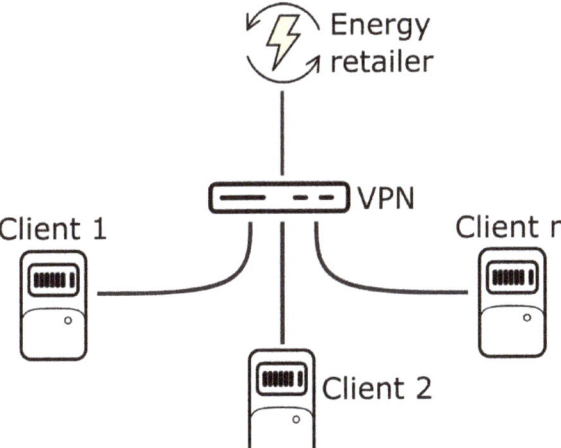

Figure 3. Proposed network scheme.

In the case of the testbed, this structure is simulated with three nodes connected in a private network. As described above, these nodes are based on low-performance devices. Thus, the network is finally as follows in Figure 4, where there are three nodes in the network, composed of a JNano, a RPi3, and a RPi4.

Figure 4. Marketplace-tested devices: up: switch; bottom-left: JNano; bottom-middle: RPi3; bottom-right: RPi4.

These three devices hold the following software:

- Raspberry Pi 3 and 4
  Both RPi3 and RPi4 contain analogue software. Both are built on an Ubuntu Server operating system (OS). Ubuntu is a general-purpose Linux-based OS, which has versions for many different architectures, among them ARM architectures such as those deployed on these devices. A server version of the OS has been installed to save as many computing resources as possible, avoiding any graphical interface.
  After that, Geth was installed on both devices. This was conducted by downloading ARM-specific compilations from the source page, after downloading GoLang packages for ARM, implemented in the native package download application for Ubuntu.
- Jetson Nano
  JNano software is built over an OS as part of the Jetson Nano Developer Kit. This is a set of tools that enables the OS to work properly on a JNano. It is a Linux-based distribution, built specifically for this device, with many similarities with other operating systems such as Ubuntu. GoLang was then installed and Geth after that, as Raspberry Pi 3 and 4 had previously been installed. In this case, Geth has a native download from the OS package manager.

Next, the necessary configurations were made to run the blockchain network. These steps are all the same for the three devices. As the network has been disconnected from the internet, it is necessary to manually set the date and hour for every system. This is performed using the *date* command. It is important to set the hour exactly: a difference of synchronization of twelve seconds is enough to avoid the nodes from connecting to each other. After that, the genesis block is generated and used to start the blockchain, where all nodes share the same network identification; the genesis block has the same Geth configuration parameters to ensure communication between nodes. Finally, the connection between all of the nodes is established, and then starts the process of validating and signing blocks.

Initially, all nodes were set as full nodes. This means that each node in the blockchain network has a complete copy of the chain. This can easily be changed in case some problems arise related to the available space on disk. In that case, the node would be set as a light node, which contains the latest blocks of the blockchain, saving disk space.

6.1. Genesis Block

The genesis block is the zero-block of any blockchain. It is the only block in every blockchain that has no reference to the hash of the previous block. Due to that, the genesis block has specific content. It is in charge of defining some parameters of the blockchain network from the point after its creation, and it is impossible to modify them once the network has started, similar to the content of any other block in a blockchain.

The genesis block is defined in Geth with a Javascript Object Notation (JSON) file called genesis.json. It has many parameters used to configure the entire blockchain network, such as chain ID, the type of block used, use of any hard fork in the network, or the consensus protocol to be implemented, among others. All these specifications are in accordance with Ethereum Improvement Proposals (EIP) [57].

According to the objectives of this paper, the genesis block must be the lightest possible. This is due to the fact that it is needed to avoid all unnecessary configurations, to reduce communication between nodes as much as possible and to configure the consensus protocol to be the one that consumes fewer computational resources of the devices. The genesis block is also responsible for other parameters, such as the gas limit for transactions, the first account of the network, which correlates with the first signer node of the network, and the initial accounts and initial amount of currency for each account.

6.2. Smart Contracts

Smart Contracts are pieces of runnable code inside the blockchain [58]. With SCs, any user of the blockchain could execute a computer program inside the blockchain network. Thus, the applications of the blockchain increase. Before SCs, any blockchain could only store transactions inside the blockchain network. Now, SCs allows any user to send a transaction to the account of an SC, which triggers the execution of the code to perform some operations. As the blockchain network is decentralized, any node in the blockchain network could write and store an SC in a block. After that, the SC must be deployed to be used by a node of the network. Once the SC is deployed, the address of the SC is public, and any node of the blockchain could access it. Fortunately, the SC could have some restrictions in its code, so that no node would be allowed to execute it. The SC source code is not automatically public to the network, so it is possible to keep the source code of the SC private, with access to the SC source code only available for the node that creates it.

For Ethereum-based blockchains such as Geth, SCs are written in a programming language called Solidity. As these types of blockchain network are Turing-complete, the way an SC is programmed is quite similar conceptually to any other code in high-level programming languages. It has some limitations due to restrictions of the blockchain and the way it is managed, such as loops in code or handling data outside of the blockchain.

For this project, a SC is proposed as the best way to deploy a marketplace inside the blockchain. However, SCs are not the only alternative to deploy a marketplace. Alternatives, such as the use of oracles [59] to create a marketplace outside the blockchain and then introduce the data into the network, could be also available. This option has been discarded for various reasons:

- It is not on approach to storing the data closest to the place where it was generated or obtained.
- It is possible to tamper with the data, so that, in the blockchain, different data could be stored than in the primitive one.
- It requires more complexity to afford it, as it implies adding an oracle between the data and the blockchain network.

For the proposed marketplace, two main SCs have been deployed along with an auxiliary one:

- Register
  This contract allows the registration of the actors involved in the marketplace. Thus, it is possible to register a prosumer, identified by its Universal Supply Point Code

(USPC), or a retailer, identified by its VAT number. It has enabled some restrictions, such as the prohibition of an external actor from participating in the marketplace, as will be discussed in the following section. It also has some extra functions to retrieve some information in the contract, such as the number of actors in the marketplace, or the count of the prosumers.

- Operation

  The second SC has the register of any operation inside the marketplace. Therefore, in the blockchain, there will be information about any generation and offer that occurs in the marketplace. The producer of this generation, the price of this generation, and the generated power are all stored in the generation register. In the case of an offer register, the blockchain stores the information of the producer and, as in the case of generation, the price and the amount of power. The SC is in charge of storing this information in the blockchain with a corresponding timestamp.

- Auxiliary contracts

  For the proper functioning of the SCs above mentioned, it is also necessary to create some more auxiliary contracts, which ensure the register and operation contract will work as desired. Therefore, the option to delete a registered actor in the system was developed, but only restricted to its own actor. For this, it was necessary to create other contracts that ensure the owner of the contract. Finally, an auxiliary contract was created to ensure that any actor in the marketplace had enough funds to execute it.

As described above, all these SCs have been written using Solidity, and it is fully compatible with the Ethereum blockchain and its derivatives. This was a possibility because of the Ethereum Virtual Machine, the environment where all Ethereum-based blockchains work. This allows the execution of the SCs in the same blockchain where the blocks are signed. This implies that the Geth compilation that is running on the nodes has native compatibility with the SCs. An example of the code of a SC function is shown in Figure 5.

```
function registerProsumer(uint256 _id, string memory _cups) public costs(msg.value) payable {
    prosumers.push(_id);
    reg[msg.sender].id = _id;
    reg[msg.sender].cid = _cups;
    reg[msg.sender].registered = true;
    cprosumers+=1;
}
```

Figure 5. Example of the *registerProsumer* function inside one of the Smart Contracts.

7. Performance Results Obtained

As described above, the proposed blockchain marketplace has been deployed on a testbed. That is, the blockchain has been executed from scratch on three nodes on these devices: a RPi4, a RPi3, and a JNano. According to the specifications of the device, the genesis block was designed with the following roles: the RPi4 will act as a validator node and as a signer node, as it is slightly more powerful in terms of CPU processing. Therefore, RPi3 and JNano will act as regular nodes on the blockchain. This is a sample of the proposed blockchain, where a few nodes would act as signers, based on their computing capacity and network connections.

The blockchain was left running for several days, while simulating the marketplace. i.e., the simulation of transactions required each 15 min per actor. After that, some key metrics were acquired, which have particular relevance to demonstrate some properties of the blockchain marketplace. These metrics have been acquired for the three nodes and compared between them. To avoid problems related to the specific time of data collection, relative metrics have been acquired, whether as the hourly average of data or in percentages.

The description of the metrics in Table 2 is as follows:

- Disk_ReadData refers to the amount of data read on the disk in bytes.

The data read from the disk are used to evaluate the performance of the node. It is desirable that the node should read the least data possible, to avoid processing time wasted in reading data operations.
- Disk_WriteData describes the amount of data written onto the disk in bytes.
This metric is directly related to the necessary disk size in case the node was a full node of the blockchain
- Network_InboundTraffic indicates the inbound traffic in the p2p network in bytes.
- Network_OutboundTraffic indicates the outbound traffic in the p2p network in bytes.
These two metrics are relevant in the case of a network failure. The more traffic inside or outside the node, the more synchronization problems that would exist. It also refers to the network speed needed to run the blockchain successfully.
- Memory_Allocated refers to the amount of memory assigned to the blockchain in bytes.
- Memory_AllocatedTotal indicates the Memory_Allocated metric related to the total RAM each device has.
- CPU_ProcessesTime describes the total time the CPU has taken during the blockchain processes.
Finally, these two metrics are crucial to evaluate whether the node could be used for other tasks at the same time as running the blockchain network.

The results of these metrics obtained at all three nodes are shown in Table 2.

Table 2. Metrics of the implemented blockchain by node.

Metric	RPi4	RPi3	JNano	Units
Disk_ReadData	9745.44	1604.88	2058.76	bytes per hour
Disk_WriteData	11,495.96	2534.52	353.64	bytes per hour
Network_InboundTraffic	1573.60	68.13	68.17	bytes per hour
Network_OutboundTraffic	2123.52	68.24	68.44	bytes per hour
Memory_Allocated	3290.40	316.84	241.67	bytes per hour
Memory_AllocatedTotal	4.11×10^{-5}	3.17×10^{-5}	6.04×10^{-6}	%
CPU_ProcessesTime	1.55	3.58	2.03	%

It is possible to obtain some conclusions by referring to these data:

- The blockchain network is technically feasible. It has been deployed, started, and run for hours without any error or strange behavior.
- Each device has different performance. In fact, the device with lower specifications in terms of CPU speed and RAM generates modest metrics compared to other devices. However, it is more than enough to run the proposed blockchain network.
- There is virtually no limit on disk usage. If we take the worst case, which is 11.5 KB per hour for RPi4, it would take roughly 10 years to occupy 1 GB of disk space.
- The blockchain does not need much network speed. A simple General Packet Radio Service (GPRS) connection has enough speed to obtain a successful connection between nodes.
- Nodes may have enough features to run other tasks while the blockchain is running. Over 90% of free CPU time in any node, and less than 1% of RAM allocated by the blockchain, make these low-performance nodes suitable for executing some data acquisition tasks. Thus, it is possible to obtain data and upload them to the blockchain from the same device, without any other connection or device that could manipulate the data quality.

To contrast the results given by the marketplace, it will be compared with a blockchain with similar characteristics, deployed over a PC instead. The PC has an Intel Core i7-10510U @1.80GHz CPU with a x86_64 architecture and 8 GB RAM, running Ubuntu Server OS. Specifically, it was deployed on three Geth nodes with a Docker container each. It has used the same genesis block and deployed the same SCs, and started and simulated the same transactions as the marketplace. The definition of the same genesis block implies that it uses the same clique PoA-based consensus protocol. Finally, the metrics have been obtained in the same way as the proposed marketplace.

The comparison has been made using the worst node for each case. Those are:

- The most data read and written in Figure 6 corresponds to RPi4.
- The most memory allocated refers to the total available in Figure 7 corresponds to RPi3.
- The highest CPU percentage in Figure 8 corresponds to RPi3.

The proposed blockchain marketplace was labeled as Marketplace and the deployed blockchain nodes over a PC as PC_blockchain.

From these results, it can be determined that powerful devices will perform better in the proposed blockchain, as the three graphs show better results for the PC_blockchain alternative. Despite this, the marketplace proposed in this paper has enough performance to run the blockchain under sufficient conditions to work as IoT nodes at the same time.

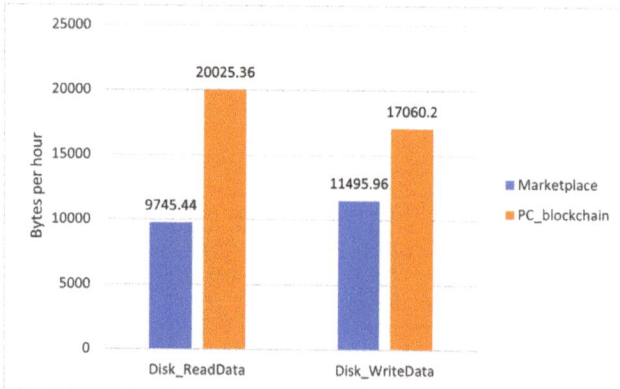

Figure 6. Comparison of disk usage metric between the proposed marketplace and an analogue blockchain over a PC.

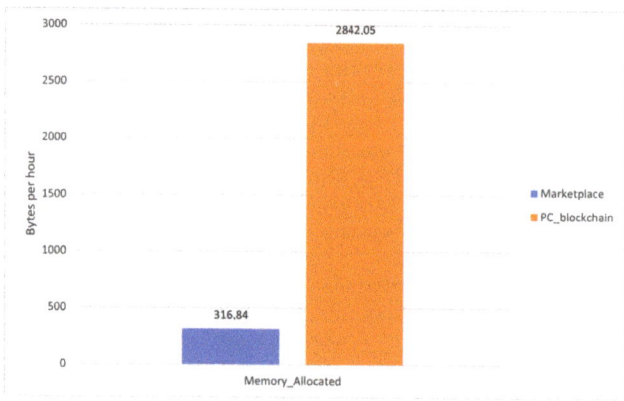

Figure 7. Comparison of the memory allocation metric between the proposed marketplace and an analogue blockchain over a PC.

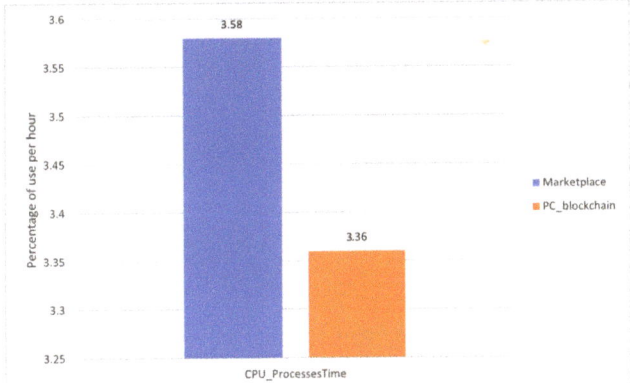

Figure 8. Comparison of CPU usage metric between the proposed marketplace and an analogue blockchain over a PC.

To evaluate the scalability in the market, some benchmarks were performed on the weaker performing device, i.e., the RPi3, using the open-source benchmarking tool Hyperledger Caliper [60]. The following benchmarks were executed using the Operation SC:

- 1000 operations at 25 transactions per second (TPS).
- 1000 operations at 50 TPS.
- 1000 operations at 100 TPS.

We obtained the results shown in the next two tables with all transactions sent successfully in each round of the benchmark. Table 3 shows the send rate and latency of the transactions, while Table 4 focuses on the node throughput and the CPU in terms of average (avg) and peak use.

Table 3. Benchmarks on RPi3: send rate and latencies in seconds (s).

Name	Send Rate (TPS)	Max Latency (s)	Min Latency (s)	Avg Latency (s)
25 TPS	25.0	7.43	2.09	4.72
50 TPS	50.1	7.71	2.15	4.90
100 TPS	99.0	9.44	2.71	6.50

Table 4. Benchmarks over RPi3: throughput, CPU usage peak, and CPU usage average.

Name	Throughput (TPS)	CPU% (Peak)	CPU% (Avg)
25 TPS	21.8	29.33	21.03
50 TPS	41.1	50.33	23.84
100 TPS	73.3	100	42.6

To reach that number of nodes, the RPi3 could support sending transactions to the blockchain, and the average CPU % used is inferred, as shown in Figure 9.

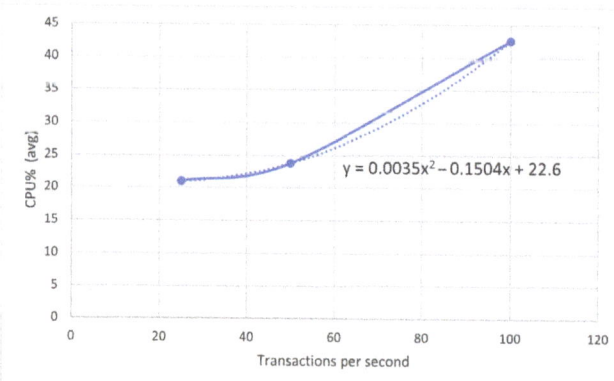

Figure 9. Percentage of CPU used for RPi3 and number of transactions per second.

Thus, if each node in the blockchain marketplace sends only one transaction per second, it is possible to reach over 128 TPS before the RPi3 reaches 100% of the CPU usage time, which means it is possible to support 128 nodes in the blockchain. Moreover, the data were registered every 15 min, so the number of real transactions was at least a third of the number of calculated transactions, and some time was needed for processing. Therefore, it is possible to say that the blockchain would support approximately 200 nodes without blockchain congestion, using only a device with lower CPU performance. Moreover, if a sidechain reaches its limit, a new sidechain can be added. This is why the proposed architecture can be used with an arbitrary number of prosumers.

With respect to CPU usage in these compact devices, it is necessary to consider if the device would suffer from high-temperature issues. These devices are known to work at high temperatures, and high CPU usage would cause temporary or permanent undesirable effects, such as throttling [61]. Due to that, CPU temperatures were monitored every half second in order to control any possible risk if the device reaches high CPU temperature. This temperature, as shown in Figure 10 has been measured in three different scenarios: before running the marketplace or idle state, running the marketplace in a normal execution, and executing the benchmarks mentioned above.

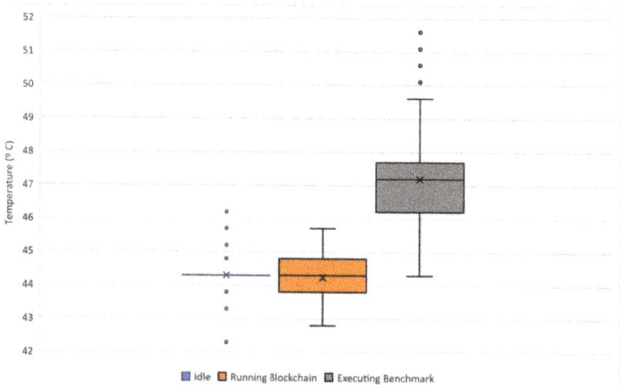

Figure 10. RPi3 CPU temperatures in different scenarios.

Finally, the power consumption of the RPi3 is measured in the same scenarios mentioned above, to visualize if the device reaches its peak consumption.

As shown in Figure 11, RPi3 power consumption is near the 5W peak consumption when it is used to execute the benchmarks. However, it remains in a safe range of power consumption when running the marketplace, which has an impact on its durability and reliability over time.

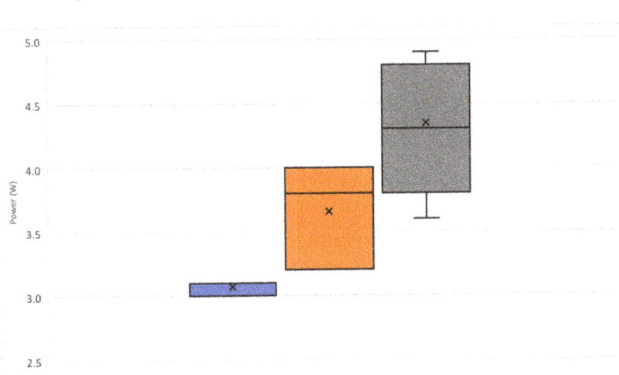

Figure 11. RPi3 power consumption in different scenarios.

8. Conclusions and Future Work

This paper has outlined a power marketplace, where blockchain and IoT are integrated to take advantage of its features, registering power offers and power demands from any actor in the system without third-party devices. Thus, this marketplace is suitable for the power industry, where different actors are involved in the smart grid.

Integration of blockchain and IoT in the same device can be possible for many different scenarios, one of which has been demonstrated in this paper. A marketplace to manage the power surplus could have actors in the network that are suited for most of the advantages that blockchain technology provides. Moreover, its implementation is made for low-performance, general-purpose devices by adapting the blockchain to its specifications. We have demonstrated that this can be achieved in a real network with a lightweight genesis block and a PoA consensus protocol, while still having enough performance to deploy some SCs in the network, at the same time as acting as an IoT node, with no performance problems in the devices.

When evaluating some key metrics on the nodes, the proposed marketplace results are light in terms of disk usage or CPU load. Therefore, it is a blockchain network that fulfills the purpose for which it was designed. This corroborates that this paper shows a valid integration of a blockchain in IoT devices, where the same device could have obtained data from some sensors and, in a different time window, register it to a blockchain network.

The following research on this topic refers to the optimization of IoT and blockchain integration. One of them is related to minimizing communications between nodes, as it is known that the SCs are triggered at specific points in time, so it would be desirable to send information between nodes only in case these data have been recorded in a block. In the same line, optimizing the consensus protocol, by changing or forking it, would be appropriate to reduce even more CPU load and RAM allocation.

This paper aims to be the start of a research line about optimizing IoT and blockchain integration in the same devices using GPU. As Geth, by how it is programmed, only uses the CPU for blockchain execution, the GPU is free for other purposes. Using both GPU and CPU at the same time, it would be possible to execute IoT algorithms at the same time. Furthermore, it would be possible to create a Geth fork that executes the blockchain code, taking advantage of the GPU, which would allow us to use other kinds of consensus algorithms in these devices. Other research lines would concern the specification of the tokenization of the energy market, in case the marketplace was a connection with real marketplaces, allowing trading among tokens.

Author Contributions: Conceptualization, J.A.G., J.I.G. and S.G.; methodology, J.A.G., J.I.G. and C.L.; software, J.A.G.; validation, S.D.-C., D.F.L. and C.L.; formal analysis, J.I.G.; investigation, J.A.G.; resources, J.A.G. and J.I.G.; data curation, J.A.G. and S.G.; writing—original draft preparation, J.I.G., S.G. and D.F.L.; writing—review and editing, J.A.G., J.I.G., S.G. and D.F.L.; visualization, S.D.-C. and C.L.; supervision, D.F.L. and C.L.; project administration, J.I.G. and C.L.; funding acquisition, C.L. All authors have read and agreed to the published version of the manuscript.

Funding: This research was supported by the Block of Things for Factory (BoT4F) project funded by the Fondo Europeo de Desarrollo Regional FEDER a través del Programa Interreg V-A España-Portugal (POCTEP) 2014–2020 and by the Ethernal Energy Project, a research project funded by the Spanish company THE SOUTH ORACLE, S.L.

Institutional Review Board Statement: Not applicable.

Informed Consent Statement: Not applicable.

Data Availability Statement: Not applicable.

Conflicts of Interest: The authors declare no conflict of interest.

References

1. Dileep, G. A survey on smart grid technologies and applications. *Renew. Energy* **2020**, *146*, 2589–2625. [CrossRef]
2. Personal, E.; Guerrero, J.; Garcia, A.; Peña, M.; Leon, C. Key performance indicators: A useful tool to assess Smart Grid goals. *Energy* **2014**, *76*, 976–988. [CrossRef]
3. Guerrero, J.; Personal, E.; Garcia Caro, S.; Parejo, A.; Rossi, M.; Garcia, A.; Perez Sanchez, R.; Leon, C. Evaluating Distribution System Operators: Automated Demand Response and Distributed Energy Resources in the Flexibility4Chile Project. *IEEE Power Energy Mag.* **2020**, *18*, 64–75. [CrossRef]
4. Das, U.; Tey, K.; Seyedmahmoudian, M.; Mekhilef, S.; Idris, M.; Van Deventer, W.; Horan, B.; Stojcevski, A. Forecasting of photovoltaic power generation and model optimization: A review. *Renew. Sustain. Energy Rev.* **2018**, *81*, 912–928. [CrossRef]
5. Dobrea, M.A.; Bichiu, S.; Opris, I.; Vasluianu, M. The Energy Efficiency of a Prosumer in a Photovoltaic System. In Proceedings of the 2020 IEEE 26th International Symposium for Design and Technology in Electronic Packaging (SIITME), Pitesti, Romania, 21–24 October 2020. [CrossRef]
6. Blaabjerg, F.; Teodorescu, R.; Liserre, M.; Timbus, A. Overview of control and grid synchronization for distributed power generation systems. *IEEE Trans. Ind. Electron.* **2006**, *53*, 1398–1409. [CrossRef]
7. Jogunola, O.; Wang, W.; Adebisi, B. Prosumers matching and least-cost energy path optimisation for peer-to-peer energy trading. *IEEE Access* **2020**, *8*, 95266–95277. [CrossRef]
8. Azghiou, K.; El Mouhib, M.; Bikrat, Y.; Benlghazi, A.; Benali, A. Guidelines for scalable and reliable photovoltaic wireless monitoring system: A case of study. *Lect. Notes Electr. Eng.* **2021**, *681*, 183–191. [CrossRef]
9. Li, Z.; Kang, J.; Yu, R.; Ye, D.; Deng, Q.; Zhang, Y. Consortium blockchain for secure energy trading in industrial internet of things. *IEEE Trans. Ind. Inform.* **2018**, *14*, 3690–3700. [CrossRef]
10. Mengelkamp, E.; Notheisen, B.; Beer, C.; Dauer, D.; Weinhardt, C. A blockchain-based smart grid: Towards sustainable local energy markets. *Comput. Sci. Res. Dev.* **2018**, *33*, 207–214. [CrossRef]
11. Li, H.; Xiao, F.; Yin, L.; Wu, F. Application of Blockchain Technology in Energy Trading: A Review. *Front. Energy Res.* **2021**, *9*, 130. [CrossRef]
12. Wang, N.; Zhou, X.; Lu, X.; Guan, Z.; Wu, L.; Du, X.; Guizani, M. When energy trading meets blockchain in electrical power system: The state of the art. *Appl. Sci.* **2019**, *9*, 1561. [CrossRef]
13. Gai, K.; Wu, Y.; Zhu, L.; Qiu, M.; Shen, M. Privacy-preserving energy trading using consortium blockchain in smart grid. *IEEE Trans. Ind. Inform.* **2019**, *15*, 3548–3558. [CrossRef]
14. Munsing, E.; Mather, J.; Moura, S. Blockchains for decentralized optimization of energy resources in microgrid networks. In Proceedings of the 2017 IEEE Conference on Control Technology and Applications (CCTA), Mauna Lani Resort, HI, USA, 27–30 August 2017. [CrossRef]
15. Alladi, T.; Chamola, V.; Rodrigues, J.; Kozlov, S. Blockchain in smart grids: A review on different use cases. *Sensors* **2019**, *19*, 4862. [CrossRef]
16. Yousefpour, A.; Fung, C.; Nguyen, T.; Kadiyala, K.; Jalali, F.; Niakanlahiji, A.; Kong, J.; Jue, J. All one needs to know about fog computing and related edge computing paradigms: A complete survey. *J. Syst. Archit.* **2019**, *98*, 289–330. [CrossRef]
17. Miglani, A.; Kumar, N.; Chamola, V.; Zeadally, S. Blockchain for Internet of Energy management: Review, solutions, and challenges. *Comput. Commun.* **2020**, *151*, 395–418. [CrossRef]
18. Erturk, E.; Lopez, D.; Yu, W. Benefits and risks of using blockchain in smart energy: A literature review. *Contemp. Manag. Res.* **2019**, *15*, 205–225. [CrossRef]

19. Konda, S.; Al-Sumaiti, A.; Panwar, L.; Panigrahi, B.; Kumar, R. Impact of Load Profile on Dynamic Interactions between Energy Markets: A Case Study of Power Exchange and Demand Response Exchange. *IEEE Trans. Ind. Inform.* **2019**, *15*, 5855–5866. [CrossRef]
20. Espe, E.; Potdar, V.; Chang, E. Prosumer communities and relationships in smart grids: A literature review, evolution and future directions. *Energies* **2018**, *11*, 2528. [CrossRef]
21. Zafar, R.; Mahmood, A.; Razzaq, S.; Ali, W.; Naeem, U.; Shehzad, K. Prosumer based energy management and sharing in smart grid. *Renew. Sustain. Energy Rev.* **2018**, *82*, 1675–1684. [CrossRef]
22. Harighi, T.; Bayindir, R.; Padmanaban, S.; Mihet-Popa, L.; Hossain, E. An Overview of Energy Scenarios, Storage Systems and the Infrastructure for Vehicle-to-Grid Technology. *Energies* **2018**, *11*, 2174. [CrossRef]
23. Lwin, M.; Yim, J.; Ko, Y.B. Blockchain-based lightweight trust management in mobile ad-hoc networks. *Sensors* **2020**, *20*, 698. [CrossRef]
24. Kim, M.; Park, K.; Yu, S.; Lee, J.; Park, Y.; Lee, S.W.; Chung, B. A Secure Charging System for Electric Vehicles Based on Blockchain. *Sensors* **2019**, *19*, 3028. [CrossRef]
25. Andoni, M.; Robu, V.; Flynn, D.; Abram, S.; Geach, D.; Jenkins, D.; McCallum, P.; Peacock, A. Blockchain technology in the energy sector: A systematic review of challenges and opportunities. *Renew. Sustain. Energy Rev.* **2019**, *100*, 143–174. [CrossRef]
26. Mengelkamp, E.; Gärttner, J.; Rock, K.; Kessler, S.; Orsini, L.; Weinhardt, C. Designing microgrid energy markets: A case study: The Brooklyn Microgrid. *Appl. Energy* **2018**, *210*, 870–880. [CrossRef]
27. Lu, Y. Implementing blockchain in information systems: A review. *Enterp. Inf. Syst.* **2021**, 1–32. [CrossRef]
28. Tsang, Y.; Wu, C.; Ip, W.; Shiau, W.L. Exploring the intellectual cores of the blockchain–Internet of Things (BIoT). *J. Enterp. Inf. Manag.* **2021**, *34*, 1287–1317. [CrossRef]
29. Dai, H.N.; Zheng, Z.; Zhang, Y. Blockchain for Internet of Things: A Survey. *IEEE Internet Things J.* **2019**, *6*, 8076–8094. [CrossRef]
30. Zhang, Y.; Kasahara, S.; Shen, Y.; Jiang, X.; Wan, J. Smart contract-based access control for the internet of things. *IEEE Internet Things J.* **2019**, *6*, 1594–1605. [CrossRef]
31. Hang, L.; Kim, D.H. Design and Implementation of an Integrated IoT Blockchain Platform for Sensing Data Integrity. *Sensors* **2019**, *19*, 2228. [CrossRef]
32. Chen, X.; Nguyen, K.; Sekiya, H. An experimental study on performance of private blockchain in IoT applications. *Peer-to-Peer Netw. Appl.* **2021**, *14*, 3075–3091. [CrossRef]
33. Pustišek, M.; Dolenc, D.; Kos, A. LDAF: Low-Bandwidth Distributed Applications Framework in a Use Case of Blockchain-Enabled IoT Devices. *Sensors* **2019**, *19*, 2337. [CrossRef]
34. Gonzalez-Amarillo, C.; Cardenas-Garcia, C.; Mendoza-Moreno, M.; Ramirez-Gonzalez, G.; Corrales, J. Blockchain-iot sensor (Biots): A solution to iot-ecosystems security issues. *Sensors* **2021**, *21*, 4388. [CrossRef] [PubMed]
35. Song, J.; Kang, E.; Shin, H.; Jang, J. A smart contract-based p2p energy trading system with dynamic pricing on ethereum blockchain. *Sensors* **2021**, *21*, 1985. [CrossRef] [PubMed]
36. Berdik, D.; Otoum, S.; Schmidt, N.; Porter, D.; Jararweh, Y. A Survey on Blockchain for Information Systems Management and Security. *Inf. Process. Manag.* **2021**, *58*, 102397. [CrossRef]
37. Casino, F.; Dasaklis, T.; Patsakis, C. A systematic literature review of blockchain-based applications: Current status, classification and open issues. *Telemat. Inform.* **2019**, *36*, 55–81. [CrossRef]
38. Bach, L.M.; Mihaljevic, B.; Zagar, M. Comparative analysis of blockchain consensus algorithms. In Proceedings of the 2018 41st International Convention on Information and Communication Technology, Electronics and Microelectronics (MIPRO), Opatija, Croatia, 21–25 May 2018. [CrossRef]
39. Bashir, I. *Mastering Blockchain*; Packt Publishing: Birmingham, UK, 2017.
40. Supreet, Y.; Vasudev, P.; Pavitra, H.; Naravani, M.; Narayan, D. Performance Evaluation of Consensus Algorithms in Private Blockchain Networks. In Proceedings of the 2020 International Conference on Advances in Computing, Communication & Materials (ICACCM), Dehradun, India, 21–22 August 2020; pp. 449–453. [CrossRef]
41. Zheng, Z.; Xie, S.; Dai, H.; Chen, X.; Wang, H. An Overview of Blockchain Technology: Architecture, Consensus, and Future Trends. In Proceedings of the 2017 IEEE International Congress on Big Data (BigData Congress), Honolulu, HI, USA, 25–30 June 2017; pp. 557–564. [CrossRef]
42. Deirmentzoglou, E.; Papakyriakopoulos, G.; Patsakis, C. A survey on long-range attacks for proof of stake protocols. *IEEE Access* **2019**, *7*, 28712–28725. [CrossRef]
43. Singh, P.; Singh, R.; Nandi, S.; Nandi, S. Managing smart home appliances with proof of authority and blockchain. *Commun. Comput. Inf. Sci.* **2019**, *1041*, 221–232. [CrossRef]
44. An, A.C.; Diem, P.T.X.; Lan, L.T.T.; Toi, T.V.; Binh, L.D.Q. Building a Product Origins Tracking System Based on Blockchain and PoA Consensus Protocol. In Proceedings of the 2019 International Conference on Advanced Computing and Applications (ACOMP), Nha Trang, Vietnam, 26–28 November 2019. [CrossRef]
45. Androulaki, E.; Barger, A.; Bortnikov, V.; Cachin, C.; Christidis, K.; De Caro, A.; Enyeart, D.; Ferris, C.; Laventman, G.; Manevich, Y.; et al. Hyperledger Fabric: A Distributed Operating System for Permissioned Blockchains. In Proceedings of the Thirteenth EuroSys Conference (EuroSys '18), Porto, Portugal, 23–26 April 2018; Association for Computing Machinery: New York, NY, USA, 2018. [CrossRef]

46. Ahmad, R.; Hasan, H.; Jayaraman, R.; Salah, K.; Omar, M. Blockchain applications and architectures for port operations and logistics management. *Res. Transp. Bus. Manag.* **2021**, *41*, 100620. [CrossRef]
47. Shabandri, B.; Maheshwari, P. Enhancing IoT Security and Privacy Using Distributed Ledgers with IOTA and the Tangle. In Proceedings of the 2019 6th International Conference on Signal Processing and Integrated Networks (SPIN), Noida, India, 7–8 March 2019; pp. 1069–1075. [CrossRef]
48. Chandra, P.; Vamsi, G.; Manoj, Y.; Mary, G. Automated energy meter using WiFi enabled raspberry Pi. In Proceedings of the 2016 IEEE International Conference on Recent Trends in Electronics, Information & Communication Technology (RTEICT), Bangalore, India, 20–21 May 2016; pp. 1992–1994. [CrossRef]
49. Pamulaparthy, M.; Jeevana Jyothi, K. Autonomous Smart Energy Meter over Internet of Things using Raspberry Pi. *IOP Conf. Ser. Mater. Sci. Eng.* **2020**, *981*, 042012. [CrossRef]
50. Bourhnane, S.; Abid, M.; Zine-Dine, K.; Elkamoun, N.; Benhaddou, D. Cluster of single-board computers at the edge for smart grids applications. *Appl. Sci.* **2021**, *11*, 10981. [CrossRef]
51. Xhafa, F.; Kilic, B.; Krause, P. Evaluation of IoT stream processing at edge computing layer for semantic data enrichment. *Future Gener. Comput. Syst.* **2020**, *105*, 730–736. [CrossRef]
52. Al Qundus, J.; Dabbour, K.; Gupta, S.; Meissonier, R.; Paschke, A. Wireless sensor network for AI-based flood disaster detection. *Ann. Oper. Res.* **2020** . [CrossRef]
53. Kanagachidambaresan, G. Introduction to Internet of Things and SBCs. In *Role of Single Board Computers (SBCs) in Rapid IoT Prototyping*; Springer: Cham, Switzerland, 2021; pp. 1–18. [CrossRef]
54. Suzen, A.A.; Duman, B.; Sen, B. Benchmark Analysis of Jetson TX2, Jetson Nano and Raspberry PI using Deep-CNN. In Proceedings of the 2020 International Congress on Human-Computer Interaction, Optimization and Robotic Applications (HORA), Ankara, Turkey, 26–28 June 2020. [CrossRef]
55. Goranovic, A.; Meisel, M.; Wilker, S.; Sauter, T. Hyperledger Fabric Smart Grid Communication Testbed on Raspberry PI ARM Architecture. In Proceedings of the 2019 15th IEEE International Workshop on Factory Communication Systems (WFCS), Sundsvall, Sweden, 27–29 May 2019; Volume 2019. [CrossRef]
56. Mahony, A.; Popovici, E. A Systematic Review of Blockchain Hardware Acceleration Architectures. In Proceedings of the 2019 30th Irish Signals and Systems Conference (ISSC), Maynooth, Ireland, 17–18 June 2019. [CrossRef]
57. Ethereum Improvement Proposals. Available online: https://eips.ethereum.org/ (accessed on 4 September 2021).
58. Bistarelli, S.; Mazzante, G.; Micheletti, M.; Mostarda, L.; Sestili, D.; Tiezzi, F. Ethereum smart contracts: Analysis and statistics of their source code and opcodes. *Internet Things* **2020**, *11*, 100198. [CrossRef]
59. Lo, S.; Xu, X.; Staples, M.; Yao, L. Reliability analysis for blockchain oracles. *Comput. Electr. Eng.* **2020**, *83*, 106582. [CrossRef]
60. Lohachab, A.; Garg, S.; Kang, B.; Amin, M. Performance evaluation of Hyperledger Fabric-enabled framework for pervasive peer-to-peer energy trading in smart Cyber–Physical Systems. *Future Gener. Comput. Syst.* **2021**, *118*, 392–416. [CrossRef]
61. Benoit-Cattin, T.; Velasco-Montero, D.; Fernández-Berni, J. Impact of Thermal Throttling on Long-Term Visual Inference in a CPU-Based Edge Device. *Electronics* **2020**, *9*, 2106. [CrossRef]

Article

Decision Support System to Classify and Optimize the Energy Efficiency in Smart Buildings: A Data Analytics Approach

Manuel Peña *, Félix Biscarri, Enrique Personal and Carlos León

Electronic Technology Department, School of Computer Science and Engineering, University of Seville, Av. Reina Mercedes S/N, 41012 Seville, Spain; fbiscarri@us.es (F.B.); epersonal@us.es (E.P.); cleon@us.es (C.L.)
* Correspondence: manuelpm@dte.us.es

Abstract: In this paper, an intelligent data analysis method for modeling and optimizing energy efficiency in smart buildings through Data Analytics (DA) is proposed. The objective of this proposal is to provide a Decision Support System (DSS) able to support experts in quantifying and optimizing energy efficiency in smart buildings, as well as reveal insights that support the detection of anomalous behaviors in early stages. Firstly, historical data and Energy Efficiency Indicators (EEIs) of the building are analyzed to extract the knowledge from behavioral patterns of historical data of the building. Then, using this knowledge, a classification method to compare days with different features, seasons and other characteristics is proposed. The resulting clusters are further analyzed, inferring key features to predict and quantify energy efficiency on days with similar features but with potentially different behaviors. Finally, the results reveal some insights able to highlight inefficiencies and correlate anomalous behaviors with EE in the smart building. The approach proposed in this work was tested on the BlueNet building and also integrated with Eugene, a commercial EE tool for optimizing energy consumption in smart buildings.

Keywords: smart building; energy efficiency; data analytics; energy optimization; decision support system

1. Introduction

The growth of energy consumption, energy resource exhaustion and significant environmental impacts [1,2] have raised concerns in most countries, which have entered international agreements for the benefit of society, such as the Paris Agreement in 2015 [3]. With a total energy consumption of 41% [4,5] in residential, public service and commercial sectors, which represents 24% of the world's CO_2 emissions [6], these sectors constitute one of the key areas of interest to address, where most action plans are focused on improving Energy Efficiency (EE) through the promotion of renewable energies and also evolving systems to minimize energy consumption. Given this scenario, this work tackles the EE challenge by proposing a new approach to support these initiatives through the optimization of EE in smart buildings using data analytics techniques with the aim of reducing energy consumption and, therefore, reducing CO_2 emissions.

Because of the relevance of these initiatives for the future of society, many international organizations are sponsoring these initiatives [7,8], such as the International Energy Agency (IEA), the U.S. Department of Energy—Energy Information Administration (DOE/EIA), the Energy European Commission (EEC) and the Organisation for Economic Co-operation and Development (OECD), with directives such as Energy Efficiency Directive 2012/27/EU (EU) or Executive Order (EO) 13514, Federal Leadership in Environmental, Energy, and Economic Performance, October 2009 (DOE/EIA). Additionally, substantial opportunities to improve energy efficiency have been expressed by the International Energy Agency (IEA), as described in [9]. These international initiatives, along with new technologies and their related research proposals, are actively contributing to tackling the above-mentioned challenges. Accordingly, with rising concerns about the exhaustion of energy resources and significant environmental impacts around the world, EE consumption is a rapidly increasing field.

From this perspective, in the last decade, huge efforts have been dedicated to the development of smart cities [10–12] in an attempt to design more sustainable cities by applying green energies and a high level of smart services [13–21] with the support of new technologies such as IoT [22–26] and 5G [27,28] and the progress made in AI [29–37], transport [38], etc. Within this context, efforts have been dedicated to increasing EE in smart buildings with more intelligent behavior, a high level of comfort and environmentally friendly operation [34,37,39–44].

There is considerable research focus on energy optimization in buildings. The increasingly sophisticated Building Automation System (BAS) [45] has become the cornerstone of intelligent modern buildings, integrating energy supply and demand factors, often known as Demand-Side Management (DSM), as energy efficiency policies that predict demand, as presented in [46]. Accordingly, most of the progress has been made in Energy Management and Control Systems (EMCSs) [47], which can manage energy policies in real time with the aim of maintaining a high level of comfort with minimum power consumption in different operating conditions [18,48,49]. In EMCSs, one of the most complex problems is optimization according to real-time environmental variables of the building. Current research approaches try to solve this problem from different points of view and with diverse techniques [50] and in different potential scenarios [51–53]. These approaches started with basic actions to improve EE [51] until more sophisticated energy management models were introduced, such as those based on a predictive controller in a Supervisory Control and Data Acquisition (SCADA) system [54], multi-objective models through a multi-criteria decision analysis [55–57], a more complex approach with Multi-Agent Systems (MAS) [58] or MAS based on occupant behaviors [59], a model based on a Markov Decision Model (MDM) [60] or a predictive control model combined with weather forecasting [61].

Within the above context, the current work is part of the KnoHolEM project, associated with FP7 (FP7-285229 KnoHolEM). The main purposes of this project are mainly focused on achieving the following seven objectives: (1) a functional energy-oriented building model complemented by a corresponding generic building ontology, (2) a specific building behavior model completed by a building-specific ontology, (3) data-mining procedures for detailed real-time energy consumption analysis, (4) algorithms derived from the building-specific ontology running in real time to acquire energy efficiency measures, (5) software for the synthesis and validation of real-time control algorithms, (6) the definition and engineering of hardware and firmware for real-time communication and optimization of energy in buildings and (7) and an interactive virtual reality smart building simulator.

In this sense and following the third and fourth objectives of the KnoHolEM project, the aim of the present paper is to optimize EE in smart buildings through a data analytics approach based on applying Data-Mining (DM) techniques, reducing energy consumption, maintaining a high degree of comfort and being environmentally friendly. Several research efforts have been dedicated to studying energy optimization in buildings with different techniques [62]. The efficiency of a building depends heavily on the way that it is used and how it is managed. In [63], the authors provided an interesting analysis of energy-efficient building design through DM techniques. In [64], the semantic modeling of building systems to support advanced data analytics for EE improvements was described. The authors of [65] presented an advanced DA framework for EE in buildings. However, the analysis of energy consumption at the use stage is essential, and it is clear that the construction characteristics of buildings strongly affect their energy consumption during their life cycle. In [66], the authors presented a review of unsupervised analytics techniques applied to EE enhancement. Finally, [67] proposed a classification approach of energy consumption in buildings.

In this work, a system to classify and measure EE in a smart building is presented. This system is implemented after an analytical process is performed to extract the knowledge hidden in historical building data. In addition, the systems are based on a set of models and algorithms corroborated by EE experts and historical data. This knowledge is used to classify each day analyzed based on its features to then predict the EE based on the insights extracted from historical data and experts. The purpose of this study is the optimization of EE based on measuring and predicting EE for each day from historical and real-time data of the BlueNet building. Subsequently, from the results obtained, it is easy to detect patterns that correlate days with poor EE with a high probability of anomalies. Finally, thanks to the progress made in this field and the results obtained, the system is in the test phase. A future goal is to integrate this framework with commercial EE software called Eugene, owned by the company Isotrol (Seville, Spain).

The paper is organized as follows: In Section 2, the smart building description and data sources of this work are presented. The description of data sources is divided into four subsections: indoor sensors, outdoor sensors, energy analyzers and, finally, EEIs. In Section 3, the proposed method, a data-mining-based DSS to measure and increase energy efficiency in the building, is detailed. It is divided into three subsections: data preprocessing (cleaning, integration, reduction and selection, and transformation), description of the classification module and, finally, the energy efficiency prediction module. In Section 4, the results of the work are presented.

2. Case Study: BlueNet Smart Building

Usually, smart buildings are made up of several data-metering units that are managed by an EMCS. This EMCS is able to act as a powerful tool for increasing EE in the building. For EE optimization, it is necessary to know how patterns of energy consumption in the building have been behaving. This knowledge is essential for understanding the consumption patterns of the BlueNet building and, therefore, for optimizing the energy efficiency of the consumption process. Thus, in this section, the BlueNet building and the technology employed in the smart building are described in detail. In addition, a description of data sources used for the classification and prediction of EE in the building is presented.

The BlueNet building is situated in the south of Spain (Seville, 37°24′29.97″ N, 6°00′18.63″ W). This building is used as an office building, and its main activity occurs from 8:00 to 17:30, although there is some variability in the schedule. The data obtained from the BlueNet building are from several data sources, such as indoor sensors, outdoor sensors, energy analyzers and EEIs, covering the main four areas of the building. The indoor sensors are made up of 16 ZigBee sensors per floor that collect indoor variables; ZigBee motes are centralized by a master ZigBee mote used as a gateway that gathers all measurements and sends them to a centralized server. Occupancy data can be obtained by many approaches, such as Passive Infrared (PIR) sensors, ultrasonic systems, camera-based systems, radar systems, CO_2 sensing, Electromagnetic (EM) signal detection systems, energy measurement devices, computer activity or sensor fusion, chair sensors or the use of multiple technologies to enhance the results [68,69]. In our case, the measurements were obtained through a combination between Radio Frequency Identity (RFID) technology, which is used by proximity cards at the entrance and exit of each room for exhaustive control for staff identity purposes, and Passive Infrared (PIR) sensors incorporated in Zigbee motes positioned strategically throughout the building space. Outdoor sensors comprise a set of sensors placed on the BlueNet building to extract environmental variables and send all of their measurements to a centralized server. Finally, the energy analyzers measure all power consumption variables of the building (HVAC, air mixers, splits, lighting, power plugs and other consumptions). These analyzers are centralized through a master Modbus device. The master Modbus device collects and transmits all consumption measurements to a centralized server that collects all of the BlueNet building information. Besides the information on these areas, a set of useful EEIs and information concerning the schedule of working and non-working days are shared, as this is quite valuable information to understand building consumption patterns and apply the knowledge extracted in this work.

2.1. Indoor Sensors

These sets of ZigBee sensors are placed in strategic locations in the BlueNet building and coordinated by a master sensor. The master sensor collects the measurements of the other sensors and sends the data to a master device that acts as a gateway. The purpose of this gateway is to collect sensor information and send it to the BlueNet centralized server. Indoor sensors are programmed to send meter readings every 30 s. These sensors collect the following measurements:

- Temperature: ZigBee sensors read the temperature in every room of the building. Generally, at least 2 or 3 sensors are used in every room to apply rules that ensure that the correct temperature is obtained.
- Humidity: ZigBee sensors read the relative humidity in every room of the BlueNet building. It is important to calculate the real feeling of the temperature in each zone.
- Lux: ZigBee sensors check the real effect of the lighting system by measuring the lux values in every room of the building.
- Presence: These sensors obtain data that indicate the presence of all occupants in the BlueNet building by identifying every person with a unique id and obtaining information about the time that every person is in BlueNet building rooms.

2.2. Outdoor Sensors

In addition, there is a set of sensors that collect metering data of the BlueNet environment. These sensors are responsible for collecting environmental data. Outdoor sensors are programmed to measure metering data every 10 min. The main measurements are:

- Temperature: These sensors take the environmental temperature every 10 min and are also able to provide the maximum, minimum and mean temperature of each day.
- Humidity: The outdoor sensors measure the environmental humidity and the amount of rain fallen to calculate the feeling of environmental temperature.
- Sunshine: These sensors obtain the amount of sunshine that irradiates onto the building every day.
- Wind: The outdoor sensors also obtain the amount and direction of the wind in the building environment.

2.3. Energy Analyzers

Energy analyzers are placed in an energy distribution panel inside of the BlueNet building. They are connected to a Programmable Logic Controller (PLC), which acts as a master Modbus device of the energy analyzers. These analyzers meter the four typologies of energy consumption in the building (HVAC, lights, power plugs and other consumptions). These analyzers are programmed to send measurements every 5 min to the master Modbus device. The principal features of every typology of consumption are described below.

2.3.1. HVAC

In the BlueNet building, HVAC is based on a VRV system. This system is made up of a set of indoor units (Daikin FXSQ-M7V1B, Daikin AC Spain S.A, Madrid, Spain) and a set of outdoor units (Daikin RXYQ-MY1B VRV II inverter—Daikin AC Spain S.A, Madrid, Spain—with heat pump). All units are connected by a DIII-Net. These connections are centralized in a DMS504B51 Daikin Lonworks Interface, which is in turn connected to a centralized server through a Lonworks/Modbus gateway (IntesisBox—HMS Industrial Networks AB S.L.U., Barcelona, Spain) using the communication protocol I3E.

HVAC was the primary area of consumption in this study due to its strong influence on EE [50–52]. Specifically, HVAC systems are appliances with the largest consumption in the building and are also the most controllable. HVAC management has the largest margin for EE improvement, consuming 143,876.7 MWh in 509 days, which represents 40.11% of the total energy consumption in the building.

2.3.2. Lights

The BlueNet building is made up of a luminary system, Philips Light Master (Luminary with ballast HF-R TD 318), connected through a DALI bus. This bus is centralized and managed by an LRC5141 controller, and it is connected to a centralized server through a Lonworks/Modbus gateway (IntesisBox—HMS Industrial Networks AB S.L.U., Barcelona, Spain). This server is responsible for sharing the data and applying certain energy management policies.

The lighting only accounted for 19.03% of total energy consumption in the BlueNet building, consuming 68,281.7 MWh throughout the 509 studied days. Thus, it is difficult to reduce this consumption (lighting is a necessary building function, and it is associated with occupancy in rooms). Nonetheless, the study of this field provides interesting information on EE, such as occupation patterns, anomalous consumptions or group behaviors.

2.3.3. Power

The consumption at power points in the BlueNet building is mainly the result of ICT equipment (PCs, servers, media, etc.). The majority of power consumption usually occurs during working hours, with the exception of some services that provide support 24 h a day.

This area is the second most relevant for this study, accounting for 120,368.7 MWh of the total consumption in the study period. Although it constitutes the second largest consumption in the building, accounting for 33.56% of the total, this energy consumption is hard to improve from the EE point of view because computers perform scheduled tasks outside of working hours, making it difficult to reduce this type of energy consumption.

2.3.4. Others

This area accounts for the minority of energy consumption, with only 26,170 MWh of total consumption. In addition, it has a strongly fixed consumption that is difficult to manage. Thus, with only 7.3% of total energy consumption, it is the least relevant field in EE.

2.4. Energy Efficiency Indicators (EEIs)

A set of EEIs of the BlueNet building were analyzed. These EEIs are based on experts' knowledge and historical data behavior of each relevant area in the BlueNet building, as described in [21]. The EE behavior in the BlueNet building is evaluated on the basis of these EEIs. In addition, each EEI provides the knowledge required to detect EE behaviors and anomalies. The EEIs are described further below.

2.4.1. Operational Changes in HVAC Compressor (OCC) Indicator

This EEI counts the number of daily on–off operations in the compressor. A large number of daily on–off operations are considered anomalous or inefficient, and it can cause one of the largest energy leaks and high inefficiency, greatly increasing energy consumption. A high on–off operation variance could indicate a possible anomaly in HVAC management (the HVAC is poorly dimensioned for this room, the HVAC is too powerful for this room, or there is a possible malfunction in the temperature sensor). Moreover, a compressor with a high OCC is prone to break down and have a shorter lifetime.

2.4.2. Number of Operational Regime Changes in the HVAC Compressor (ORCC) Indicator

This EEI counts the number of daily ORCC periods and the number of minutes in which the daily ORCC periods occurred. An ORCC is defined as a change in the compressor power consumption greater than 0.5 kW with respect to the previous measurement (10 min). These parameters were specified based on the results of DM techniques under the consensus of Isotrol HVAC experts. Thus, a large number of operational regime changes in the HVAC compressor (ORCC) is considered abnormal or inefficient.

This EEI can indicate that the HVAC system is not properly calibrated (HVAC is too powerful for this requirement) or that the temperature in the room is forcing the HVAC compressor to constantly change its operating mode. Furthermore, this EEI can denote a possible improvement in HVAC management, softening the HVAC consumption curve.

2.4.3. Switch on HVAC Compressor and Abnormal Changes in Indoor Temperature (SONCCIT) Indicator

This EEI counts the number of total minutes per day with a SONCCIT anomaly. In addition, this EEI averages the daily active power in the HVAC compressor during the compressor anomaly and the number of periods per day in which a switched-on compressor anomaly is observed (an anomalous period is specified as the aggregation of consecutive anomalous data points). A SONCCIT anomaly is defined when the HVAC system is turned on (the HVAC compressor consumption is higher than 1.7 kWh) and produces a change in indoor temperature greater than or equal to 1 °C between samples (10 min).

It is considered abnormal or inefficient when the room temperature decreases sharply (winter) or increases sharply (summer) while the compressor is running. This could be due to a sudden leakage of heat (winter) or a sudden influx of heat (summer) in the room, which counteracts the effect of the HVAC system (i.e., opened windows or doors).

2.4.4. Switch off HVAC Compressor and Abnormal Changes in Indoor Temperature (SOFFCCIT) Indicator

This EEI counts the total minutes per day with a SOFFCCIT anomaly. In addition, this EEI averages the daily active power in the HVAC compressor and the number of periods per day in which a switched-off compressor anomaly is observed (as in SONCCIT, an anomalous period is defined as the aggregation of consecutive anomalous data points).

It is considered abnormal or inefficient when the room temperature rises sharply (winter) while the compressor is not running. This could be due to a heat source (electric heater) replacing the HVAC system and can indicate the inefficient use of power energy. From historical data, a 1 °C increase between samples (10 min) during the winter period or a 1 °C decrease during the summer months is considered anomalous.

2.4.5. No Persons in BlueNet Building and Switch on HVAC Compressors (NPSONC) Indicator

This EEI counts the total daily minutes in which an NPSONC anomaly is detected, the number of periods per day with an NPSONC anomaly (an anomalous period is defined as the aggregation of consecutive anomalous data points) and the average active power consumption per day by the HVAC compressor during the anomaly. An anomalous function of the compressor is identified when there is an absence of occupants or the lights are switched off and the compressor is switched on (NPSONC). This could indicate that the air conditioner is switched on accidentally, considering that there are no building occupants if the lights are not switched on.

In this study, five data sources were analyzed: indoor sensors, outdoor sensors, schedule of working days, energy analyzers and EEIs. These data sources were employed during this work, in which every area was exhaustively and carefully examined to detect any possible improvement in energy management.

Some important information for this first analysis of historical data behaviors yielded the following results: 40.11% of building consumption was attributable to HVAC, 33.56% of energy consumption was due to power, 19.03% was spent on lighting, and another 7.3% was due to other activities, with an accumulated energy consumption of 358,697.1 MWh in the BlueNet building during the analyzed period. The most relevant consumption of the BlueNet building was due to HVAC operation with 143,876.7 MWh, and specifically, the major HVAC consumption was attributable to the compressor motor engine consumption, accounting for 46.38% of the total HVAC consumption.

Thus, a system to measure and optimize EE in the BlueNet building through DA techniques was developed. The objectives of this system were to: extract the knowledge hidden in BlueNet building data, develop a classification module and build an EE prediction module that helps to predict EE for each day. This EE classification is able to compare EE on days with similar characteristics, regardless of the season and other factors, which would be difficult to compare without this information, and also allows the presence of anomalous energy consumptions and other possible problems to be identified. The EE prediction module is able to quantify the energy efficiency every day, comparing days with similar energy efficiency conditions based on clusters.

The architecture of the EE optimization system is depicted in Figure 1.

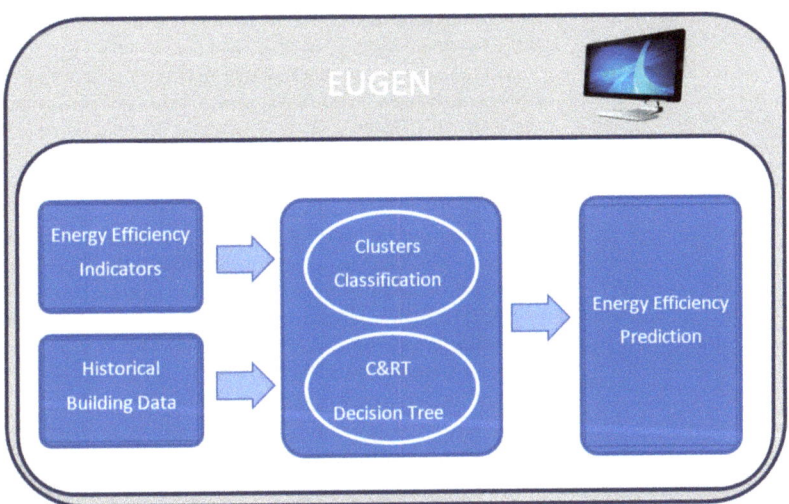

Figure 1. Prototype architecture.

3. The Data-Mining-Based Decision Support System to Optimize EE in the Smart Building

This section describes the approach selected to optimize the EE in the BlueNet building. This approach is based on a hybrid decision support system that combines a search based on the historical data to classify the energy consumption on days with similar features and a prediction module for additional EE interpolations. This approach takes advantage of the knowledge extracted from the historical data of the building, making it possible to reveal information about behavioral patterns in energy management, as well as knowledge provided by HVAC experts to optimize EE. Therefore, this approach comprises a module for the classification of energy consumption, which is based on CR&T decision tree [70] and cluster classification, and a module for EE prediction, which is based on metrics. With both modules, the days are analyzed with the EEI results to discover patterns and detect anomalies or other possible problems.

Data can provide some insights hidden in the behavior of historical data. Normally, the expert's knowledge is hidden in the collected dataset. Knowledge Discovery in Databases (KDD) refers to the overall data-mining process of discovering useful knowledge from large amounts of data. The DM process consisted of 6 essential phases: understanding the business, understanding the data, data preprocessing, modeling, evaluation and deployment. Once the phases of understanding the business and data were carried out, the next phase was the preprocessing of data. In this phase, the data were cleaned, and the different data sources were integrated, reduced and selected, and finally, transformed [71,72] (described in Section 3.1). After these phases, data were prepared for the modeling phase, which is explained in Sections 3.2 and 3.3 and evaluated in Section 4.

On the one hand, the first study aimed to extract the knowledge of energy consumption behaviors of the BlueNet building and determine how the energy has been consumed (e.g., regime changes in the compressor, operational changes in the compressor, lighting patterns and others). Secondly, the influence of each BlueNet building variable in every area (indoor sensors, outdoor sensors, HVAC, lighting, power and other consumption measurements, and EEIs) was quantified.

On the other hand, after extracting the knowledge hidden in BlueNet building data, an EE classification of the historical data was carried out through a hybrid system of classification. This EE classification allows the establishment of a relationship between days with similar features and also enables experts to compare the behavior on these days in EE terms. The classification module is based on a decision tree, the clustering of historical building data and the use of a set of energy efficiency indicators (Section 3.2). As a result, this classification identifies the presence of anomalous energy consumptions and other possible problems. In addition, a prediction system to quantify EE every day was developed (Section 3.3). This prediction is based on the features of each cluster classification, with the aim of assisting in the quantification of energy efficiency every day.

In order to carry out this work, a powerful and extended tool in the analytics area was used, namely, the SPSS Modeler tool (originally Statistical Package for Social Sciences Inc., an IBM company). This tool was used to perform the preprocessing and modeling tasks, as well as to evaluate the models. In addition, SPSS Modeler includes IA libraries used by the classification and prediction systems through DM techniques.

Currently, the decision support system to optimize EE in the BlueNet building is in the testing phase, and it will be connected to Eugene, an EE tool owned by Isotrol. These modules provided all of the knowledge extracted from the large amount of data provided by the BlueNet building with the aim of improving the results of the EE tool.

3.1. Data Preprocessing

The DM process requires an initial phase of data preprocessing, in which the data are analyzed, filtered and formatted [71]. BlueNet data comprise different data sources: energy consumption, environmental sensors inside of the building, external climate sensors, EEIs and other sources of data, all of which have their own temporal frequency. Thus, to manage different timestamps among the recorded data, the frequency was synchronized with a period of 10 min. The time interval for these data sources is between January 2011 and March 2013 (509 days).

The type of data strongly depends on the source of the data. The types of data and their time bases are as follows:

- Indoor sensors (30 s basis): mote_id, timestamp (YYYY/MM/DD hh:dd:ss), temperature (Celsius degrees), percentage_humidity, CO_2 and lux (lumens).
- Indoor sensors (30 s basis): mote_id, timestamp (YYYY/MM/DD hh:dd:ss) and employee_id (presence).
- Outdoor sensors (10 min basis): sensor_id, latitude, longitude, timestamp (yyyy/mm/dd hh:dd:ss), wind_direction (degrees), max_wind_speed (m/s), min_wind_speed (m/s), ave_wind_speed (m/s), UV_index, max_humidity, min_humidity, ave_humidity, precipitation (l/m^2) and sunshine_radiation (W/m^2).
- Energy Analyzers—HVAC (5 min basis): timestamp (YYYY/MM/DD hh:mm:ss), AP_CLI_FASE1 (kW), AP_CLI_FASE2 (kW) and AP_CLI_FASE3 (kW).
- Energy Analyzers—Lights (5 min basis): timestamp (YYYY/MM/DD hh:mm:ss), AP_LIG_FASE1 (kW), AP_LIG_FASE2 (kW) and AP_LIG_FASE3 (kW).
- Energy Analyzers—Power (5 min basis): timestamp (YYYY/MM/DD hh:mm:ss), AP_POW_FASE1 (kW), AP_POW_FASE2 (kW) and AP_POW_FASE3 (kW).
- Energy Efficiency Indicators—OCC, ORCC, SONCCIT, SOFFCCIT and NPSONC (10 min basis): timestamp (YYYY/MM/DD hh:mm:ss) and anomaly (true/false).
- Energy Efficiency Indicators—OCC (10 min basis): timestamp (YYYY/MM/DD hh:mm:ss) and anomaly (true/false).
- Energy Efficiency Indicators—ORCC (10 min basis): timestamp (YYYY/MM/DD hh:mm:ss) and anomaly (true/false).
- Energy Efficiency Indicators—SONCCIT (10 min basis): timestamp (YYYY/MM/DD hh:mm:ss) and anomaly (true/false).

- Energy Efficiency Indicators—SOFFCCIT (10 min basis): timestamp (YYYY/MM/DD hh:mm:ss) and anomaly (true/false).
- Energy Efficiency Indicators—NPSONC (10 min basis): timestamp (YYYY/MM/DD hh:mm:ss) and anomaly (true/false).

In this regard, the techniques used during the entire preprocessing stage were the following:

- Data cleaning: removing all missing and null values, as well as inconsistencies found in each data source;
- Data transformation: normalizing all data to the same period (10 min) to facilitate the aggregation of data and their analysis;
- Data reduction: simplifying the data with the aim of providing meaningful data.

Once the data were normalized and standardized, the data sample consisted of 5,088,765 records corresponding to EEI data and the set of variables of the BlueNet building. From all of these variables, only the most relevant ones were selected for use in the study. A list of these variables is summarized in Appendix A.

Initially, the 5,088,765 records of the dataset were analyzed. From these records, 636,848 missing data records were found, corresponding to null values. These records were filtered and removed from the data sample. After the exclusion of outliers, the next requirement was ensuring the data quality. Thus, a set of rules were established to guarantee the data quality:

- Detect every outlier in the data sample and fix it with the average value of data dispersion through DM techniques.
- Data consistency validation: every sample of data requires at least one value for each 30 min period.

As a consequence of this first phase of data quality control, 636,848 records were filtered in the preprocessing task, corresponding to null values. After applying basic rules to ensure data consistency, 34 records were filtered. Finally, the data sample was reduced from 5,088,765 records to 4,451,883 records.

Once the preprocessing phase was carried out, a set of 25 relevant variables was selected from the total of variables measured in the BlueNet building for the tasks of data classification and data prediction (described in Appendix A). This selection was carefully analyzed by studying the degree of the influence of each selected variable through Principal Component Analysis (PCA) techniques [73]. At first, the standardization of continuous variables was carried out, so each one contributed equally to the analysis. Second, a covariance matrix was established to identify correlations between the selected variables; pairs that had a positive score were correlated, and pairs that had a negative score were inversely correlated. Third, the eigenvectors and eigenvalues of the covariance matrix were computed to identify the principal components. These principal components are new variables constructed as linear combinations or mixtures of the initial variables; principal components are uncorrelated and provide the maximum information based on their variance. On the basis of the results, the most meaningful variables were selected, providing 25 relevant variables. Most of the variables analyzed were used in other similar research studies [74,75].

3.2. Classification Module

The classification module was developed through DM techniques with the aim of classifying days with similar features and comparing these days to quantify the EE. These clusters were developed with the aim of extracting insights about how different features affected days in EE terms. The second aim of these comparisons is to determine the clusters in which the detected anomalies are concentrated and infer key features that can provide some insights to reveal anomalies. Once the classification and detection are carried out, the main objective is to predict days with anomalies and then apply policies to improve the EE. This classification module uses the most influential information supplied by EEIs;

HVAC, lighting, indoor and outdoor environmental information; BlueNet occupation; and holiday/work schedule.

In the first instance, data for the classification module were analyzed through Generalized Rule Induction (GRI) [76] and the Apriori algorithm [77] to study the correlation between the selected variables and consumption. GRI begins with the original set S as a root node. It iterates through every unused attribute of the set S and calculates the entropy H(S) or the information gain IG(S) of that attribute. It selects the attribute with the smallest entropy (or largest IG), and the set S is partitioned by the selected attribute. Finally, the algorithm continues recursion on each subset.

Entropy H(S) is a measure of the amount of uncertainty in the dataset S:

$$H(S) = \sum_{x \in X} p(x) \log_2 p(X)$$

where S is the current dataset for which entropy is being calculated, X is the set of classes in S, and p(x) is the proportion of the number of elements in class x to the number of elements in set S.

Information Gain IG(A) is the measure of the difference in entropy from before to after the set S is split on the basis of an attribute A. In other words, it measures how much uncertainty in S was reduced after splitting set S on the basis of attribute A:

$$IG(S, A) = H(S) - \sum_{t \in T} p(t) H(t) = H(S) - H(S|A)$$

where H(S) is the entropy of set S; T is the subsets created by splitting the set S by attribute A, such as $S = \bigcup_{t \in T} t$; p(t) is the proportion of the number of elements in t to the number of elements in set S; and H(t) is the entropy of subset t.

Additionally, the Apriori algorithm is used for frequent item set mining and association rule learning, and it uses breadth-first search and a hash tree structure (Algorithm 1).

Algorithm 1. A Priori with breadth-first search.

Apriori(T,ε)
    $L_1 \longleftarrow \{large\ 1 - itemsets\}$
    $k \longleftarrow 2$
    while $L_{k-1} \neq \emptyset$
        $C_k \longleftarrow \{a \cup \{b\} \mid a \in L_{k-1} \wedge b \notin a\} - \{c \mid \{s \mid s \subseteq c \wedge |s| = k-1\} \nsubseteq L_{k-1}\}$
        for transformations $t \in T$
            $C_t \longleftarrow \{c \mid c \in C_k \wedge c \subseteq t\}$
            for candidates $c \in C_t$
                $count[c] \longleftarrow count[c + 1]$
        $L_k \longleftarrow \{c \mid c \in C_k \wedge count[c] \geq \varepsilon\}$
        $k \longleftarrow k + 1$
    return $\bigcup_k L_k$

where
T—The set of data;
ε—Confidence threshold;
k—Size of the set of candidate items;
C_k—Candidate set at level k;
c—Candidate c;
count[c]—Pointer to the candidate set c.

As a result, 24 features were selected from the initial set of 25 variables that were previously filtered. These features provided more detailed and useful information to develop the classification module through DM techniques with the aim of supplying the most accurate results possible. The features selected for the classification module are shown in Table 1.

Table 1. Selected features.

Feature	Unit
AEMET_AT	°C
AP_COMPRESSOR	kW
AP_COMPRESSOR_MINUTES	minutes
AP_COMPRESSORS_MEAN	kW
AP_LIGTH	kW
DATE	DD/MM/YY hh:min:ss
HOLIDAY	DD/MM/YY
LIG_MINUTES	minutes
NPSONC_MINUTES	minutes
NPSONC_PERIODNUMBER	integer
NPSONC_TIMES	integer
OCC_STOPPOINTS	integer
ORCC_PERIODNUMBER	integer
ORCC_MINUTES	minutes
ORCC_TIMES	integer
PRESENCE_ID_ENT	DD/MM/YY hh:min:ss
PRESENCE_ID_EXI	DD/MM/YY hh:min:ss
SOFFCCIT_MINUTES	minutes
SOFFCCIT_PERIODNUMBER	integer
SOFFCCIT_TIMES	integer
SONCCIT_AP	kW
SONCCIT_MINUTES	minutes
SONCCIT_PERIODNUMBER	integer
SONCCIT_TIMES	integer

Subsequently, the resultant variables, also called features, were analyzed and classified according to outdoor temperature (AEMET_AT), occupant presence (PRESENCE_ID_ENT and PRESENCE_ID_EXI) and HVAC compressor consumption (AP_COMPRESSORS_MEAN).

Firstly, a filter was applied to exclude all data without compressor measurements. On the one hand, after obtaining the sample without null compressor function values, an analysis of this sample was carried out to realize the segmentation of the data according to outdoor temperature (AEMET_AT). As a result of applying a binning algorithm, 8 groups of outdoor temperature (TE_Mean_BIN) were defined with a range from 3.56 °C to 39.52 °C in 5 °C intervals, as is shown in Table 2. The binning algorithm is used to reduce the effects of minor observation errors, and it carried out bucketing, where bins have an equal frequency of 5 degrees Celsius following the formula:

$$L = \frac{\max(x) - \min(x)}{n} \quad (1)$$

where L is the length of the bucket, and n is the number of buckets.

Table 2. External temperature binning.

Bin	Lower	Upper
1	≥3,562,762	<8,562,762
2	≥8,562,762	<13,562,762
3	≥13,562,762	<18,562,762
4	≥18,562,762	<23,562,762
5	≥23,562,762	<28,562,762
6	≥28,562,762	<33,562,762
7	≥33,562,762	<38,562,762
8	≥38,562,762	<39,529,412

On the other hand, the distribution of the building occupancy was analyzed by applying statistical methods and defining 3 clear groups: a group with fewer than 50 persons, which mainly represents holidays; a group with between 50 and 110 persons, which indicates days with medium occupancy; and finally, a group with a range between 110 and 160 persons corresponding to working days with high occupancy, which contains the majority of the sample, as is shown in Figure 2.

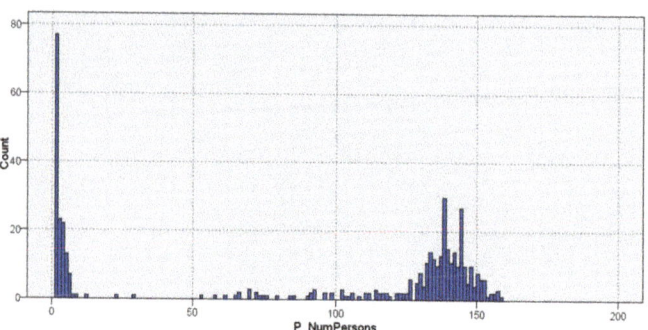

Figure 2. Occupancy distribution.

Finally, the data were represented in terms of compressor consumption (AP_COMPRESS ORS_MEAN), external temperature (AEMET_AT) and occupancy (PRESENCE_ID_ENT and PRESENCE_ID_EXI) to be further analyzed. Furthermore, a discriminant analysis between workdays and non-workdays was performed to provide more depth to our model. The sample distribution is illustrated in Figure 3.

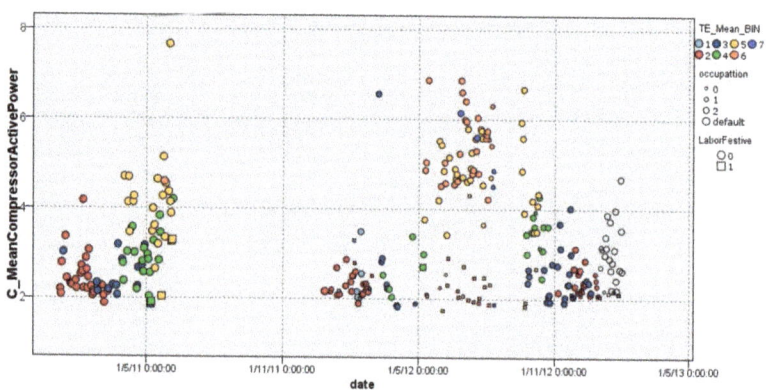

Figure 3. Sample compressor consumption distribution.

After representing and studying the data distribution through the different features, a classification model based on a C&RT decision tree using Gini impurity measures was carried out, with the aim of modeling the sample based on the mean daily active power of the compressors (AP_COMPRESSORS_MEAN). As a result, a set of rules were obtained. C&RT is used for both classification and regression, and it uses the Gini Index (*GI*) criterion to split a node into subnodes. It starts with the training set as a root node, and after splitting the root node in two, it splits the subsets using the same logic recursively until it finds that further splits will not result in any pure subnodes or reaches the maximum number of leaves in a growing tree. The Gini Index is expressed as follows:

$$GI = \sum_{i=0}^{c} P_i(1 - P_i)$$

where c is total classes, and P_i is the probability of class i.

This set of rules was the basis for substantiating our clustering model. The scheme of this classification algorithm is shown in Figure 4.

```
TE_Mean_BIN in [ 1 2 3 4 ]  [Ave: 2,639, Effect: -0,581 ] (192)
  TE_Mean_BIN in [ 1 2 3 ]  [Ave: 2,532, Effect: -0,106 ] (154)
    LaboralFestive in [ 0 ]  [Ave: 2,572, Effect: 0,04 ] (135)
      TE_Mean_BIN in [ 2 ]  [Ave: 2,566, Effect: -0,006 ]  ⇨ 2,566 (86)
      TE_Mean_BIN in [ 1 3 ]  [Ave: 2,583, Effect: 0,011 ]  ⇨ 2,583 (49)
    LaboralFestive in [ 1 ]  [Ave: 2,247, Effect: -0,285 ] (19)            Cluster 1
      TE_Mean_BIN in [ 1 2 ]  [Ave: 2,295, Effect: 0,048 ] (15)
        TE_Mean_BIN in [ 1 ]  [Ave: 2,409, Effect: 0,114 ]  ⇨ 2,409 (4)
        TE_Mean_BIN in [ 2 ]  [Ave: 2,253, Effect: -0,041 ]  ⇨ 2,253 (11)
      TE_Mean_BIN in [ 3 ]  [Ave: 2,067, Effect: -0,18 ]  ⇨ 2,067 (4)
  TE_Mean_BIN in [ 4 ]  [Ave: 3,07, Effect: 0,431 ] (38)
    LaboralFestive in [ 0 ]  [Ave: 3,086, Effect: 0,016 ]  ⇨ 3,086 (34)     Cluster 2
    LaboralFestive in [ 1 ]  [Ave: 2,934, Effect: -0,136 ]  ⇨ 2,934 (4)
TE_Mean_BIN in [ 5 6 7 ]  [Ave: 4,182, Effect: 0,962 ] (116)
  LaboralFestive in [ 0 ]  [Ave: 4,834, Effect: 0,652 ] (86)
    TE_Mean_BIN in [ 5 ]  [Ave: 4,538, Effect: -0,296 ]  ⇨ 4,538 (53)
    TE_Mean_BIN in [ 6 7 ]  [Ave: 5,31, Effect: 0,476 ]  ⇨ 5,31 (33)        Cluster 3
  LaboralFestive in [ 1 ]  [Ave: 2,315, Effect: -1,868 ] (30)
    TE_Mean_BIN in [ 5 ]  [Ave: 2,545, Effect: 0,231 ]  ⇨ 2,545 (13)
    TE_Mean_BIN in [ 6 7 ]  [Ave: 2,138, Effect: -0,177 ]  ⇨ 2,138 (17)
```

Figure 4. Set of rules for the clustering of the sample with the objective of average active power compressor versus temperature and work or non-workdays.

As can be seen at the first level of the decision tree, the sample was classified into 2 groups distinguished by external temperature (*AEMET_AT*). On the one hand, groups with lower external temperature ranging from 3.5 °C to 23.5 °C correspond to groups 1, 2, 3 and 4. On the other hand, groups with higher external temperature ranging from 23.5 °C to 39.5 °C correspond to groups 5, 6 and 7. Once this first classification was carried out, it was possible to observe a division between days with lower temperature: one group has external temperature ranges from 3.5 °C to 18.5 °C (1, 2 and 3) with very little compressor use, and another group has a comfortable temperature range ranging from 18.5 °C to 23.5 °C (4) with slight compressor use.

These 3 groups were derived from the following clusters analyzed to study EE in the BlueNet building.

3.2.1. Cluster 1

In cluster 1 (Figure 5), it was possible to observe a classification of days with lower external temperature ranging from 3.5 °C to 18.5 °C. This group (158 days) had an average compressor consumption ranging from 2.566 kW/h to 2.583 kW/h on working days with an effect of −0.006 and 0.011, respectively. On non-working days (*LaborFestive* equals 1), the average consumption of the compressor ranged from 2.067 kW/h to 2.409 kW/h with an effect of −0.18 and 0.114, respectively, which indicates the effect on the entropy—average—when removing a value from the cluster. Furthermore, it should be noted that most of the days followed this distribution, while scattered days usually coincided with individual cases of high consumption over a short period of time. In addition, non-working days had a homogeneous distribution with compressors having quite low average active power consumption, very close to 2 kW/h.

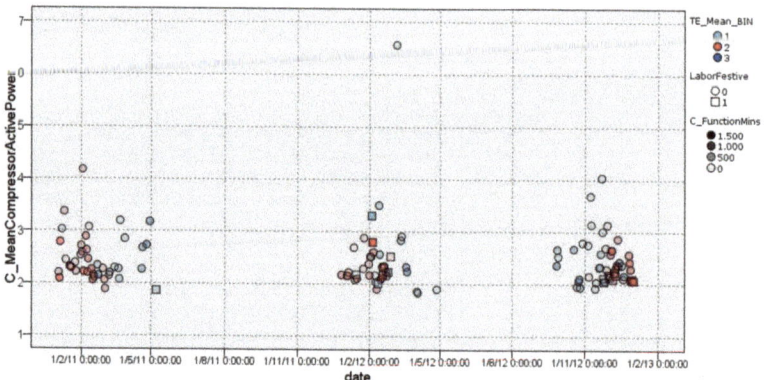

Figure 5. Graphical representation of distribution of cluster 1.

3.2.2. Cluster 2

In cluster 2 (Figure 6), it was possible to observe a classification of days with a comfortable external temperature that ranged between 18.5 °C and 23.5 °C. This group (38 days) had an average compressor consumption of 3.086 kW/h on workdays with an effect of −0.016. On non-working days, the compressor average was 2.934 kW/h with an effect of −0.136. In addition, it was possible to observe a wider dispersion and high similarity between workdays and non-workday cases. This is caused by a comfortable external temperature ranging between 18.5 °C and 23.5 °C. In addition, in Figure 6, it is possible to observe the number of minutes in which the HVAC system was switched on ($C\_FunctionMins$).

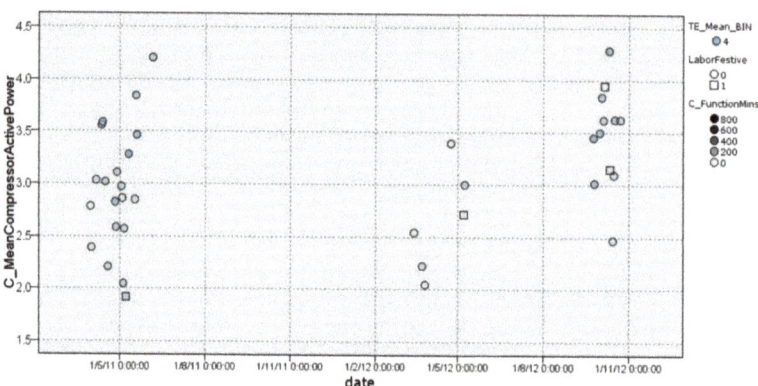

Figure 6. Graphical representation of distribution of cluster 2.

3.2.3. Cluster 3

After analyzing cluster 1 and cluster 2, cluster 3 was the most relevant and interesting, and it covered the majority of cases with high external temperature. This cluster, with 116 days, has a somewhat complex distribution because it was analyzed carefully and split into 3 well-defined groups, as is shown in Figure 7.

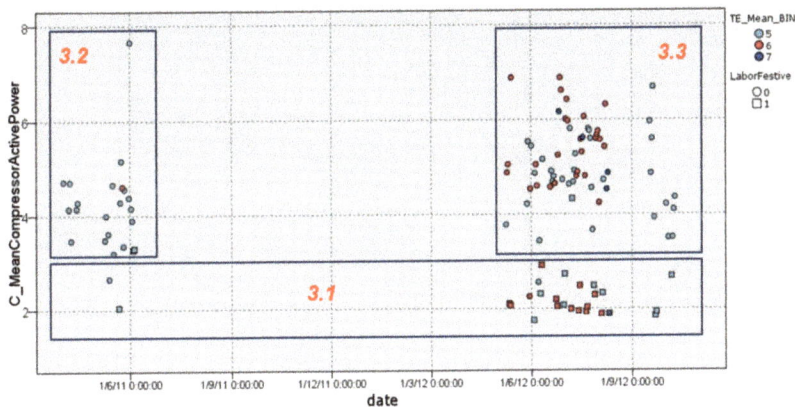

Figure 7. Graphical representation of distribution of cluster 3.

Cluster 3.1

Cluster 3.1 contains the majority of non-working days, in which the compressor had a low range of average energy consumption, with values between 1.7 kW/h and 3 kW/h for the entire period covered in the data sample (26 days) (Figure 8).

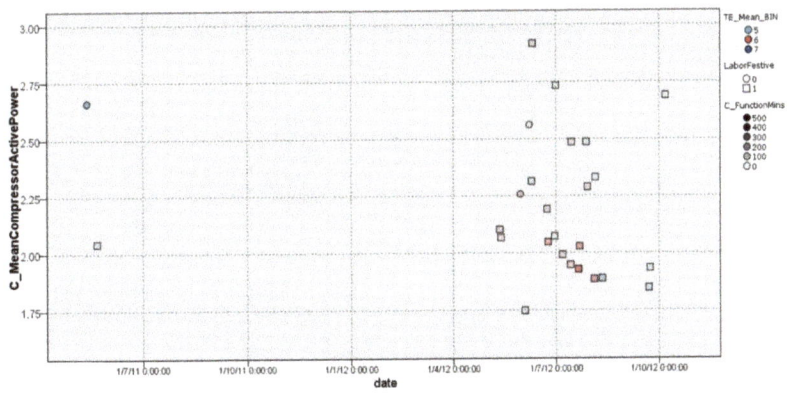

Figure 8. Graphical representation of distribution of cluster 3.1.

Cluster 3.2

Cluster 3.2 includes the majority of working days during the year 2011 until the months of July–August (22 days), during which consumption was higher, with values of 3 kW/h and 5 kW/h, except for a day on which data were dispersed to almost 8 kW/h. As it is possible to observe, this cluster corresponds to a season with more moderate temperatures, as is illustrated in Figure 9. Furthermore, in the illustration in Figure 9, it is possible to observe the number of minutes in which the HVAC system was switched on (C_FunctionMins), providing further detail.

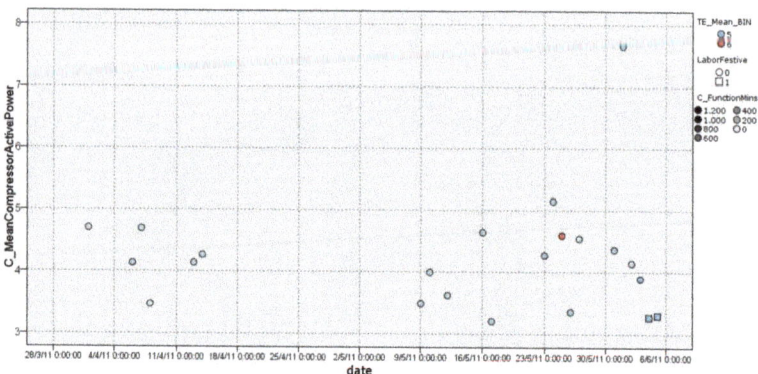

Figure 9. Graphical representation of distribution of cluster 3.2.

Cluster 3.3

Finally, cluster 3.3 corresponds to most of the working days during 2012 from April to October. This cluster (68 days) is characterized by days with a large dispersion, caused by a hotter season with higher temperatures, and an average compressor consumption ranging between 3 kW/h and 7 kW/h in a regular manner, as is illustrated in Figure 10.

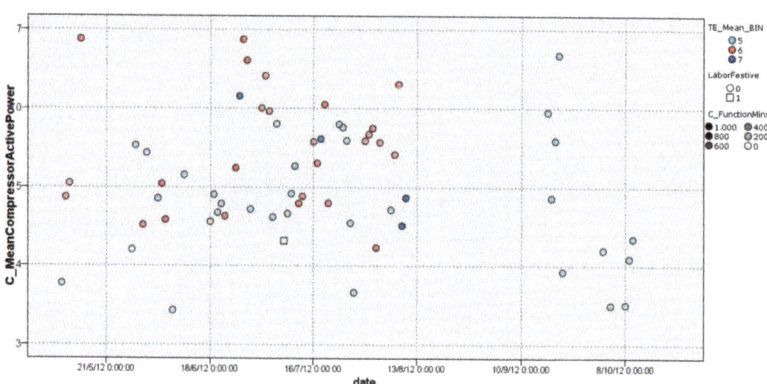

Figure 10. Graphical representation of distribution of cluster 3.3.

In this case, the behavior of all days is characterized by a high compressor consumption and a high number of changes in compressor consumption. Thus, on days with similar environmental characteristics, the features that indicate the grades of efficiency comprise lower compressor consumption, uniform compressor consumption and a small number of slight changes in compressor consumption.

3.3. Energy Efficiency Prediction Module

Once the developed classification module was implemented, the results of this module were carefully studied. As a result of the study, an energy efficiency prediction module was developed. The purpose of this module was mainly to obtain a metric of EE for this study. Therefore, clusters were studied independently, and the EE estimation was based on historical behavior inferred through statistical methods.

First, every cluster was selected, and a distribution analysis for each sample was carried out. The distribution analysis was based on an energy consumption histogram. This distribution curve follows a distribution that is similar to a Gaussian distribution; thus, the EE categories were fitted with a function of the sample distribution, as is shown in Figure 11.

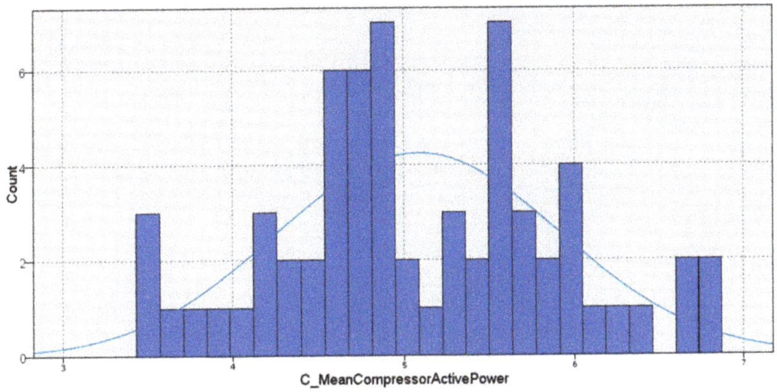

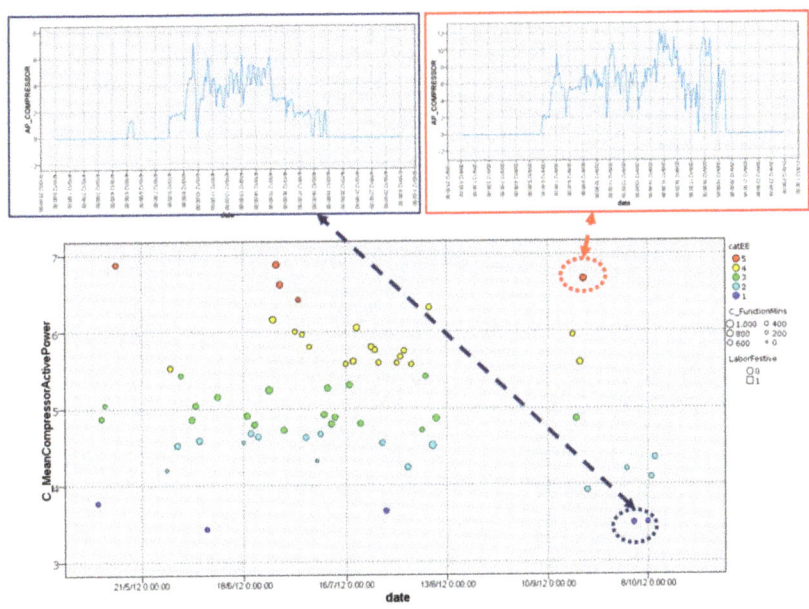

Figure 11. Gaussian distribution of energy consumption in cluster 3.3.

Thus, the days were categorized into 5 EE categories based on cluster dispersion. These categories were established on the basis of the normal distribution, which approximates a Gaussian distribution. The limits for each category comprise the following segments: $\mu - 3/2\,\sigma$, $\mu - 1/2\,\sigma$, $\mu + 1/2\,\sigma$ and $\mu + 3/2\,\sigma$, where μ is the mean, and σ is covariance. As a result, the consumption on each day was categorized according to every cluster characteristic following the classification based on the data distribution shown in Figure 12.

Figure 12. EE category prediction for cluster 3.3 and energy consumption behaviors.

Besides EE categories for each cluster, this module is able to provide some insights that can help to reveal and detect the patterns and anomalies observed on inefficient days. For example, the consumption behaviors on 4 October 2012 (cat. 1) and 20 September 2012 (cat. 5) had the same features (cluster 3.3), which comprise the same range of outdoor temperatures, number of persons and other environmental conditions, but the energy consumption behavior was very different, as it was highlighted through the EE category carried out by the EE prediction module. Figure 12 shows a wide range of consumption (0–6 kWh vs. 0–12 kWh), a greater number of stop points (5 OCC vs. 3 OCC) and regime changes (43 ORCC with 293 min of anomalous function vs. 77 ORCC with 532 min of anomalous function), a less softened energy consumption curve with a larger variation in energy consumption changes at the same time, etc. All of these results can indicate an anomaly, such as HVAC breakdown, poorly dimensioned HVAC, a possible isolation problem in the room and other issues.

In summary, the EE prediction module indicates the EE category for every day as a function of the historical behavior for the cluster sample. In addition, the EE prediction module, together with the classification module, can indicate possible anomalous consumption patterns that occur in the building and other possible anomalies. These cases were studied, and the results show a tight correlation between the detected OCC and ORCC anomalies and the index of EE, as shown in Tables 3–7. At the same time, it is observed that SOFFCCIT, SONCCIT and NPSONC anomalies are not correlated with inefficient consumption behavior.

Table 3. Cluster 1 anomaly distribution. Each EE Category is associated with a color as depicted in Figure 12.

Cluster 1	Days	OCC	ORCC	SOFFCCIT	SONCCIT	NPSONC	TOTAL
EE Categ. 1	0	-	-	-	-	-	0
EE Categ. 2	9	0	1	1	1	9	12
EE Categ. 3	8	0	5	9	6	8	28
EE Categ. 4	2	0	1	1	0	2	4
EE Categ. 5	0	0	1	0	0	0	1
	19	0	8	11	7	19	45

Table 4. Cluster 2 anomaly distribution. Each EE Category is associated with a color as depicted in Figure 12.

Cluster 1	Days	OCC	ORCC	SOFFCCIT	SONCCIT	NPSONC	TOTAL
EE Categ. 1	3	0	0	0	0	1	1
EE Categ. 2	8	0	0	0	0	1	1
EE Categ. 3	13	0	0	0	0	3	3
EE Categ. 4	12	0	2	0	0	0	2
EE Categ. 5	2	0	2	0	0	1	3
	38	0	4	0	0	6	10

Table 5. Cluster 3.1 anomaly distribution. Each EE Category is associated with a color as depicted in Figure 12.

Cluster 1	Days	OCC	ORCC	SOFFCCIT	SONCCIT	NPSONC	TOTAL
EE Categ. 1	0	-	-	-	-	-	0
EE Categ. 2	11	0	1	0	0	7	8
EE Categ. 3	8	0	1	0	0	4	5
EE Categ. 4	4	0	0	0	0	2	2
EE Categ. 5	3	0	0	0	0	1	1
	26	0	2	0	0	14	16

Table 6. Cluster 3.2 anomaly distribution. Each EE Category is associated with a color as depicted in Figure 12.

	Cluster 1	Days	OCC	ORCC	SOFFCCIT	SONCCIT	NPSONC	TOTAL
	EE Categ. 1	0	-	-	-	-	-	0
	EE Categ. 2	7	0	3	0	0	3	6
	EE Categ. 3	12	0	3	0	0	3	6
	EE Categ. 4	2	0	0	0	0	0	0
	EE Categ. 5	1	0	1	0	0	0	1
		22	0	7	0	0	6	13

Table 7. Cluster 3.3 anomaly distribution. Each EE Category is associated with a color as depicted in Figure 12.

	Cluster 1	Days	OCC	ORCC	SOFFCCIT	SONCCIT	NPSONC	TOTAL
	EE Categ. 1	5	0	0	0	0	0	0
	EE Categ. 2	16	8	12	0	0	1	21
	EE Categ. 3	20	13	16	0	0	1	30
	EE Categ. 4	18	14	14	0	0	2	30
	EE Categ. 5	5	4	5	0	0	0	9
		64	39	47	0	0	4	90

The above-mentioned correlation between detected OCC and ORCC anomalies versus the index of EE highlights not only the number of anomalies but also the period (in minutes) of each anomaly for ORCC and the number of stop points for OCC, as shown in Table 8. This insight is even more relevant when one of the principles of EE consists in softening the consumption curve, aiming to avoid peaks and large changes in consumption. Furthermore, OCC and ORCC anomalies denote large changes and peaks in consumption, which means that the HVAC system is poorly dimensioned or calibrated, the HVAC system is too powerful, or a possible breakdown in a temperature sensor has occurred. On the other hand, a compressor with high OCC is prone to break and to have a shortened lifetime. Table 8 highlights the correlation between an increasing number of stop points for OCC and an increasing number of minutes with ORCC in each period when EE decreases. Additionally, in Table 9 are detailed the average of OCC and ORCC anomalies per day and its correlation with each EE Category.

Table 8. Stop points and minutes for OCC and ORCC anomalies vs. EE category. Each EE Category is associated with a color as depicted in Figure 12.

Cluster	Anomaly	Days	Categ. 1	Categ. 2	Categ. 3	Categ. 4	Categ. 5
Cluster 1	OCC	5	0	0	0	0	0
	ORCC	16	0	313 (1)	1769 (5)	418 (1)	544 (1)
Cluster 2	OCC	20	0	0	0	0	0
	ORCC	18	0	0	0	724 (2)	658 (2)
Cluster 3.1	OCC	20	0	0	0	0	0
	ORCC	18	0	340 (1)	315 (1)	0	0
Cluster 3.2	OCC	20	0	0	0	0	0
	ORCC	18	0	1020 (3)	945 (3)	0	543 (1)
Cluster 3.3	OCC	20	0	64 (8)	142 (13)	158 (14)	39 (4)
	ORCC	18	0	5026 (12)	7278 (16)	5881 (14)	2671 (5)
Total		64	0	64/6699	142/10307	158/7023	39/4416

261

Table 9. Average stop points and minutes per day for OCC and ORCC anomalies. Each EE Category is associated with a color as depicted in Figure 12.

Cluster	Anomaly	Days	Categ. 1	Categ. 2	Categ. 3	Categ. 4	Categ. 5
Cluster 1	OCC	5	0	0	0	0	0
	ORCC	16	0	313	353.8	418	544
Cluster 2	OCC	20	0	0	0	0	0
	ORCC	18	0	0	0	362	329
Cluster 3.1	OCC	20	0	0	0	0	0
	ORCC	18	0	340	315	0	0
Cluster 3.2	OCC	20	0	0	0	0	0
	ORCC	18	0	462.7	437.7	0	543
Cluster 3.3	OCC	20	0	8	10.9	11.3	9.8
	ORCC	18	0	418.8	454.9	420.1	534.2
Total		64	0	8/394.1	10.9/412.3	11.3/413.1	9.8/490.7

4. Summary and Experimental Results

The objective of the present work was to present a decision support system to optimize Energy Management and Control Systems (EMCSs) in smart buildings for any energy-consuming operations. Thus, in order to achieve the main purpose of this work, a model to extract knowledge to realize the efficient management of energy consumption was developed. This model was based on a classification module and a prediction module developed through DM techniques and statistical inference. All BlueNet variables were considered for this EE classification study (indoor, outdoor, energy analyzers, EEIs and work schedule). In addition, the knowledge extracted from data and EE experts was considered for this purpose. Once the study was carried out, five clusters were identified. Each cluster was defined by the following set of conditions:

- Cluster 1: This cluster represents days with lower external temperature ranging from 3.5 °C to 23.5 °C. This cluster does not discriminate between energy consumption or between work and non-workdays.
- Cluster 2: This cluster represents days with a softer curve of intermediate temperature ranging from 23.5 °C to 28.5 °C. This cluster does not discriminate between energy consumption or between work and non-workdays.
- Cluster 3.1: This cluster groups days with higher external temperature ranging from 28.5 °C to 39.6 °C, in which most of the days are non-working days with low energy consumption.
- Cluster 3.2: This cluster represents days with higher external temperature ranging mainly from 28.5 °C to 33.5 °C. Most of the selected days are working days with high energy consumption.
- Cluster 3.3: This cluster groups days with higher external temperature ranging mainly from 28.5 °C to 39.6 °C, in which most of the selected days are working days with high energy consumption.

Once clusters were defined, days that belong to the same cluster were compared to observe differences among behaviors and, consequently, the effects of the different variables on each day. In order to validate and quantify the results, the results were analyzed and corroborated by EE experts of the BlueNet building, and additionally, a module for EE prediction was developed. The aim of the EE prediction module was to evaluate EE according to the type of cluster analyzed. The results obtained were applied to the sample of 509 days with the following results.

All of the results in this study were corroborated by EE experts. In summary, only 3.25% of the total days classified were considered to be very efficient, 29% were efficient and

41.36% were classified with normal EE. The remaining 20.29% of the days were considered to be energy inefficient, and 6.36% of days were classified as very inefficient, indicating a large margin for EE improvement, as detailed in Table 10.

Moreover, in this study, days that represent anomalous behavior were detected with the support of the classification module and prediction module. These anomalous behaviors usually correspond to days that were very inefficient. In addition, these anomalous behaviors can indicate possible problems that are affecting the BlueNet building (HVAC breakdown, incorrectly dimensioned and other problems), as shown in Tables 2–8. With the results in this study, system integration with EE software could yield substantial benefits: detection of anomalous behaviors in energy consumption, facilitation of energy savings, large profits due to consumption reduction and environmentally friendly management of the building.

Table 10. Distribution of days for each cluster based on EE category. Each EE Category is associated with a color as depicted in Figure 12.

Cluster	Days	Categ. 1	Categ. 2	Categ. 3	Categ. 4	Categ. 5
Cluster 1	171	0%	32.35%	48.87%	13.53%	5.25%
Cluster 2	91	7.89%	21.05%	34.21%	31.58%	5.27%
Cluster 3.1	26	0%	42.31%	30.77%	15.39%	11.53%
Cluster 3.2	22	0%	31.82%	54.55%	9.09%	4.54%
Cluster 3.3	64	7.81%	25%	31.25%	28.13%	7.81%
Total	374	3.25%	29%	41.36%	20.29%	6.36%

5. Conclusions

After a bibliographical review, an intensive field of research with strong interest in EE was identified. This review revealed that similar approaches have been considered, but no other works have been performed with the aim of rendering support in EE classification and prediction. Specifically, systems able to measure EE and support the detection of anomalies have not been investigated with similar approaches or results.

This paper presents a system for the optimization of energy consumption in smart buildings. The main purposes of this work are the following: extract the knowledge hidden in building data, develop a classification module and build an EE prediction module that helps to predict EE for each day. This proposal was tested in the BlueNet building scenario, but it is applicable to any other building and is not specific to BlueNet.

The process of energy optimization was carried out through a hybrid system whose cornerstone is a classification module and an EE prediction module. The classification module is made up of a hybrid system based on a CR&T decision tree and a clustering model. As a result, this module is able to compare EE on days with similar characteristics, regardless of the season and other factors, which are difficult to compare without this information, and it is also able to highlight the presence of anomalous energy consumptions and other possible problems. This module provides an objective point of view that is key to measuring, comparing and predicting the EE for each day.

The EE prediction module is able to quantify the energy efficiency of each day, comparing days with similar EE conditions supported by previous cluster classifications. This module is able to measure and predict the efficiency for each day based on the knowledge extracted from historical data by applying statistical analysis. Furthermore, this module is able to unveil insights that highlight correlations between inefficiencies and anomalous behaviors.

Finally, this work presents a classification for each day of the historical data and their respective EE categories. These results were compared and corroborated by experts to, first, understand how the energy consumption is behaving and, second, understand the reasons for this behavior and how to enhance the efficiency. Furthermore, our approach highlights some evidence that days with less efficiency (6.36% and 20.29% of days respectively) contain more anomalies and that these anomalies also occurred over a greater amount of time. Furthermore, this work unveils an exponential correlation between OCC and ORCC anomalies and EE.

Based on the results obtained in the BlueNet building and aligned with this research area, interesting future research lines that could be explored include the automation of the full process through AI techniques that are able to classify and predict EE in an analytical manner in real time and, based on anomalous behavior, determine how to apply actionable measures to correct and improve EE in the building automatically. On the other hand, this work aimed to support the gap in the understanding of building behavior, as cited in [39]. A future goal is the integration of this module with the commercial EE software called Eugene, whose owner is the company Isotrol.

Author Contributions: Conceptualization, M.P., F.B., E.P. and C.L.; methodology, M.P., F.B. and C.L.; software, M.P.; validation, M.P.; formal analysis, M.P. and F.B.; investigation, M.P., F.B., E.P. and C.L.; resources, M.P., F.B. and C.L.; data curation, M.P.; writing—original draft preparation, M.P.; writing—review and editing, M.P., F.B., E.P. and C.L.; visualization, M.P. and F.B.; supervision, M.P., F.B., E.P. and C.L.; project administration, C.L.; funding acquisition, C.L. All authors have read and agreed to the published version of the manuscript.

Funding: This research was funded by the European Commission—Framework Program 7, FP7-285229 KnoHolEM.

Institutional Review Board Statement: Not applicable.

Informed Consent Statement: Not applicable.

Data Availability Statement: All data and code will be made available on request to the correspondent author's email with appropriate justification.

Acknowledgments: The authors thank FP7-285229 KnoHolEM for providing funds for this project. The authors thank Isotrol for their collaboration in this work by providing knowledge and cooperation to extract the data from the BlueNet building database.

Conflicts of Interest: The authors declare no conflict of interest.

Appendix A. Main Variables for Optimization in BlueNet Building

AEMET_AT (°C): Outdoor air temperature in Celsius degrees outside of the BlueNet building.
AP_AIR_MIXERS (kW): Active power in air mixers of the HVAC system.
AP_COMPRESSOR_MEAN (kW): Average daily active power of the HVAC compressor.
AP_COMPRESSOR_MINUTES (min): Total number of daily minutes of HVAC compressor use.
AP_COMPRESSOR (kW): Active power of the HVAC compressor.
AP_SPLITS (kW): Active power in splits of the HVAC system.
AP_LIGTH (kW): Active power in the lighting system.
LABOR/FESTIVE (time): Date of a non-working day.
LIG_MINUTES (min): Time in minutes in which lighting was working during a day.
LIG_PERIODNUMBER: Total number of periods of lighting during a day.
NPSONC_ MINUTES (min): Time in minutes in which lighting was working and an NPSONC anomaly was detected.
NPSONC_ PERIODNUMBER: Total number of periods with an NPSONC anomaly during a day. A period constitutes the number of consecutive points with NPSONC anomalies detected.
NPSONC_ TIMES: Number of times with an NPSONC anomaly during a day.
OCC_STOPPOINTS: Number of daily On-Off operations in HVAC compressor (OCC).
ORCC_PERIODNUMER: Total number of periods with an operational regime change (ORCC) in HVAC compressor during a day.

ORCC_MINUTES (min): Time in minutes in which an operational regime change in HVAC compressor (ORCC) anomaly was detected.

ORCC _ TIMES: Number of times with an operational regime change in HVAC compressor (ORCC) anomaly during a day.

SOFFCCIT _MINUTES (min): Time in minutes in which the HVAC compressor was switched off and an abnormal change in interior temperature (SOFFCCIT) was detected.

SOFFCCIT_PERIODNUMBER: Total number of periods in which the HVAC compressor was switched off and an abnormal change in interior temperature (SOFFCCIT) is detected during a day.

SOFFCCIT_TIMES: Number of times in which the HVAC compressor was switched off and an abnormal change in interior temperature (SOFFCCIT) was detected during a day.

SONCCIT_AP (kW): Active power waste in the HVAC system while the HVAC compressor was switched on and an abnormal change in interior temperature (SONCCIT) was detected.

SONCCIT_MINUTES (min): Time in minutes in which the HVAC compressor was switched off and an abnormal change in interior temperature (SONCCIT) was detected.

SONCCIT_PERIODNUMBER: Total number of periods in which the HVAC compressor was switched off and an abnormal change in interior temperature (SONCCIT) was detected during a day.

SONCCIT_TIMES: Number of times in which the HVAC compressor was switched off and an abnormal change in interior temperature (SONCCIT) was detected during a day.

ZIGBEE_AT ($°C$): Indoor air temperature in Celsius degrees inside the BlueNet building.

References

1. Rashid, Y.R.; Sulaiman, M.S.; Aziz, A.; Selamat, H.; Yani, A.H.M.; Kandar, M.Z. Greening government's office buildings: PWD Malaysia experiences. *Procedia Eng.* **2011**, *21*, 1056–1060. [CrossRef]
2. Pérez-Lombard, L.; Ortiz, J.; Pout, C. A review on buildings energy consumption information. *Energy Build.* **2008**, *40*, 394–398. [CrossRef]
3. Peters, G.P.; Andrew, R.M.; Canadell, J.G.; Fuss, S.; Jackson, R.B.; Korsbakken, J.I.; Le Quéré, C.; Nakicenovic, N. Key indicators to track current progress and future ambition of the Paris Agreement. *Nat. Clim. Chang.* **2017**, *7*, 118. [CrossRef]
4. EIA, U.S. *Key Energy Statistics*; Energy Information Administration: Washington, DC, USA, 2017.
5. EIA, U.S. *Annual Energy Outlook*; Energy Information Administration: Washington, DC, USA, 2018.
6. Day, A.; Jones, P.; Turton, J. Development of a UK Centre for Efficient and Renewable Energy in Buildings (CEREB). *Renew. Energy* **2012**, *49*, 166–170. [CrossRef]
7. de Alegría Mancisidor, I.; de Basurto Uraga, P.D.; de Arbulo López, P.R. European Union's renewable energy sources and energy efficiency policy review: The Spanish perspective. *Renew. Sustain. Energy Rev.* **2009**, *13*, 100–114. [CrossRef]
8. EEA. *Financial Support for Energy Efficiency in Buildings 2018*; EEA: Copenhagen, Denmark, 2018.
9. DOE/EIA. *Annual Energy Outlook 2021*; DOE/EIA: Washington, DC, USA, 2021.
10. Yamagata, Y.; Seya, H. Simulating a future smart city: An integrated land use-energy model. *Appl. Energy* **2013**, *112*, 1466–1474. [CrossRef]
11. Lazaroiu, G.C.; Roscia, M. Definition methodology for the smart cities model. *Energy* **2012**, *47*, 326–332. [CrossRef]
12. Personal, E.; Guerrero, J.I.; Garcia, A.; Peña, M.; Leon, C. Key performance indicators: A useful tool to assess Smart Grid goals. *Energy* **2014**, *76*, 976–988. [CrossRef]
13. Vadiee, A.; Martin, V. Thermal energy storage strategies for effective closed greenhouse design. *Appl. Energy* **2013**, *109*, 337–343. [CrossRef]
14. Singh, R.; Tiwari, G. Energy conservation in the greenhouse system: A steady state analysis. *Energy* **2010**, *35*, 2367–2373. [CrossRef]
15. Djevic, M.; Dimitrijevic, A. Energy consumption for different greenhouse constructions. *Energy* **2009**, *34*, 1325–1331. [CrossRef]
16. Hoseini, A.G.; Dahlan, N.D.; Berardi, U.; Hoseini, A.G.; Makaremi, N.; Hoseini, M.G. Sustainable energy performances of green buildings: A review of current theories, implementations and challenges. *Renew. Sustain. Energy Rev.* **2013**, *25*, 1–17. [CrossRef]
17. Hoseini, A.G.; Dahlan, N.; Berardi, U.; Hoseini, A.G.; Makaremi, N. The essence of future smart houses: From embedding ICT to adapting to sustainability principles. *Renew. Sustain. Energy Rev.* **2013**, *24*, 593–607. [CrossRef]
18. Dounis, A.I.; Caraiscos, C. Advanced control systems engineering for energy and comfort management in a building environment—A review. *Renew. Sustain. Energy Rev.* **2009**, *13*, 1246–1261. [CrossRef]
19. Omer, A.M. Renewable building energy systems and passive human comfort solutions. *Renew. Sustain. Energy Rev.* **2008**, *12*, 1562–1587. [CrossRef]
20. Xia, C.; Zhu, Y.; Lin, B. Renewable energy utilization evaluation method in green buildings. *Renew. Energy* **2008**, *33*, 883–886. [CrossRef]

21. Peña, M.; Biscarri, F.; Guerrero, J.I.; Monedero, I.; León, C. Rule-based system to detect energy efficiency anomalies in smart buildings, a data mining approach. *Expert Syst. Appl.* **2016**, *56*, 242–255. [CrossRef]
22. Zanella, A.; Bui, N.; Castellani, A.; Vangelista, L.; Zorzi, M. Internet of Things for Smart Cities. *IEEE Int. Things J.* **2014**, *1*, 22–32. [CrossRef]
23. Singh, D.; Tripathi, G.; Jara, A.J. A survey of Internet-of-Things: Future vision, architecture, challenges and services. *Int. Things (WF-IoT)* **2014**, *1*, 287–292. [CrossRef]
24. Al-Fuqaha, A.; Guizani, M.; Mohammadi, M.; Aledhari, M.; Ayyash, M. Internet of Things: A Survey on Enabling Technologies, Protocols, and Applications. *IEEE Commun. Surv. Tutor.* **2015**, *17*, 2347–2376. [CrossRef]
25. Rathore, M.M.; Paul, A.; Hong, W.-H.; Seo, H.; Awan, I.; Saeed, S. Exploiting IoT and big data analytics: Defining Smart Digital City using real-time urban data. *Sustain. Cities Soc.* **2018**, *40*, 600–610. [CrossRef]
26. Pawar, P.; Kumar, M.T.; Vittal, K.P. An IoT based Intelligent Smart Energy Management System with accurate forecasting and load strategy for renewable generation. *Measurement* **2020**, *152*, 107187. [CrossRef]
27. Palattella, M.; Dohler, M.; Grieco, A.; Rizzo, J.; Torsner, T.; Engel, L.; Ladid, L. Internet of Things in the 5G era: Enablers architecture and business models. *IEEE J. Sel. Areas Commun.* **2016**, *34*, 510–527. [CrossRef]
28. Skouby, K.E.; Lynggaard, P. Smart home and smart city solutions enabled by 5G, IoT, AAI, and CoT services. In Proceedings of the 2014 International Conference on Contemporary Computing and Informatics (IC3I), Mysuru, India, 27–29 November 2014; pp. 874–878.
29. Goraczek, M.; Pereira, G.; Falco, E.; Kleinhans, R.; Parycek, P. Using Fuzzy Cognitive Maps as Decision Support Tool for Smart Cities. In *CeDEM Asia 2016: Proceedings of the International Conference for E-Democracy and Open Government, Asia 2016, 7–9 Decemer 2016, Deagu, Korea*; Edition Donau-Universität Krems: Krems an der Donau, Austria, 2017.
30. Somu, N.; Raman, G.; Ramamritham, K. A hybrid model for building energy consumption forecasting using long short term memory networks. *Appl. Energy* **2020**, *261*, 114131. [CrossRef]
31. Ikhide, M.; Egaji, A.; Ahmed, A. Developing Energy Control and Optimisation Methodology for Built Environment of the Future. *Renew. Energy Sustain. Build.* **2020**, *45*, 567–577. [CrossRef]
32. Fan, C.; Sun, Y.; Xiao, F.; Ma, J.; Lee, D.; Wang, J.; Tseng, Y.C. Statistical investigations of transfer learning-based methodology for short-term building energy predictions. *Appl. Energy* **2020**, *262*, 114499. [CrossRef]
33. Hosseinnezhad, V.; Shafie-Khah, M.; Siano, P.; Catalao, J.P.S. An Optimal Home Energy Management Paradigm with an Adaptive Neuro-Fuzzy Regulation. *IEEE Access* **2020**, *8*, 19614–19628. [CrossRef]
34. Broujeny, R.S.; Madani, K.; Chebira, A.; Amarger, V.; Hurtard, L. Data-Driven Living Spaces' Heating Dynamics Modeling in Smart Buildings using Machine Learning-Based Identification. *Sensors* **2020**, *20*, 1071. [CrossRef]
35. Pallonetto, F.; De Rosa, M.; Milano, F.; Finn, D.P. Demand response algorithms for smart-grid ready residential buildings using machine learning models. *Appl. Energy* **2019**, *239*, 1265–1282. [CrossRef]
36. Fan, C.; Xiao, F.; Yan, C.; Liu, C.; Li, Z.; Wang, J. A novel methodology to explain and evaluate data-driven building energy performance models based on interpretable machine learning. *Appl. Energy* **2018**, *235*, 1551–1560. [CrossRef]
37. Reynolds, J.; Ahmad, M.W.; Rezgui, Y.; Hippolyte, J.L. Operational supply and demand optimisation of a multi-vector district energy system using artificial neural networks and a genetic algorithm. *Appl. Energy* **2019**, *235*, 699–713. [CrossRef]
38. Olszewski, R.; Pałka, P.; Turek, A. Solving "Smart City" Transport Problems by Designing Carpooling Gamification Schemes with Multi-Agent Systems: The Case of the So-Called "Mordor of Warsaw". *Sensors* **2018**, *18*, 141. [CrossRef] [PubMed]
39. Yoshino, H.; Hong, T.; Nord, N. IEA EBC annex 53: Total energy use in buildings—Analysis and evaluation methods. *Energy Build.* **2017**, *152*, 124–136. [CrossRef]
40. Noailly, J. Improving the energy efficiency of buildings: The impact of environmental policy on technological innovation. *Energy Econ.* **2012**, *34*, 795–806. [CrossRef]
41. Andrews, C.J.; Krogmann, U. Explaining the adoption of energy-efficient technologies in U.S. commercial buildings. *Energy Build.* **2009**, *41*, 287–294. [CrossRef]
42. Colmenar-Santos, A.; de Lober, L.N.T.; Borge-Diez, D.; Castro-Gil, M. Solutions to reduce energy consumption in the man-agement of large buildings. *Energy Build.* **2013**, *56*, 66–77. [CrossRef]
43. Biyik, E.; Kahraman, A. A predictive control strategy for optimal management of peak load, thermal comfort, energy storage and renewables in multi-zone buildings. *J. Build. Eng.* **2019**, *25*, 100826. [CrossRef]
44. Chumnanvanichkul, P.; Chirapongsananurak, P.; Hooncharoen, N. Three-level Classification of Air Conditioning Energy Consumption for Building Energy Management System Using Data Mining Techniques. In Proceedings of the 2019 IEEE PES GTD Grand International Conference and Exposition Asia (GTD Asia), Bangkok, Thailand, 20 March–23 March 2019; pp. 611–615. [CrossRef]
45. Marinakis, V.; Karakosta, C.; Doukas, H.; Androulaki, S.; Psarras, J. A building automation and control tool for remote and real time monitoring of energy consumption. *Sustain. Cities Soc.* **2013**, *6*, 11–15. [CrossRef]
46. Azadeh, A.; Saberi, M.; Ghaderi, S.; Gitiforouz, A.; Ebrahimipour, V. Improved estimation of electricity demand function by integration of fuzzy system and data mining approach. *Energy Convers. Manag.* **2008**, *49*, 2165–2177. [CrossRef]
47. Altwies, J.E.; Nemet, G.F. Innovation in the U.S. building sector: An assessment of patent citations in building energy control technology. *Energy Pol.* **2013**, *52*, 819–831. [CrossRef]

48. Pang, X.; Wetter, M.; Bhattacharya, P.; Haves, P. A framework for simulation-based real-time whole building performance assessment. *Build. Environ.* **2012**, *54*, 100–108. [CrossRef]
49. Kolokotsa, D.; Kalaitzakis, K.; Antonidakis, E.; Stavrakakis, G. Interconnecting smart card system with PLC controller in a local operating network to form a distributed energy management and control system for buildings. *Energy Convers. Manag.* **2002**, *43*, 119–134. [CrossRef]
50. Nguyen, T.A.; Aiello, M. Energy intelligent buildings based on user activity: A survey. *Energy Build.* **2013**, *56*, 244–257. [CrossRef]
51. Escrivá-Escrivá, G. Basic actions to improve energy efficiency in commercial buildings in operation. *Energy Build.* **2011**, *43*, 3106–3111. [CrossRef]
52. D'Agostino, D.; Cuniberti, B.; Bertoldi, P. Energy consumption and efficiency technology measures in European non-residential buildings. *Energy Build.* **2017**, *153*, 72–86. [CrossRef]
53. Koezjakov, A.; Urge-Vorsatz, D.; Crijns-Graus, W.; van den Broek, M. The relationship between operational energy demand and embodied energy in Dutch residential buildings. *Energy Build.* **2018**, *165*, 233–245. [CrossRef]
54. Figueiredo, J.; Sá da Costa, J. A SCADA system for energy management in intelligent buildings. *Energy Build.* **2012**, *49*, 85–98. [CrossRef]
55. Diakaki, C.; Grigoroudis, E.; Kolokotsa, D. Towards a multi-objective optimization approach for improving energy efficiency in buildings. *Energy Build.* **2008**, *40*, 1747–1754. [CrossRef]
56. Diakaki, C.; Grigoroudis, E.; Kabelis, N.; Kolokotsa, D.; Kalaitzakis, K.; Stavrakakis, G. A multi-objective decision model for the improvement of energy efficiency in buildings. *Energy* **2010**, *35*, 5483–5496. [CrossRef]
57. Wu, W.; Guo, J.; Li, J.; Hou, H.; Meng, Q.; Wang, W. A multi-objective optimization design method in zero energy building study: A case study concerning small mass buildings in cold district of China. *Energy Build.* **2018**, *158*, 1613–1624. [CrossRef]
58. Doukas, H.; Nychtis, C.; Psarras, J. Assessing energy-saving measures in buildings through an intelligent decision support model. *Build. Environ.* **2009**, *44*, 290–298. [CrossRef]
59. Yang, R.; Wang, L. Development of multi-agent system for building energy and comfort management based on occupant behaviors. *Energy Build.* **2013**, *56*, 1–7. [CrossRef]
60. Klein, L.; Kwak, J.-Y.; Kavulya, G.; Jazizadeh, F.; Becerik-Gerber, B.; Varakantham, P.; Tambe, M. Coordinating occupant behavior for building energy and comfort management using multi-agent systems. *Autom. Constr.* **2012**, *22*, 525–536. [CrossRef]
61. Oldewurtel, F.; Parisio, A.; Jones, C.; Gyalistras, D.; Gwerder, M.; Stauch, V.; Lehmann, B.; Morari, M. Use of model predictive control and weather forecasts for energy efficient building climate control. *Energy Build.* **2012**, *45*, 15–27. [CrossRef]
62. Stamatescu, G.; Stamatescu, I.; Arghira, N.; Drăgana, C.; Făgărăşan, I. Data-driven methods for smart building AHU subsystem modelling. In Proceedings of the 2017 9th IEEE International Conference on Intelligent Data Acquisition and Advanced Computing Systems: Technology and Applications, Bucharest, Romania, 21–23 September 2017; pp. 617–621.
63. Kim, H.; Stumpf, A.; Kim, W. Analysis of an energy efficient building design through data mining approach. *Autom. Constr.* **2011**, *20*, 37–43. [CrossRef]
64. Petrushevski, F.; Gaida, S.; Beigelböck, B.; Sipetic, M.; Zucker, G.; Schiefer, C.; Schachinger, D.; Kastner, W. Semantic Building Systems Modeling for Advanced Data Analytics for Energy Efficiency. In Proceedings of the 15th IBPSA Conference, San Francisco, CA, USA, 7–9 August 2017; pp. 622–627.
65. Schachinger, D.; Gaida, S.; Kastner, W.; Petrushevski, F.; Reinthaler, C.; Sipetic, M.; Zucker, G. An advanced data analytics framework for energy efficiency in buildings. In Proceedings of the 2016 IEEE 21st International Conference on Emerging Technologies and Factory Automation (ETFA), Berlin, Germany, 6–9 September 2016; pp. 1–4. [CrossRef]
66. Fan, C.; Xiao, F.; Li, Z.; Wang, J. Unsupervised data analytics in mining big building operational data for energy efficiency enhancement: A review. *Energy Build.* **2018**, *159*, 296–308. [CrossRef]
67. Li, X.; Bowers, C.; Schnier, T. Classification of Energy Consumption in Buildings with Outlier Detection. *IEEE Trans. Ind. Electron.* **2009**, *57*, 3639–3644. [CrossRef]
68. Cardillo, E.; Li, C.; Caddemi, A. Embedded heating, ventilation, and air-conditioning control systems: From traditional technologies toward radar advanced sensing. *Rev. Sci. Instrum.* **2021**, *92*, 061501. [CrossRef]
69. Labeodan, T.; Zeiler, W.; Boxem, G.; Zhao, Y. Occupancy measurement in commercial office buildings for demand-driven control applications—A survey and detection system evaluation. *Energy Build.* **2015**, *93*, 303–314. [CrossRef]
70. Dhurandhar, A.; Dobra, A. Probabilistic Characterization of Random Decision Trees. *J. Mach. Learn. Res.* **2008**, *9*, 2321–2348.
71. Maimon, O.; Rokach, L. *Data Mining and Knowledge Discovery Handbook*, 2nd ed.; Springer: Berlin/Heidelberg, Germany, 2010.
72. Han, J.; Kamber, M. *Data Mining. Concepts and Techniques*, 1st ed.; Morgan Kaufmann: Burlington, MA, USA, 2001.
73. Sehgal, S.; Singh, H.; Agarwal, M.; Bhasker, V. Data analysis using principal component analysis 2014. In Proceedings of the 2014 International Conference on Medical Imaging, m-Health and Emerging Communication Systems (MedCom), Greater Noida, India, 7–8 November 2014; pp. 45–48.
74. Hou, Z.; Lian, Z.; Yao, Y.; Yuan, X. Data mining-based sensor fault diagnosis and validation for building air conditioning system. *Energy Conv. Manag.* **2006**, *47*, 2479–2490. [CrossRef]
75. Wu, S.; Clements-Croome, D. Understanding the indoor environment through mining sensory data—A case study. *Energy Build.* **2007**, *39*, 1183–1191. [CrossRef]

76. Zhang, C.; Zhang, S. *Association Rule Mining: Models and Algorithms*; Springer: Berlin/Heidelberg, Germany, 2002; ISBN 3-540-43533-6.
77. Dongre, J.; Prajapati, G.L.; Tokekar, S.V. The role of Apriori algorithm for finding the association rules in Data mining. In Proceedings of the 2014 International Conference on Issues and Challenges In Intelligent Computing Techniques, Ghaziabad, India, 7–8 February 2014; pp. 657–660. [CrossRef]

Article

Convolutional Neural Networks for Segmenting Cerebellar Fissures from Magnetic Resonance Imaging

Robin Cabeza-Ruiz [1,*], Luis Velázquez-Pérez [2,3], Alejandro Linares-Barranco [4,5,6] and Roberto Pérez-Rodríguez [1,2]

1. CAD/CAM Study Centre, University of Holguín, Holguín 80100, Cuba; roberto.perez@uho.edu.cu
2. Cuban Academy of Sciences, Havana 10200, Cuba; velazq63@gmail.com
3. Centre for the Research and Rehabilitation of Hereditary Ataxias, Holguín 80100, Cuba
4. Robotics and Tech. of Computers Lab, University of Seville, 41012 Seville, Spain; alinares@us.es
5. Escuela Politécnica Superior (EPS), University of Seville, 41011 Seville, Spain
6. Smart Computer Systems Research and Engineering Lab (SCORE), Research Institute of Computer Engineering (I3US), University of Seville, 41012 Seville, Spain
* Correspondence: robbinc91@uho.edu.cu

Abstract: The human cerebellum plays an important role in coordination tasks. Diseases such as spinocerebellar ataxias tend to cause severe damage to the cerebellum, leading patients to a progressive loss of motor coordination. The detection of such damages can help specialists to approximate the state of the disease, as well as to perform statistical analysis, in order to propose treatment therapies for the patients. Manual segmentation of such patterns from magnetic resonance imaging is a very difficult and time-consuming task, and is not a viable solution if the number of images to process is relatively large. In recent years, deep learning techniques such as convolutional neural networks (CNNs or convnets) have experienced an increased development, and many researchers have used them to automatically segment medical images. In this research, we propose the use of convolutional neural networks for automatically segmenting the cerebellar fissures from brain magnetic resonance imaging. Three models are presented, based on the same CNN architecture, for obtaining three different binary masks: fissures, cerebellum with fissures, and cerebellum without fissures. The models perform well in terms of precision and efficiency. Evaluation results show that convnets can be trained for such purposes, and could be considered as additional tools in the diagnosis and characterization of neurodegenerative diseases.

Keywords: convolutional neural network; cerebellum segmentation; neurodegenerative disease; cerebellar fissures; magnetic resonance imaging

1. Introduction

The human cerebellum plays an essential role in critical tasks, like motor coordination and cognition, and is related to other functions, e.g., language and emotions [1,2]. Diseases like spinocerebellar ataxias (SCAs), multiple sclerosis (MD), or Alzheimer's disease (AD), are known to cause damage in the cerebellum, conducting patients to progressive loss in such functions and, in some cases, to premature death [3]. Cerebellar damage caused by such diseases occurs in the form of degeneration, reducing the cerebellar volume. The damage can be seen as large fissures, and grows with the progression of the disease. Knowing how to observe such fissures allows specialists to obtain some important characteristics from the patients, like volume loss related to the specific disease.

Segmentation of magnetic resonance imaging (MRI) is often performed, and clinicians make research with several patients, with the goal of learning more about the disease, and how to treat it better. However, manual segmentation of MRIs is a complex and time-consuming task, and becomes impractical as the number of images increases. For that reason, computational tools are required for performing those processes automatically.

Automated cerebellum processing from MRIs has been addressed by several authors, in studies mainly oriented to the delineation and volume calculation of the whole organ and its lobules [1,4–8], deep nuclei segmentation [9], and gray/white matter segmentation [10]. Diedrichsen et al. [7] proposed a probabilistic atlas of the human cerebellum, and performed automatic cerebellum parcellation by combining it with registered images. Weier et al. [8] parcellated cerebellum using patch-based label-fusion and a template library composed of manually labelled images. Romero et al. [5] proposed CERES, which is currently one state-of-the-art pipeline for cerebellar segmentation and parcellation, based on atlas templates and several registration steps for each image to be processed. Manjón and Coupé [10] proposed VolBrain as a tool for subcortical structure segmentation, based on multi-atlas label-fusion. Dolz, Desrosiers and Ben Ayed [11] used a fully convolutional neural network which has been tested in [6] for cerebellar parcellation, obtaining good results. Han et al. [1] proposed the ACAPULCO pipeline, which relies on convolutional neural networks, for performing cerebellar parcellation from MRIs. Kim et al. [9] performed deep cerebellar nuclei segmentation using a fully connected densenet. Thyreau and Taki [12] used convolutional neural networks for brain cortical tissue parcellation from an initial brain mask.

Currently, two of the top-most ranked applications on cerebellar segmentation and parcellation are CERES and ACAPULCO. CERES is based on multi-atlas segmentation, and consists of a pipeline which includes several registration stages, inhomogeneity corrections, and intensity normalizations. It has outperformed all other solutions in the study made by Carass et al. [6]. ACAPULCO is based on convolutional neural networks. The system uses a first CNN to find a bounding box of the cerebellum, and a second, deeper CNN to divide the organ into 28 regions. As reported by Han et al. [1] it surpassed an improved version of CERES in the segmentation of various cerebellar lobules.

In the last decade, convolutional neural networks [13] have experimented a rapid development, as the number of researchers using them for medical image processing grows, in systems where performance is an important factor [14–17]. Specifically, for brain MRI processing, convnets have been successfully applied in segmentation and classification tasks, predicting the stage of Alzheimer's disease [18], cerebellum [4] and brain parcellation [19], and tumor detection and segmentation [20].

Despite the excellence of the existing methods and the reported results, none of this research is oriented to correctly segment and determine all important fissures in cerebellum of patients with neurodegenerative diseases. Figure 1 shows a comparison between segmentations produced by CERES and ACAPULCO for one magnetic resonance from a SCA2 patient with severe cerebellar atrophy. It can be seen that CERES made a better recognition of increased fissures than ACAPULCO, however, some of them have been incorrectly classified as cerebellar tissue. This phenomenon must be related to the training images and labels for both methods, but it should have great impact on the calculation of volumes for the affected parts. As the fissures are classified as cerebellar tissue, the resulting volumes should be larger than the actual ones, giving an incorrect idea of the atrophy produced in the patient's cerebellum. Images were generated with ITK-Snap software [21], CERES segmentation was obtained through the web portal (https://www.volbrain.upv.es/, accessed on 5 December 2021), and ACAPULCO segmentation was obtained by using a docker container shared by the authors in the original paper [1].

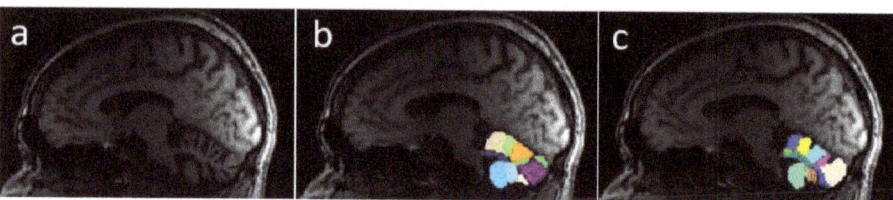

Figure 1. Comparison between segmentations on MRI of SCA2 patient. In (**a**) the original imaging, in (**b**) segmentation produced by ACAPULCO, and in (**c**) segmentation obtained by CERES.

This article proposes the use of convolutional neural networks for segmenting the cerebellum and its fissures. The study comprises analysis over three CNN models, based in the same architecture, for obtaining binary masks of the whole cerebellum without fissures, the cerebellum with its fissures, and the fissures mask itself. Our analysis demonstrates the feasibility of convnets for such tasks. We think that the existence of tools for recognizing the cerebellar fissures from brain MRIs of patients with cerebellar disorders should improve the automated volume estimation currently applied by the aforementioned research, bringing the calculations closer to the real values. Produced segmentations might give an idea of the total volume loss in patients, as well as the stage and progression of the disease itself. As part of the performed analysis, our system is compared with ACAPULCO and CERES, demonstrating an improvement in the segmentation of cerebellar tissue with a correct estimate of the fissures. Additionally, a simple procedure is proposed to help in the construction of similar datasets, relying on an existing mask of the structure to be segmented.

2. Materials and Methods

2.1. Models and Implementation Details

Our three models are built upon the same U-Net architecture. The only differences between models are the labels used for training. Table 1 shows the difference between the three models. The proposed structure is based on U-Net [22], a well-known CNN architecture which takes advantage of feature maps created in previous steps. This characteristic gives the network the ability of processing more complex images while reducing the computational requirements. The system consists of four down- and up-sample steps, composed of inception modules [23] and instance normalization layers, and two chained inception modules as a bottleneck. Each inception module is composed of four convolutional layers, one max pooling operation, and a final concatenation. After each inception module, an instance normalization [1,24] layer processes the produced features. All the activation layers (one per convolution) are Rectified Linear Units (ReLU) [25]. Figure 2 shows the main architecture. The total number of inception modules used was 10, and the number of filters passed to them, in sequence, were 16, 16, 32, 64, and 128 for the contracting path. For the decoding section, the number of parameters were 128, 64, 32, 16, 16. Note that, for each inception module, the output size is four times the input size; e.g., a module with an input of size 128 will return an output with 512 feature maps. The final layer of the architecture consists of a convolutional layer with one single filter, returning the segmented mask from the input.

Table 1. Differences between the three used models.

Model Name	Desired Output	Reference Figure
M1	Binary mask with only cerebellar fissures	Figure 3g
M2	Binary mask of the cerebellum with its fissures	Figure 3h
M3	Binary mask of the cerebellum without fissures	Figure 3f

Implementation was made with Keras [26] and TensorFlow backend [27], using the Python 3.7 programming language, and the training was done on a 16 GB Tesla P100-SXM2 GPU, available through a Jupyter notebook on Google Colab (https://colab.research.google.com/, accessed on 5 December 2021). The used optimizer was Adam [28], with its default values. To avoid overfitting, a dropout of 0.3 was established after the last convolutional layer of each model. Rather than preparing a single model for predicting the three desired features, we trained separated ones for simplicity, making our task a single label segmentation problem. Finally, image cropping was done for reducing computational cost of algorithms. All images were cropped to a volume containing only the cerebellum.

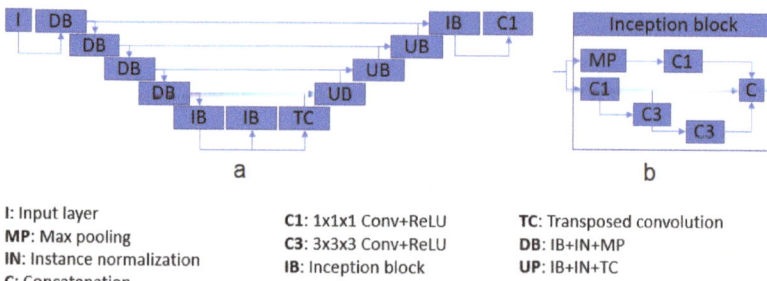

I: Input layer
MP: Max pooling
IN: Instance normalization
C: Concatenation

C1: 1x1x1 Conv+ReLU
C3: 3x3x3 Conv+ReLU
IB: Inception block

TC: Transposed convolution
DB: IB+IN+MP
UP: IB+IN+TC

Figure 2. Architecture diagram (**a**), and inception module pipeline (**b**).

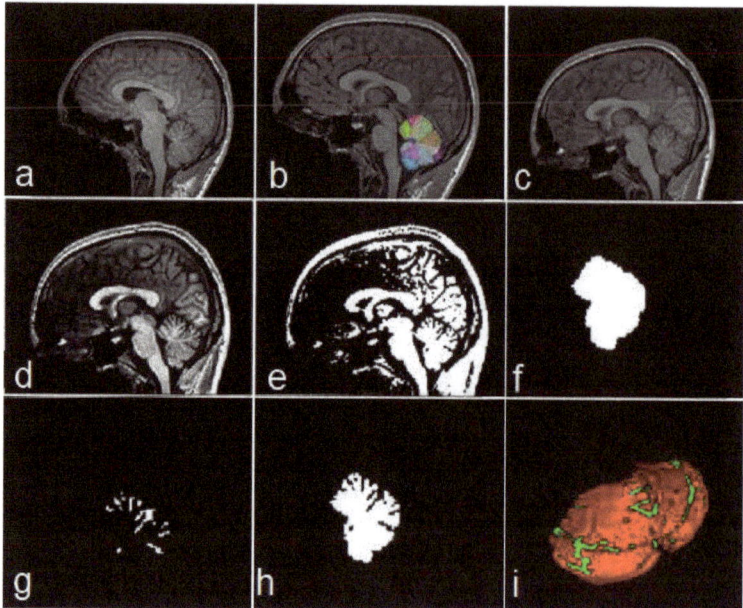

Figure 3. Steps of data construction procedure. Sagittal views of the original image (**a**), cerebellar mask obtained with ACAPULCO (**b**), result of BFC and registration (**c**), contrast-enhanced image (**d**), binary image obtained (**e**), feature map containing the whole cerebellar tissue (**f**), obtained fissures mask (**g**), and cerebellum with fissures (**h**). In (**i**) a 3D view of the union of (**g**,**h**); red color represents the cerebellar tissue, and green color shows the fissures.

2.2. Data Preparation and Dataset Construction

The used dataset consists of 24 magnetic resonances retrieved from the Cuban Neurosciences Center. The images belong to 15 patients, divided into three categories: five healthy controls, five presymptomatic carriers, and five patients diagnosed with spinocerebellar ataxia type 2 (SCA2). Presymptomatic carriers in this research are treated as patients, as it is well known that cerebellar atrophy due to SCA2 may be present long before the disease onset [29–31].

Building a manually labelled dataset from 3D images is a very difficult task. For this reason, we created a simple procedure for the preparation of our dataset. For each MRI, the following steps were applied:

1. Obtain a cerebellar mask, using any existent technique. See Figure 3b.
2. Bias Field Correction (BFC) for reducing intensity inhomogeneities. The algorithm used in this research was the N4 method [32].

3. Image registration to the 1 mm isotropic ICBM 2009c template [33] in MNI space. See Figure 3c.
4. Obtain a contrast-enhanced image (Figure 3d).
5. Binarize equalized image using any existent technique (Figure 3e).
6. Build a mask containing the cerebellar segmentation obtained in step 1 (output 1). See Figure 3f.
7. Build a feature map containing cerebellar fissures, by applying binary xor operation to outputs from steps 5 and 6 (output 2, Figure 3g,i).
8. Build a feature map containing the cerebellar tissue, with all its fissures, by subtracting output 1 from output 2 (output 3, Figure 3h,i).
9. Imaging cropping for reducing computational cost.

For the original cerebellum mask, any available tool can be used, but we highly recommend using ACAPULCO [1] or CERES [5], which are state-of-the-art pipelines for cerebellum parcellation. For this study, ACAPULCO was used, accessed through a docker image shared by the authors in the original paper. The segmented masks have been manually corrected, in order to eliminate any errors than can occur. Manual correction was done with the software ITK-Snap [21].

N4 bias field correction and rigid registration were performed with the ANTS suit [34], available at http://stnava.github.io/ANTs/ (accessed on 5 December 2021).

Enhanced-contrast images were obtained by following a pipeline of intensity normalization (Equation (1)), rescaling to range [1; 255], and histogram equalization. This contrast-enhanced image will serve as the input for the three segmentation models.

$$i = \frac{i - mean(i)}{std(i)} \quad (1)$$

To obtain the binary maps, we computed the Otsu threshold [35], and kept only those voxels with an intensity higher than the calculated threshold. If the original imaging contains a high contrast, some errors may be carried through this procedure, obtaining an incorrect binary map (i.e., several parts of the cerebellar tissue can be removed). For that reason, the binary images must be visually inspected and corrected.

For creating the mask parting from the original cerebellum segmentation, we used the Morphological Snakes algorithm [36]. We applied this step as it improves border smoothness, and may be used to regularize segmentations created/corrected by different raters. The original implementation can be found at https://github.com/pmneila/morphsnakes (accessed on 20 November 2021).

Steps 6, 7 and 8 from the algorithm (outputs 1, 2 and 3), are used as the output maps for the system training. They correspond to the whole cerebellar mask, cerebellar fissures, and cerebellum tissue with its fissures, respectively. The last step is optional, but recommendable if low computational resources are available.

By following the procedure, the construction of an entire dataset may be significantly reduced, since user interaction is limited only to correcting errors, which in some cases are minimal.

2.3. Analysis Description

From the 24 images composing our dataset, 17 were used for training, two for validation, and five for testing purposes. To avoid overfitting, data augmentation was applied to those images on the train/validation partition. The images were augmented using combinations of rotations in the range [−10°, 10°] and shifts on random axes, in the range [−10, 10]. For every training/validation image, 40 new augmented images were created. The three models were trained during 120 epochs, and evaluations were made on the five unseen images.

For testing the robustness of trained models, we tested on subsets of other three datasets:
1. Ten magnetic resonances from the Hammers 2017 dataset (Hammers) [37]. The dataset contains 30 MRIs from healthy subjects, manually segmented by experts into

95 regions [38–40]. From the 95 labels, we used only 17 and 18, corresponding to left and right cerebellum, respectively.
2. Ten magnetic resonances from the Dallas Lifespan Brain Study dataset (DLBS) [41–43]. The dataset contains 315 MRIs of healthy people, some of them are healthy carriers of APOE gene. The initial cerebellar maps for this dataset were obtained by combining the labels from the output of ACAPULCO.
3. Seven magnetic resonances obtained from BrainWeb [44,45]. The site allows the construction of simulated MRIs from healthy people and MS patients, based on templates. The images used in this study were constructed simulating mild, moderate and severe MS lesions (http://www.bic.mni.mcgill.ca/brainweb/, accessed on 5 December 2021).

As a preprocessing stage, steps 2–4 and 9 from the described procedure were applied. Therefore, our preprocess comprises bias field correction, registration to MNI space, contrast enhancement, and image cropping.

For the model predicting cerebellar fissures (M1), no postprocessing technique was applied. The evaluations were carried on the untouched outputs. In the case of the models responsible for segmenting cerebellum with and without fissures (M2 and M3, respectively), a selection of longest connected component was done, classifying only the biggest structure as cerebellar tissue.

Finally, for evaluating the impact of the current research, segmentations of model M2 were compared with the results of ACAPULCO and CERES.

2.4. Evaluation Metrics

Dice Score (DSC, F1-score), overlap coefficient (OC), specificity (SP, True negative rate, TNR), sensitivity (SN, True positive rate, TPR), and area under the ROI curve (AUC), are used as the evaluation metrics for the three models. DSC allows comparison of two volumes of the same dimensions through Equation (2) [46]:

$$DSC = \frac{2 \times \sum_i^N p_i g_i}{\sum_i^N p_i^2 + \sum_i^N g_i^2} \quad (2)$$

where N represents the total number of voxels in one image, p belongs to the prediction volume, and g belongs to the ground truth volume. SP allows to quantify the proportion of those voxels that do not belong to the ground truth mask, and can be obtained with Equation (3) [47]:

$$SP = \frac{TN}{FP + TN} \quad (3)$$

where TP and TN are the number of voxels which have been correctly recognized as part of the mask and part of the background, respectively, and FP, FN correspond to those incorrectly identified as mask and background, respectively. SN allows to quantify the proportion of voxels that belong to the ground truth mask, and can be obtained as in Equation (4) [48]:

$$SN = \frac{TP}{TP + FN} \quad (4)$$

OC allows to calculate how close a finite set is from the other, in terms of overlapping [49]. A perfect overlap would have a value of 1, and two images without any overlapping should obtain 0 score. It can be calculated with Equation (5).

$$OC = \frac{\sum_i^N p_i g_i}{\min\left(\sum_i^N p_i, \sum_i^N g_i\right)} \quad (5)$$

AUC is used as a measurement of a classifier's performance, being more complete than the usual overall accuracy [48,50], and can be obtained with Equation (6).

$$AUC = 1 - \frac{1}{2}\left(\frac{FP}{FP + TN} + \frac{FN}{FN + TP}\right) \quad (6)$$

The measures were selected based on the guidelines proposed by Taha and Hanbury [48], attending to the following properties and requirements on 3D medical image segmentation: outliers exist (some outsider voxels might be incorrectly classified as ground truth), complex boundary (cerebellar fissures present very complex shapes and boundaries), and contour is important.

3. Results

This section exposes the result of evaluations performed to the three models. Table 2 allows to analyze the mean scores for the three models in the whole test set. It can be seen that the worst results were obtained by model M1. Models M2 and M3 achieved very high scores in evaluations.

Table 2. Mean scores for models M1, M2 and M3 in the whole set of test images.

	M1	M2	M3
DSC	0.761	0.965	0.959
OC	0.826	0.982	0.978
SP	0.997	0.992	0.991
SN	0.749	0.977	0.969
AUC	0.871	0.985	0.980

For an easy understanding and analysis, we decided to divide into six subsections. The first four subsections correspond to results on each dataset used, the fifth presents our time analysis, and the last subsection corresponds to the comparison with segmentations produced by ACAPULCO and CERES.

3.1. Results for Our Dataset

The three models (see Table 1) were tested on five unseen magnetic resonance images. The test subset contained one healthy control (subject 1), two presymptomatic carriers (subjects 2 and 4), and two SCA2 patients (subjects 3 and 5). Figure 4 shows a comparison between the original masks and the segmentations produced by M1, M2 and M3. It can be appreciated the similarity between original and segmented images. Some errors remain, mainly in the contour of segmented masks; those errors will be covered in next investigations. Table 3 shows the result of the evaluations on model M1, segmenting cerebellar fissures only.

Table 3. Evaluation results for model M1 in our dataset.

	Subject 1	Subject 2	Subject 3	Subject 4	Subject 5
DSC	0.803	0.895	0.864	0.834	0.875
OC	0.882	0.914	0.924	0.895	0.876
SP	0.998	0.998	0.997	0.999	0.996
SN	0.737	0.877	0.924	0.780	0.875
AUC	0.868	0.938	0.960	0.889	0.935

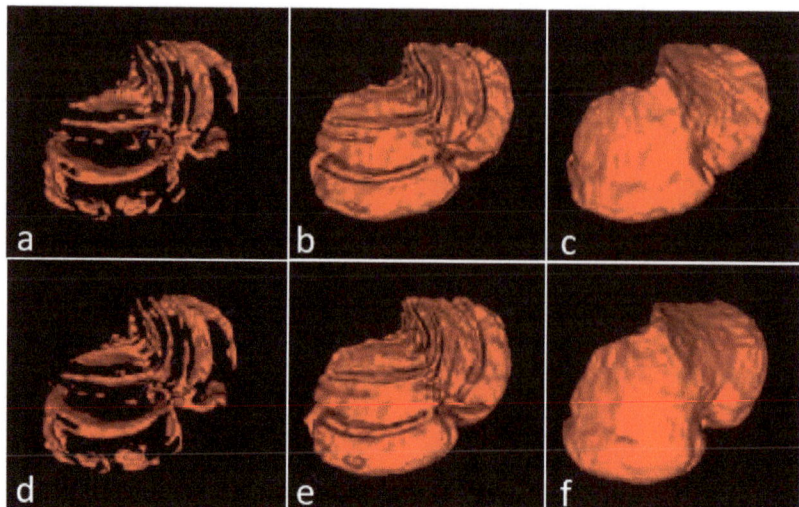

Figure 4. Masks and predictions for an MRI belonging to one of the SCA2 patients in the test subset. The top row shows the original masks, obtained with the procedure described in Section 2.2, and the bottom row displays the segmentations produced by our models. Cerebellar fissures in (**a**,**d**), cerebellum tissue with fissures in (**b**,**e**), and whole cerebellum without any fissure in (**c**,**f**).

Produced segmentations have relatively good scores. Mean DSC and OC are 0.854 and 0.898, respectively. All SP are above 0.99, which means an optimal recognition of background voxels. Low SN values represent some errors in the voxels belonging to cerebellar fissures, mainly in the MRI belonging to the healthy control (0.73, the minimum SN value). It seems that the best behavior was obtained for subject 3, one of the SCA2 patients in our dataset. Note that segmenting cerebellar fissures is a difficult task and, as such, characteristics change greatly between different people. Furthermore, no postprocessing was applied to the results of model M1. Figure 4d shows an example of the outputs produced by our model, compared against the ground truth mask in Figure 4a.

Table 4 shows the evaluation results for model M2 (segmentation of cerebellum tissue with its fissures). As observed, results for this model were much better than the previous one. This is a logical result, considering that segmenting a single, larger structure, which is always located in the same place on MRI, should be easier than segmenting smaller regions with many position changes. The best scores were achieved for the subject 4 MRI, producing better segmentations. The mean values for DSC and OC are 0.973 and 0.987, respectively. SP, SN and AUC are all above 0.98, which means a good background and foreground voxel classification. Figure 4e displays an example output from this model.

Table 4. Evaluation results for cerebellar tissue with fissures (model M2) in our dataset.

	Subject 1	Subject 2	Subject 3	Subject 4	Subject 5
DSC	0.976	0.977	0.970	0.981	0.965
OC	0.991	0.984	0.992	0.992	0.977
SP	0.993	0.995	0.995	0.995	0.994
SN	0.991	0.984	0.992	0.992	0.977
AUC	0.992	0.989	0.993	0.994	0.986

Table 5 shows the results for the model segmenting the whole cerebellum (M3). As in Table 4, all scores are above 0.95, which gives the idea of a high precision in the segmentation results. Mean DSC and OC are 0.969 and 0.982, respectively. As in evaluation for model M2, SP, SN and AUC are above 0.98, which means a high-quality segmentation. In a general way, the segmentations obtained by models M1, M2 and M3 have a good quality. Models M2 and M3 obtained better scores than M1.

Table 5. Evaluation results for whole cerebellum segmentation without fissures (model M3).

	Subject 1	Subject 2	Subject 3	Subject 4	Subject 5
DSC	0.976	0.975	0.954	0.980	0.963
OC	0.984	0.980	0.987	0.986	0.976
SP	0.994	0.994	0.991	0.995	0.991
SN	0.984	0.980	0.987	0.986	0.976
AUC	0.989	0.987	0.989	0.991	0.984

3.2. Results on Hammers Dataset

The three models were evaluated using a subset of the Hammers 2017 dataset. For this evaluation, we used the first 10 images. The images in the dataset are named from a01 to a30; we used images from a01 to a10. The images were processed with the same procedure described in Section 2.2, but manual correction of generated binary maps was not performed, as we wanted to check the possibility of automatically creating a new dataset. As a cerebellar map for the initial step, the original segmentations were conveniently corrected. Therefore, the rest of the dataset preparation was done in a fully automatic manner.

Evaluation results for model M1 on this dataset can be observed in Table A1. This time the segmentations produced were less precise. The mean DSC obtained was 0.755, while the mean overlap coefficient was 0.826. We believe that this result presents a direct relation with the fact that binary maps for each MRI were not manually corrected. A revision of those features should improve the segmentation, and it will be covered in future investigations. As in evaluation with images from our dataset, high SP and low SN and AUC were obtained, meaning that the model had some trouble identifying the tissue belonging to cerebellar fissures.

Results for model M2 are presented in Table A2. It may be observed that the scores obtained are competitive with those obtained in our dataset, as mean DSC and OC are 0.951 and 0.983, respectively. The scores in the segmentations were quite high and close to each other. Minimum DSC and OC are 0.945 and 0.975, respectively, which indicates very realistic segmentations as in previous evaluation of model M2. SP, SN and AUC are above 0.98, which demonstrates a high-quality segmentation on cerebellar tissue with fissures.

Finally, Table A3 shows the evaluations for model M3. As in Table A2, the results are very promising, giving mean DSC and OC with values of 0.947 and 0.976 respectively. The rest of calculated scores, all above 0.98, also give the notion of very good segmentations.

As in the previous case, the worst results were achieved for the model M1, in the segmentation of cerebellar fissures.

3.3. Results on DLBS Dataset

As a third set of MRIs for evaluating the methods, 10 images from the Dallas Lifespan Brain Study were used. For our purposes, we selected 10 MRIs belonging to older APOE-ε4 gene carriers.

Results of the evaluation on segmentations produced by model M1 can be observed in Table A4. As in the previous discussion on cerebellar fissure segmentation (Section 3.2), the DSCs are between 0.71 and 0.76. Mean DSC and OC are 0.745 and 0.799, respectively. The rest of the scores remain similar to analysis performed in our dataset and Hammers: low SN, which means errors in the precise classification of the fissures.

Evaluations for model M2 are presented in Table A5, and some improvement can be seen with respect to evaluations on Hammers dataset. Mean values of DSC and OC are 0.967 and 0.975, respectively, for a very good segmentation of cerebellum with its fissures. As expected, values of SN, SP and AUC are above 0.96.

Scores for model M3 are shown in Table A6. Again, the scores are quite good, with mean DSC and OC of 0.963 and 0.975, respectively.

3.4. Results on Dataset from BrainWeb

As commented in Section 2.3, seven MRIs were generated through the BrainWeb web portal, simulating multiple sclerosis. The images were created with variable parameters such as rotation, noise level, and MS severity.

Table A7 shows the scores for model M1, presenting the same situation as previous evaluations. Mean DSC and OC obtained were 0.728 and 0.81, respectively, and the sensitivity was severely affected. Table A8 shows the evaluations for model M2, with another surprising result. Achieved scores are all above 0.97, and the mean DSC and OC were 0.973 and 0.988, respectively. The same occurs with the scores for model M3 (Table A9), with 0.964 and 0.982 as mean DSC and OC, respectively. Despite the high scores achieved in this dataset, we believe that further analysis should be performed, as all images are created from two original templates: one for severe MS, and one for mild and moderate MS.

3.5. Time Analysis

An analysis was performed to evaluate the time our architecture takes to segment new images. All experiments were carried out on a Lenovo computer, equipped with an Intel Core i3-8145U processor, and 8 GB RAM. Table 6 shows the mean times for models M1, M2 and M3, as well as preprocessing and load times.

Table 6. Mean times for loading, preprocessing, and segmentation processes. From left to right column are presented: dataset names, load times, preprocessing times, and segmentation time for M1, M2 and M3. The time is expressed in seconds (s).

	Load	Preprocessing	Segmentation		
			M1	M2	M3
Ours	0.06	227.57	53.40	51.44	53.00
Hammers	0.06	263.77	49.89	50.24	50.42
DLBS	0.07	206.73	55.85	60.26	56.90
Brainweb	0.04	180.06	55.08	54.75	61.34

The load times for each dataset are small, ranging from 0.04 to 0.07 s. Preprocessing times ranged from 177.95 to 265.75 s. This is the most time-consumer phase in our pipeline, as it involves bias field correction, image registration, normalization, histogram equalization, and cropping.

For model M1, the best segmentation times were obtained over Hammers subset, with a mean processing time of 49.89 s per image. The global mean time of this model was 53.43 s. Segmentation times for M2 were slightly higher, averaging 54.55 s. The best results were also obtained for Hammers subset, with a mean time of 50.21 s. Finally, results of time analysis for model M3 were better on Hammers subset, with a mean of 50.42 s. The mean time for all the images was 55.23 s.

In a general manner, the total time needed for processing an MRI is the sum of loading, preprocessing and segmentation tasks. Since our three models work with the same cropped portion of the preprocessed MRI, the load and preprocessing operations are performed only once on each image. The total time for every image is then the sum of loading, preprocessing, and segmentations for M1, M2 and M3. The total mean time of processing for our models was 385.26 s (about six minutes for each image). Considering that manual segmentation can take several hours for each MRI, we believe that it is a remarkable advance in such task. However, finding a faster BFC/registration technique should greatly improve this result, as preprocessing is the most time-consuming phase of our process.

3.6. Comparison with Other Methods

For stablishing an improvement on cerebellar tissue segmentation with special attention to fissures, comparisons were made with ACAPULCO and CERES. We compared the results of our model M2 with the segmentations produced by these two tools. Segmentations from ACAPULCO were obtained using the docker image that the authors made available in the original paper [1], and segmentations from CERES were obtained through a web portal available to the public, also shared by the authors on their paper [6].

As these are tools for cerebellar parcellation, a binary mask of the whole cerebellum was obtained for each segmentation, constructed by combining all the labels in the segmented images. The evaluations were performed on the five test magnetic resonances of our cohort, and the 10 images from the DLBS dataset. The measures used for the comparison were dice score (DSC), overlap coefficient (OC), and specificity (SP). Table 7 shows the comparison of DSC in our images.

Table 7. DSC comparison between our M2 model, ACAPULCO and CERES.

	S.1	S.2	S.3	S.4	S.5
M2	0.976	0.977	0.970	0.981	0.965
ACAPULCO	0.910	0.905	0.894	0.909	0.900
CERES	0.935	0.927	0.911	0.924	0.926

As can be seen, our model M2 achieved higher DSC than both methods. Mean DSC were 0.973, 0.903 and 0.924 for M2, Acapulco and CERES, respectively. CERES performed better than ACAPULCO in the segmentation, but in general both methods only identify the largest fissures, and a substantial part of the small fissures is misclassified. We think that this event is related with the segmentations used in both methods as a training/knowledge base, since both methods were used without any modification. The best behavior for both methods was on segmenting the first resonance image, corresponding to a healthy control.

Figure 5 presents an example of segmentation produced by the three models for a subject in our dataset. As the figure shows, ACAPULCO (Figure 5d) only detected parts of the biggest fissures, while the smaller ones are classified as cerebellar tissue. CERES (Figure 5c) recognized fissures better than ACAPULCO, but some of them are also misclassified. Furthermore, some irregularities are present in the front of the cerebellum, leaving some holes in the mask produced by CERES. Segmentations obtained by model M2 (Figure 5b) are very close to the real ones, correctly recognizing most of the fissures.

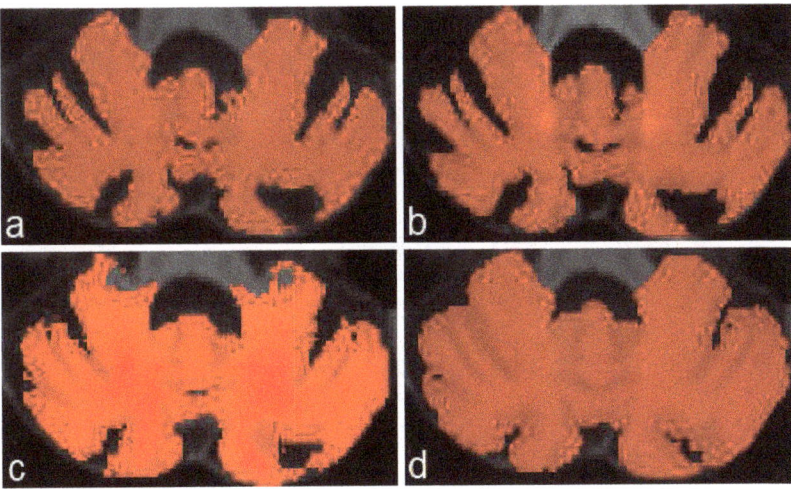

Figure 5. Example of segmentations produced by the approaches for a sample image from our dataset. Original mask (**a**), followed by segmentation produced by M2 (**b**), CERES (**c**) and ACAPULCO (**d**).

Table 8 shows a comparison for the OC scores achieved by the three methods. Mean scores for M2, ACAPULCO and CERES were 0.987, 0.994 and 0.988, respectively. Results are very close between approaches, but in general terms, ACAPULCO achieved higher

OC scores. This is a logical conclusion, as ACAPULCO tends to misclassify fissures. As a result, the original masks are almost entirely contained in segmentations produced by ACAPULCO. The same happens with segmentations produced by CERES.

Table 8. OC comparison between our M2 model, ACAPULCO and CERES.

	S.1	S.2	S.3	S.4	S.5
M2	0.991	0.984	0.992	0.992	0.977
ACAPULCO	0.999	0.990	0.990	0.996	0.999
CERES	0.991	0.989	0.981	0.991	0.988

In Table 9 are included the results of the SP analysis for the three models. It can be appreciated that M2 model achieved the higher scores, followed by CERES, and finally ACAPULCO. The mean values are 0.994, 0.971 and 0.964, respectively.

Table 9. SP comparison between our M2 model, ACAPULCO and CERES.

	S.1	S.2	S.3	S.4	S.5
M2	0.993	0.995	0.995	0.995	0.994
ACAPULCO	0.965	0.967	0.966	0.971	0.953
CERES	0.973	0.973	0.980	0.977	0.955

Tables 7–9 clearly indicate that model M2 produced better segmentations than ACAPULCO and CERES. Higher DSC and SP combined with lower OC, means that our approach correctly identifies the most of fissures on the cerebellum.

Table 10 shows the DSC comparison for the DLBS subset. The three approaches obtained close dice scores, with a mean value of 0.967, 0.931 and 0.945, respectively. The 10 images for this comparison belong to healthy controls, which means less fissures, so the scores for ACAPULCO and CERES were increased.

Table 10. DSC comparison between the approaches, in DLBS subset.

Subject No.	M2	ACAPULCO	CERES
1	0.969	0.926	0.945
2	0.965	0.939	0.950
3	0.960	0.938	0.941
4	0.966	0.936	0.950
5	0.978	0.922	0.953
6	0.962	0.930	0.946
7	0.971	0.937	0.954
8	0.960	0.911	0.921
9	0.967	0.935	0.944
10	0.974	0.936	0.954

Figure 6 shows a case of the segmentations produced for this second dataset. As in the previous example, the best segmentations were produced by model M2 (Figure 6b). There are some irregularities on borders, which we think can be corrected by applying some postprocessing technique (rather than longest connected component, which is the only postprocessing we currently apply on segmentations). In this example, ACAPULCO was capable of segmenting some fissures better than CERES (Figure 6c,d).

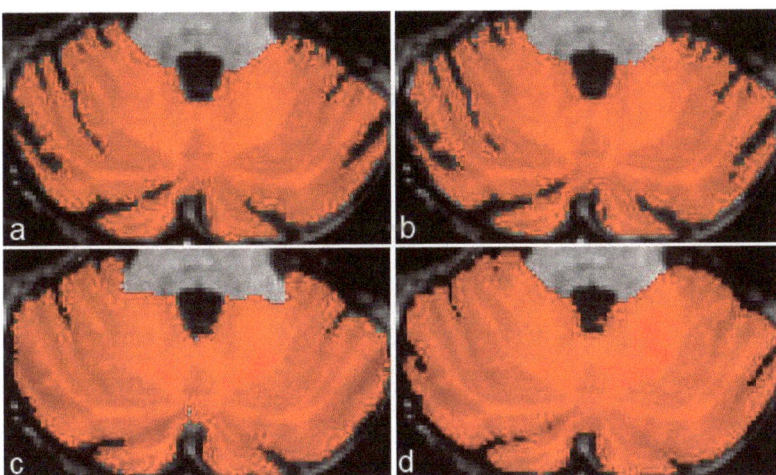

Figure 6. Segmentations produced by the three approaches for a sample image from the DLBS dataset. Original mask (**a**), followed by segmentation produced by M2 (**b**), CERES (**c**) and ACAPULCO (**d**).

Table 11 shows a comparison for the OC scores achieved in the DLBS dataset. Higher values were obtained by ACAPULCO, followed by CERES, and finally M2. The mean values were 0.998, 0.980 and 0.975, respectively. This represents the same phenomena as Table 8: segmentations produced by ACAPULCO and CERES include the original masks because of the problems when recognizing cerebellar fissures, resulting in elevated OC.

Table 11. OC comparison between the approaches, in DLBS subset.

Subject No.	M2	ACAPULCO	CERES
1	0.970	0.996	0.983
2	0.974	0.999	0.975
3	0.971	0.993	0.982
4	0.969	0.998	0.984
5	0.987	0.995	0.982
6	0.981	0.994	0.975
7	0.977	0.998	0.981
8	0.964	0.999	0.980
9	0.975	0.998	0.985
10	0.982	0.995	0.987

In Table 12 are included the SP scores achieved in the DLBS dataset. As in Table 9, model M2 presented the best behavior, which means that the classification of background voxels was better. Mean scores were 0.993, 0.970 and 0.977, respectively.

Results for this dataset were similar to those obtained in our five test MRIs. The model M2 presented higher DSC and SP, and lower OC than ACAPULCO and CERES. This means that M2 identifies cerebellar fissures better than the other approaches.

Table 12. SP comparison between the approaches, in DLBS subset.

Subject No.	M2	ACAPULCO	CERES
1	0.995	0.969	0.978
2	0.992	0.970	0.979
3	0.991	0.972	0.971
4	0.994	0.973	0.975
5	0.995	0.972	0.976
6	0.992	0.973	0.982
7	0.996	0.969	0.983
8	0.993	0.973	0.982
9	0.993	0.970	0.980
10	0.995	0.961	0.968

4. Discussion

Three models have been proposed for segmentation tasks on human cerebellum from magnetic resonance imaging: the first model (M1) segments cerebellar fissures, the second (M2) segments the cerebellum with the most of its fissures, and the third (M3) obtains the whole cerebellum without fissures. The three models were tested on a total of 32 MRIs, composed of 21 healthy controls, four SCA2 patients, and seven MRIs with multiple sclerosis.

In the case of cerebellar fissure segmentation (model M1), the best DSC obtained was 0.895 in our dataset, and the worst case presented a score of 0.707 in the Hammers dataset. We observed that the best results were achieved on the MRIs of SCA2 patients with severe atrophy, indicating that the model might not be capable of correctly find the fissures in healthy people. More tests need to be done to verify if the proposed U-Net architecture can be modified in any way, or more augmentation techniques/training epochs are necessary for improving segmentation results. A postprocessing stage could be added too, increasing the possibility of producing better segmentations. Despite the low results (minimum DSC = 0.707), we have not seen other investigations dedicated to specifically segmenting and quantifying the cerebellar fissures, and we consider this to be a good starting point for future researches on this kind of study.

The model for segmentation of the cerebellum with its fissures (M2) presented very precise results, with DSC ranging from 0.946 to 0.981 among the four subsets used for testing. This result implies that volumetric calculations might be performed in the human cerebellum, with a higher grade of precision. We think that the model could be integrated in some greater pipeline for characterizing neurodegenerative diseases. The model performed well on MRIs of healthy people and patients, making it suitable for the task.

The model for segmenting the whole cerebellum (M3) also obtained very good results, with dice scores ranging from 0.946 to 0.980, demonstrating very precise segmentations in the 32 test images. Obtained scores highly reduce the chance of overfitting during training process, and allow the affirmation that models have sufficient generalization for working with images from different origins.

Segmentations produced by the models M2 and M3 could be used to improve current cerebellar segmentation/parcellation methods, obtaining more accurate volumetric estimations on patients with cerebellar degeneration caused by SCAs or other neurodegenerative diseases. Furthermore, the procedure proposed in Section 2.2 for the creation of our dataset can be adapted to any research with the same interests, always providing the correct mask at the beginning.

The three models present good performance in terms of efficiency, as total time needed when processing a new image is about six minutes (less than three minutes if the loading and preprocessing stages are not considered).

The model M2 was compared with two state-of-the-art approaches, obtaining better scores in all cases. The comparison was only made with 15 resonance images, and deeper comparisons will be performed in future researches.

Based on analysis results, we may conclude that convolutional neural networks can be applied on segmenting complicated features from brain magnetic resonances. Not only well-defined organs such as cerebellum, but also fissures can be obtained, always providing the correct dataset and adequate training. Our model trained for cerebellar fissures did not obtain such high scores as expected, but we think that fissures can be obtained by combining outputs of models M2 and M3.

The outcomes of this study should provide a comprehensive set of tools to specialists in neurodegenerative diseases. Digital tools can be generated and incorporated into existing visualization applications, increasing the speed and precision in diagnosis and characterization.

For more in-depth evaluation of the proposed method, larger datasets must be tried, as well as other CNN architectures, with different grades of complexity, and a higher number of features. In future research we aim to integrate the models described here into more complex architectures and pipelines, such as cerebellum parcellation.

5. Conclusions

This article has evaluated the possibility of applying convolutional neural networks for automatically segmenting the cerebellum and its fissures from brain magnetic resonance imaging. Three models, built upon the same U-Net based architecture, have been proposed for segmenting cerebellar fissures, cerebellum with all fissures, and cerebellum without any fissures. Analysis has been performed on 32 MRIs, including healthy controls, presymptomatic carriers, SCA2 patients, and multiple sclerosis patients. The best dice scores achieved were 0.895, 0.981 and 0.98 on each task, respectively. The proposed architecture is highly efficient, since segmentations can be carried on in less than a minute after preprocessing. Analysis results indicate that convnets are capable of segmenting the human cerebellum with high precision. The model prepared for segmenting the cerebellum with its fissures was compared with two existent methods, achieving better results than both in all tests. The images resulting from the segmentations could be incorporated into higher pipelines, dedicated to diagnosing or characterizing any disease that affects the cerebellum, and could help to improve the estimation of volume loss and general damage to the cerebellum. Furthermore, a simple method has been proposed for facilitating the construction of similar datasets. The use of the procedure should help to quickly construct datasets, saving time and efforts.

Author Contributions: Conceptualization, L.V.-P., R.P.-R. and R.C.-R.; methodology, R.C.-R.; software, R.C.-R.; validation, R.C.-R., L.V.-P., R.P.-R. and A.L.-B.; formal analysis, L.V.-P. and R.P.-R.; investigation, R.C.-R.; resources, L.V.-P.; data curation, R.C.-R.; writing—original draft preparation, R.C.-R.; writing—review and editing, L.V.-P., R.P.-R. and A.L.-B.; visualization, R.C.-R.; supervision, L.V.-P., R.P.-R. and A.L.-B. All authors have read and agreed to the published version of the manuscript.

Funding: This research received no external funding.

Institutional Review Board Statement: Not applicable.

Informed Consent Statement: Not applicable.

Data Availability Statement: At the time of writing this paper, the original images are being uploaded to https://github.com/robbinc91/cerebellar_fissures_segmentation_cnn (accessed on 8 December 2021). The computer codes for creating the dataset are also being shared, allowing other researchers to replicate our study. The rest of datasets used in this research are publicly available on the internet. The hammers 2017 dataset can be accessed from http://brain-development.org/ (accessed on 5 December 2021), simulated multiple sclerosis images can be accessed from http://www.bic.mni.mcgill.ca/brainweb/ (accessed on 5 December 2021), and DLBS dataset may be obtained from https://fcon_1000.projects.nitrc.org/indi/retro/dlbs.html (accessed on 5 December 2021).

Conflicts of Interest: The authors declare no conflict of interest.

Appendix A

This appendix contains the tables for the evaluation results on the 10 first images from Hammers, DLBS and BrainWeb datasets.

Table A1. Evaluation results for cerebellar fissures (model M1) on Hammers subset.

Subject No.	DSC	OC	SP	SN	AUC
1	0.740	0.908	0.999	0.624	0.812
2	0.752	0.864	0.998	0.665	0.832
3	0.816	0.828	0.998	0.805	0.901
4	0.789	0.828	0.998	0.753	0.875
5	0.726	0.729	0.996	0.722	0.859
6	0.838	0.855	0.997	0.855	0.926
7	0.724	0.773	0.998	0.773	0.886
8	0.735	0.799	0.996	0.799	0.897
9	0.732	0.847	0.998	0.645	0.821
10	0.707	0.833	0.998	0.614	0.806

Table A2. Evaluation results for cerebellar tissue with fissures (model M2) on Hammers subset.

Subject No.	DSC	OC	SP	SN	AUC
1	0.945	0.986	0.988	0.986	0.987
2	0.966	0.976	0.993	0.976	0.985
3	0.945	0.990	0.986	0.990	0.988
4	0.954	0.982	0.988	0.982	0.985
5	0.952	0.981	0.987	0.981	0.984
6	0.947	0.989	0.984	0.989	0.987
7	0.949	0.992	0.986	0.992	0.989
8	0.953	0.975	0.986	0.975	0.980
9	0.951	0.988	0.988	0.988	0.988
10	0.950	0.976	0.987	0.976	0.982

Table A3. Evaluation results for whole cerebellum segmentation, without fissures (model M3) on Hammers subset.

Subject No.	DSC	OC	SP	SN	AUC
1	0.950	0.973	0.990	0.973	0.981
2	0.955	0.958	0.992	0.958	0.975
3	0.938	0.985	0.983	0.985	0.984
4	0.949	0.978	0.986	0.978	0.982
5	0.944	0.972	0.984	0.972	0.978
6	0.945	0.982	0.983	0.982	0.982
7	0.943	0.992	0.984	0.992	0.988
8	0.953	0.977	0.985	0.977	0.981
9	0.950	0.979	0.988	0.979	0.983
10	0.946	0.965	0.987	0.965	0.976

Table A4. Evaluation results for cerebellar fissures (model M1) on DLBS dataset.

Subject No.	DSC	OC	SP	SN	AUC
1	0.776	0.813	0.997	0.813	0.905
2	0.738	0.833	0.997	0.833	0.915
3	0.730	0.858	0.997	0.858	0.927
4	0.723	0.724	0.997	0.723	0.860
5	0.720	0.794	0.998	0.658	0.828
6	0.769	0.776	0.998	0.763	0.880
7	0.738	0.752	0.997	0.724	0.861
8	0.723	0.875	0.998	0.616	0.807
9	0.719	0.754	0.997	0.664	0.875
10	0.769	0.820	0.999	0.725	0.862

Table A5. Evaluation results for cerebellar tissue with fissures (model M2) on DLBS dataset.

Subject No.	DSC	OC	SP	SN	AUC
1	0.969	0.970	0.995	0.970	0.983
2	0.965	0.974	0.992	0.974	0.983
3	0.960	0.971	0.991	0.971	0.981
4	0.966	0.969	0.994	0.969	0.982
5	0.978	0.987	0.995	0.987	0.991
6	0.962	0.981	0.992	0.981	0.987
7	0.971	0.977	0.996	0.966	0.981
8	0.960	0.964	0.993	0.964	0.978
9	0.967	0.975	0.993	0.975	0.984
10	0.974	0.982	0.995	0.982	0.989

Table A6. Evaluation results for whole cerebellum segmentation, without fissures (model M3) on DLBS dataset.

Subject No.	DSC	OC	SP	SN	AUC
1	0.968	0.978	0.993	0.978	0.985
2	0.965	0.980	0.991	0.980	0.985
3	0.964	0.987	0.990	0.987	0.988
4	0.963	0.989	0.993	0.969	0.981
5	0.963	0.963	0.994	0.963	0.979
6	0.959	0.978	0.990	0.978	0.984
7	0.969	0.969	0.994	0.969	0.981
8	0.948	0.961	0.993	0.935	0.964
9	0.964	0.973	0.991	0.973	0.982
10	0.968	0.972	0.995	0.972	0.983

Table A7. Evaluation results for cerebellar fissures (model M1) on BrainWeb dataset.

Subject No.	DSC	OC	SP	SN	AUC
1	0.722	0.729	0.993	0.714	0.854
2	0.711	0.932	0.993	0.691	0.842
3	0.715	0.724	0.993	0.706	0.849
4	0.722	0.938	0.993	0.707	0.850
5	0.739	0.781	0.994	0.701	0.847
6	0.754	0.801	0.994	0.713	0.853
7	0.738	0.767	0.993	0.912	0.852

Table A8. Evaluation results for cerebellar tissue with fissures (model M2) on BrainWeb dataset.

Subject No.	DSC	OC	SP	SN	AUC
1	0.973	0.986	0.996	0.961	0.979
2	0.972	0.981	0.995	0.964	0.979
3	0.973	0.985	0.996	0.962	0.979
4	0.973	0.984	0.996	0.962	0.979
5	0.974	0.980	0.995	0.969	0.982
6	0.974	0.989	0.995	0.970	0.982
7	0.974	0.982	0.995	0.966	0.981

Table A9. Evaluation results for whole cerebellum segmentation, without fissures (model M3) on BrainWeb dataset.

Subject No.	DSC	OC	SP	SN	AUC
1	0.965	0.983	0.995	0.949	0.972
2	0.964	0.982	0.995	0.946	0.970
3	0.965	0.982	0.995	0.948	0.971
4	0.965	0.983	0.995	0.948	0.971
5	0.963	0.982	0.995	0.944	0.969
6	0.963	0.983	0.995	0.944	0.970
7	0.963	0.981	0.995	0.945	0.970

References

1. Han, S.; Carass, A.; He, Y.; Prince, J.L. Automatic Cerebellum Anatomical Parcellation using U-Net with Locally Constrained Optimization. *IEEE Trans. Med. Imaging* **2020**, *218*, 116819. [CrossRef] [PubMed]
2. Kansal, K.; Yang, Z.; Fishman, A.M.; Sair, H.I.; Ying, S.H.; Jedynak, B.M.; Prince, J.L.; Onyike, C.U. Structural cerebellar correlates of cognitive and motor dysfunctions in cerebellar degeneration. *Brain* **2017**, *140*, 707–720. [CrossRef] [PubMed]
3. Klockgether, T.; Mariotti, C.; Paulson, H.L. Spinocerebellar ataxia. *Nat. Rev. Dis. Primers* **2019**, *5*, 24. [CrossRef]
4. Han, S.; He, Y.; Carass, A.; Ying, S.H.; Prince, J.L. Cerebellum Parcellation with Convolutional Neural Networks. *Proc. SPIE Int. Soc. Opt. Eng.* **2019**, *10949*, 109490K. [CrossRef]
5. Romero, J.; Coupé, P.; Giraud, R.; Ta, V.; Fonov, V.; Park, M.T.; Chakravarty, M.; Voineskos, A.; Manjón, J. CERES: A new cerebellum lobule segmentation method. *Neuroimage* **2016**, *147*, 916–924. [CrossRef] [PubMed]
6. Carass, A.; Cuzzocreo, J.L.; Han, S.; Hernandez-castillo, C.R.; Rasser, P.E.; Ganz, M.; Beliveau, V.; Dolz, J.; Ayed, I.B.; Desrosiers, C.; et al. Comparing fully automated state-of-the-art cerebellum parcellation from magnetic resonance images. *Neuroimage* **2018**, *183*, 150–172. [CrossRef]
7. Diedrichsen, J.; Balsters, J.H.; Flavell, J.; Cussans, E.; Ramnani, N. A probabilistic MR atlas of the human cerebellum. *Neuroimage* **2009**, *46*, 39–46. [CrossRef] [PubMed]
8. Weier, K.; Fonov, V.; Lavoie, K.; Doyon, J.; Collins, D.L. Rapid Automatic Segmentation of the Human Cerebellum and its Lobules (RASCAL)—Implementation and Application of the Patch-based Label-fusion Technique with a Template Library to Segment the Human Cerebellum. *Hum. Brain Mapp.* **2014**, *35*, 5026–5039. [CrossRef] [PubMed]
9. Kim, J.; Patriat, R.; Kaplan, J.; Solomon, O.; Harel, N. Deep Cerebellar Nuclei Segmentation via Semi-Supervised Deep Context-Aware Learning from 7T Diffusion MRI. *IEEE Access* **2020**, *8*, 101550–101568. [CrossRef]
10. Manjón, J.V.; Coupé, P. volBrain: An Online MRI Brain Volumetry System. *Front. Neuroinform.* **2016**, *10*, 1–14. [CrossRef]
11. Dolz, J.; Desrosiers, C.; Ben Ayed, I. 3D fully convolutional networks for subcortical segmentation in MRI: A large-scale study. *Neuroimage* **2018**, *170*, 456–470. [CrossRef]
12. Thyreau, B.; Taki, Y. Learning a cortical parcellation of the brain robust to the MRI segmentation with convolutional neural networks. *Med. Image Anal.* **2020**, *61*, 101639. [CrossRef] [PubMed]
13. Zeiler, M.D.; Fergus, R. Visualizing and Understanding Convolutional Networks. *Anal. Chem. Res.* **2014**, *12*, 818–833. [CrossRef]
14. Duran-Lopez, L.; Dominguez-Morales, J.P.; Conde-Martin, A.F.; Vicente-Diaz, S.; Linares-Barranco, A. PROMETEO: A CNN-Based Computer-Aided Diagnosis System for WSI Prostate Cancer Detection. *IEEE Access* **2020**, *8*, 128613–128628. [CrossRef]
15. Duran-Lopez, L.; Dominguez-Morales, J.P.; Corral-jaime, J.; Vicente-Diaz, S.; Linares-Barranco, A. COVID-XNet: A Custom Deep Learning System to Diagnose and Locate COVID-19 in Chest X-ray Images. *Appl. Sci.* **2020**, *10*, 5683. [CrossRef]
16. Duran-Lopez, L.; Dominguez-Morales, J.P.; Rios-Navarro, A.; Gutierrez-Galan, D.; Jimenez-Fernandez, A.; Vicente Diaz, S.; Linares-Barranco, A. Performance Evaluation of Deep Learning-Based Prostate Cancer Screening Methods in Histopathological Images: Measuring the Impact of the Model's Complexity on Its Processing Speed. *Sensors* **2021**, *21*, 1122. [CrossRef]

17. Amaya-Rodriguez, I.; Duran-Lopez, L.; Luna-Perejon, F.; Civit-Masot, J.; Dominguez-Morales, J.P.; Vicente, S.; Civit, A.; Cascado, D.; Linares-Barranco, A. Glioma Diagnosis Aid through CNNs and Fuzzy-C Means for MRI. In Proceedings of the 11th International Joint Conference on Computational Intelligence, Vienna, Austria, 17–19 September 2019; pp. 528–535.
18. Farooq, A.; Anwar, S.M.; Awais, M.; Rehman, S. A Deep CNN based Multi-class Classification on Alzheimer's Disease using MRI. In Proceedings of the 2017 IEEE International Conference on Imaging Systems and Techniques (IST), Beijing, China, 18–20 October 2017; pp. 1–6.
19. Milletari, F.; Ahmadi, S.A.; Kroll, C.; Plate, A.; Rozanski, V.; Maiostre, J.; Levin, J.; Dietrich, O.; Ertl-Wagner, B.; Bötzel, K.; et al. Hough-CNN: Deep learning for segmentation of deep brain regions in MRI and ultrasound. Comput. *Vis. Image Underst.* **2016**, *164*, 92–102. [CrossRef]
20. Chen, L.; Bentley, P.; Mori, K.; Misawa, K.; Fujiwara, M.; Rueckert, D. DRINet for Medical Image Segmentation. *IEEE Trans. Med. Imaging* **2018**, *37*, 1–11. [CrossRef]
21. Yushkevich, P.A.; Gao, Y.; Gerig, G. ITK-SNAP: An interactive tool for semi-automatic segmentation of multi-modality biomedical images. In Proceedings of the 2016 38th Annual International Conference of the IEEE Engineering in Medicine and Biology Society (EMBC), Orlando, FL, USA, 16–20 August 2016; pp. 3342–3345.
22. Ronneberger, O.; Fischer, P.; Brox, T. U-net: Convolutional networks for biomedical image segmentation. In Proceedings of the International Conference on Medical Image Computing and Computer-Assisted Intervention, Munich, Germany, 5–9 October 2015; Volume 9351, pp. 234–241. [CrossRef]
23. Szegedy, C.; Liu, W.; Jia, Y.; Sermanet, P.; Reed, S.; Anguelov, D.; Erhan, D.; Vanhoucke, V.; Rabinovich, A. Going Deeper with Convolutions. In Proceedings of the IEEE Conference on Computer Vision and Pattern Recognition, Boston, MA, USA, 7–12 June 2015; pp. 1–9.
24. Ulyanov, D.; Vedaldi, A.; Lempitsky, V. Improved Texture Networks: Maximizing Quality and Diversity in Feed-forward Stylization and Texture Synthesis. In Proceedings of the IEEE Conference on Computer Vision and Pattern Recognition, Honolulu, HI, USA, 21–26 July 2017; pp. 6924–6932.
25. Nair, V.; Hinton, G.E. Rectified Linear Units Improve Restricted Boltzmann Machines. In Proceedings of the International Conference on Machine Learning, Haifa, Israel, 21–24 June 2010.
26. Chollet, F. Keras: The Python Deep Learning Library; Astrophysics Source Code Library ascl-1806. 2018. Available online: https://ui.adsabs.harvard.edu/abs/2018ascl.soft06022C (accessed on 5 December 2021).
27. Agarwal, A.; Barham, P.; Brevdo, E.; Chen, Z.; Citro, C.; Corrado, G.S.; Davis, A.; Dean, J.; Devin, M.; Ghemawat, S.; et al. TensorFlow: Large-Scale Machine Learning on Heterogeneous Distributed Systems. *arXiv* **2016**, arXiv:1603.04467.
28. Kingma, D.P.; Ba, J.L. Adam: A method for stochastic optimization. In Proceedings of the 3rd International Conference for Learning Representations (ICLR), San Diego, CA, USA, 7–9 May 2015.
29. Reetz, K.; Rodríguez, R.; Dogan, I.; Mirzazade, S.; Romanzetti, S.; Schulz, J.B.; Cruz-Rivas, E.M.; Alvarez-Cuesta, J.A.; Aguilera Rodríguez, R.; Gonzalez Zaldivar, Y.; et al. Brain atrophy measures in preclinical and manifest spinocerebellar ataxia type 2. *Ann. Clin. Transl. Neurol.* **2018**, *5*, 128–137. [CrossRef] [PubMed]
30. Inagaki, A.; Iida, A.; Matsubara, M.; Inagaki, H. Positron emission tomography and magnetic resonance imaging in spinocerebellar ataxia type 2: A study of symptomatic and asymptomatic individuals. *Eur. J. Neurol.* **2005**, *12*, 725–728. [CrossRef]
31. Seidel, K.; Siswanto, S.; Brunt, E.R.P.; Den Dunnen, W.; Korf, H.W.; Rüb, U. Brain pathology of spinocerebellar ataxias. *Acta Neuropathol.* **2012**, *124*, 1–21. [CrossRef] [PubMed]
32. Tustison, N.J.; Avants, B.B.; Cook, P.A.; Zheng, Y.; Egan, A.; Yushkevich, P.A.; Gee, J.C. N4ITK: Improved N3 Bias Correction. *IEEE Trans. Med. Imaging* **2010**, *29*, 1310–1320. [CrossRef]
33. Fonov, V.S.; Evans, A.C.; Mckinstry, R.C.; Almli, C.R.; Collins, D.L. Unbiased nonlinear average age-appropriate brain templates from birth to adulthood. *Neuroimage* **2009**, *47*, S102. [CrossRef]
34. Avants, B.B.; Tustison, N.; Johnson, H. Advanced Normalization Tools (ANTS). *Insight J.* **2009**, 2.
35. Otsu, N. A Threshold Selection Method from Gray-Level Histograms. *IEEE Trans. Syst. Man Cybern.* **1979**, *9*, 62–66. [CrossRef]
36. Márquez-neila, P.; Baumela, L.; Alvarez, L. A morphological approach to curvature-based evolution of curves and surfaces. *IEEE Trans. Pattern Anal. Mach. Intell.* **2013**, *36*, 2–17. [CrossRef] [PubMed]
37. Brain Development Webpage. Available online: https://brain-development.org/brain-atlases/ (accessed on 5 December 2021).
38. Hammers, A.; Allom, R.; Koepp, M.J.; Free, S.L.; Myers, R.; Lemieux, L.; Mitchell, T.N.; Brooks, D.J.; Duncan, J.S. Three-Dimensional Maximum Probability Atlas of the Human Brain, with Particular Reference to the Temporal Lobe. *Hum. Brain Mapp.* **2003**, *19*, 224–247. [CrossRef]
39. Gousias, I.S.; Rueckert, D.; Heckemann, R.A.; Dyet, L.E.; Boardman, J.P.; Edwards, A.D.; Hammers, A. Automatic segmentation of brain MRIs of 2-year-olds into 83 regions of interest. *Neuroimage* **2008**, *40*, 672–684. [CrossRef]
40. Faillenot, I.; Heckemann, R.A.; Frot, M.; Hammers, A. Macroanatomy and 3D Probabilistic Atlas of the Human Insula. *Neuroimage* **2017**, *150*, 88–98. [CrossRef] [PubMed]
41. Mennes, M.; Biswal, B.; Castellanos, F.X.; Milham, M.P. Making data sharing work: The FCP/INDI experience. *Neuroimage* **2013**, *15*, 683–691. [CrossRef] [PubMed]
42. Kennedy, K.M.; Rodrigue, K.M.; Bischof, G.N.; Hebrank, A.C.; Reuter-Lorenz, P.A.; Park, D.C. Age Trajectories of Functional Activation Under Conditions of Low and High Processing Demands: An Adult Lifespan fMRI Study of the Aging Brain. *Neuroimage* **2015**, *104*, 21–34. [CrossRef] [PubMed]

43. Chan, M.Y.; Park, D.C.; Savalia, N.K.; Petersen, S.E.; Wig, G.S. Decreased segregation of brain systems across the healthy adult lifespan. *Proc. Natl. Acad. Sci. USA* **2014**, *111*, E4997–E5006. [CrossRef] [PubMed]
44. Cocosco, C.A.; Kollokian, V.; Kwan, R.K.S.; Evans, A.C. BrainWeb: Online Interface to a 3D MRI Simulated Brain Database. *Neuroimage* **1997**, *5*, 425.
45. Kwan, R.K.; Evans, A.C.; Pike, G.B. MRI Simulation-Based Evaluation of Image-Processing and Classification Methods. *IEEE Trans. Med. Imaging* **1999**, *18*, 1085–1097. [CrossRef]
46. Milletari, F.; Navab, N.; Ahmadi, S.A. V-Net: Fully convolutional neural networks for volumetric medical image segmentation. In Proceedings of the 2016 Fourth International Conference on 3D Vision (3DV), Stanford, CA, USA, 25–28 October 2016; pp. 565–571. [CrossRef]
47. Fawcett, T. An introduction to ROC analysis. *Pattern Recognit. Lett.* **2006**, *27*, 861–874. [CrossRef]
48. Taha, A.A.; Hanbury, A. Metrics for evaluating 3D medical image segmentation: Analysis, selection, and tool. *BMC Med. Imaging* **2015**, *15*, 29. [CrossRef]
49. Vijaymeena, M.K.; Kavitha, K. A survey on similarity measures in text mining. *Mach. Learn. Appl. Int. J.* **2016**, *3*, 19–28.
50. Bradley, A.E. The use of the area under the ROC curve in the evaluation of machine learning algorithms. *Pattern Recognit.* **1997**, *30*, 1145–1159. [CrossRef]

MDPI
St. Alban-Anlage 66
4052 Basel
Switzerland
Tel. +41 61 683 77 34
Fax +41 61 302 89 18
www.mdpi.com

MDPI Books Editorial Office
E-mail: books@mdpi.com
www.mdpi.com/books

www.ingramcontent.com/pod-product-compliance
Lightning Source LLC
LaVergne TN
LVHW070155100526
838202LV00015B/1949